A Basic Course in Partial Differential Equations

A Basic Course in Partial Differential Equations

Qing Han

Graduate Studies in Mathematics

Volume 120

American Mathematical Society
Providence, Rhode Island

2000 *Mathematics Subject Classification.* Primary 35–01.

For additional information and updates on this book, visit
www.ams.org/bookpages/gsm-120

Library of Congress Cataloging-in-Publication Data

Han, Qing.
A basic course in partial differential equations / Qing Han.
p. cm. — (Graduate studies in mathematics ; v. 120)
Includes bibliographical references and index.
ISBN 978-0-8218-5255-2 (alk. paper)
1. Differential equations, Partial. I. Title.

QA377.H31819 2010
515′. 353—dc22

2010043189

∞ The paper used in this book is acid-free and falls within the guidelines
established to ensure permanence and durability.
Visit the AMS home page at http://www.ams.org/

10 9 8 7 6 5 4 3 2 1 16 15 14 13 12 11

To Yansu, Raymond and Tommy

Contents

Preface

Is it really necessary to classify partial differential equations (PDEs) and to employ different methods to discuss different types of equations? Why is it important to derive *a priori* estimates of solutions before even proving the existence of solutions? These are only a few questions any students who just start studying PDEs might ask. Students may find answers to these questions only at the end of a one-semester course in basic PDEs, sometimes after they have already lost interest in the subject. In this book, we attempt to address these issues *at the beginning*. There are several notable features in this book.

First, the importance of *a priori estimates* is addressed at the beginning and emphasized throughout this book. This is well illustrated by the chapter on first-order PDEs. Although first-order linear PDEs can be solved by the method of characteristics, we provide a detailed analysis of *a priori* estimates of solutions in sup-norms and in integral norms. To emphasize the importance of these estimates, we demonstrate how to prove the existence of weak solutions with the help of basic results from functional analysis. The setting here is easy, since L^2-spaces are needed only. Meanwhile, all important ideas are in full display. In this book, we do attempt to derive explicit expressions for solutions whenever possible. However, these explicit expressions of solutions of special equations usually serve mostly to suggest the correct form of estimates for solutions of general equations.

The second feature is the illustration of the necessity to classify second-order PDEs at the beginning. In the chapter on general second-order linear PDEs, immediately after classifying second-order PDEs into elliptic, parabolic and hyperbolic type, we discuss various boundary-value problems and initial/boundary-value problems for the Laplace equation, the heat equation and the wave equation. We discuss energy methods for proving uniqueness and find solutions in the plane by separation of variables. The explicit expressions of solutions demonstrate different properties of solutions of different types of PDEs. Such differences clearly indicate that there is unlikely to be a unified approach to studying PDEs.

Third, we focus on simple models of PDEs and study these equations in detail. We have chapters devoted to the Laplace equation, the heat equation and the wave equation, and use several methods to study each equation. For example, for the Laplace equation, we use three different methods to study its solutions: the fundamental solution, the mean-value property and the maximum principle. For each method, we indicate its advantages and its shortcomings. General equations are not forgotten. We also discuss maximum principles for general elliptic and parabolic equations and energy estimates for general hyperbolic equations.

The book is designed for a one-semester course at the graduate level. Attempts have been made to give a balanced coverage of different classes of partial differential equations. The choice of topics is influenced by the personal tastes of the author. Some topics may not be viewed as *basic* by others. Among those not found in PDE textbooks at a comparable level are estimates in L^∞-norms and L^2-norms of solutions of the initial-value problem for the first-order linear differential equations, interior gradient estimates and differential Harnack inequality for the Laplace equation and the heat equation by the maximum principle, and decay estimates for solutions of the wave equation. Inclusions of these topics reflect the emphasis on estimates in this book.

This book is based on one-semester courses the author taught at the University of Notre Dame in the falls of 2007, 2008 and 2009. During the writing of the book, the author benefitted greatly from comments and suggestions of many of his friends, colleagues and students in his classes. Tiancong Chen, Yen-Chang Huang, Gang Li, Yuanwei Qi and Wei Zhu read the manuscript at various stages. Minchun Hong, Marcus Khuri, Ronghua Pan, Xiaodong Wang and Xiao Zhang helped the author write part of Chapter 8. Hairong Liu did a wonderful job of typing an early version of the manuscript. Special thanks go to Charles Stanton for reading the entire manuscript carefully and for many suggested improvements.

I am grateful to Natalya Pluzhnikov, my editor at the American Mathematical Society, for reading the manuscript and guiding the effort to turn it into a book. Last but not least, I thank Edward Dunne at the AMS for his help in bringing the book to press.

Qing Han

Chapter 1

Introduction

This chapter serves as an introduction of the entire book.

In Section 1.1, we first list several notations we will use throughout this book. Then, we introduce the concept of partial differential equations.

In Section 1.2, we discuss briefly well-posed problems for partial differential equations. We also introduce several function spaces whose associated norms are used frequently in this book.

In Section 1.3, we present an overview of this book.

1.1. Notation

In general, we denote by x points in $\mathbb{R}^n$ and write $x = (x_1, \cdots, x_n)$ in terms of its coordinates. For any $x \in \mathbb{R}^n$, we denote by $|x|$ the standard *Euclidean norm*, unless otherwise stated. Namely, for any $x = (x_1, \cdots, x_n)$, we have

$$|x| = \left(\sum_{i=1}^{n} x_i^2 \right)^{\frac{1}{2}}.$$

Sometimes, we need to distinguish one particular direction as the time direction and write points in $\mathbb{R}^{n+1}$ as (x, t) for $x \in \mathbb{R}^n$ and $t \in \mathbb{R}$. In this case, we call $x = (x_1, \cdots, x_n) \in \mathbb{R}^n$ the *space variable* and $t \in \mathbb{R}$ the *time variable*. In $\mathbb{R}^2$, we also denote points by (x, y).

Let Ω be a domain in $\mathbb{R}^n$, that is, an open and connected subset in $\mathbb{R}^n$. We denote by $C(\Omega)$ the collection of all continuous functions in Ω, by $C^m(\Omega)$ the collection of all functions with continuous derivatives up to order m, for any integer $m \geq 1$, and by $C^\infty(\Omega)$ the collection of all functions with continuous derivatives of arbitrary order. For any $u \in C^m(\Omega)$, we denote by

$\nabla^m u$ the collection of all partial derivatives of u of order m. For $m = 1$ and $m = 2$, we usually write $\nabla^m u$ in special forms. For first-order derivatives, we write ∇u as a vector of the form

$$\nabla u = (u_{x_1}, \cdots, u_{x_n}).$$

This is the *gradient vector* of u. For second-order derivatives, we write $\nabla^2 u$ in the matrix form

$$\nabla^2 u = \begin{pmatrix} u_{x_1x_1} & u_{x_1x_2} & \cdots & u_{x_1x_n} \\ u_{x_2x_1} & u_{x_2x_2} & \cdots & u_{x_2x_n} \\ \vdots & \vdots & \ddots & \vdots \\ u_{x_nx_1} & u_{x_nx_2} & \cdots & u_{x_nx_n} \end{pmatrix}.$$

This is a symmetric matrix, called the *Hessian matrix* of u. For derivatives of order higher than two, we need to use multi-indices. A multi-index $\alpha \in \mathbb{Z}_+^n$ is given by $\alpha = (\alpha_1, \cdots, \alpha_n)$ with nonnegative integers $\alpha_1, \cdots, \alpha_n$. We write

$$|\alpha| = \sum_{i=1}^n \alpha_i.$$

For any vector $\xi = (\xi_1, \cdots, \xi_n) \in \mathbb{R}^n$, we denote

$$\xi^\alpha = \xi_1^{\alpha_1} \cdots \xi_n^{\alpha_n}.$$

The partial derivative $\partial^\alpha u$ is defined by

$$\partial^\alpha u = \partial_{x_1}^{\alpha_1} \cdots \partial_{x_n}^{\alpha_n} u,$$

and its order is $|\alpha|$. For any positive integer m, we define

$$|\nabla^m u| = \left(\sum_{|\alpha|=m} |\partial^\alpha u|^2 \right)^{\frac{1}{2}}.$$

In particular,

$$|\nabla u| = \left(\sum_{i=1}^n u_{x_i}^2 \right)^{\frac{1}{2}},$$

and

$$|\nabla^2 u| = \left(\sum_{i,j=1}^n u_{x_ix_j}^2 \right)^{\frac{1}{2}}.$$

A hypersurface in $\mathbb{R}^n$ is a surface of dimension $n-1$. Locally, a C^m-hypersurface can be expressed by $\{\varphi = 0\}$ for a C^m-function φ with $\nabla\varphi \neq 0$. Alternatively, by a rotation, we may take $\varphi(x) = x_n - \psi(x_1, \cdots, x_{n-1})$ for a C^m-function ψ of $n-1$ variables. A domain $\Omega \subset \mathbb{R}^n$ is C^m if its boundary $\partial\Omega$ is a C^m-hypersurface.

A *partial differential equation* (henceforth abbreviated as PDE) in a domain $\Omega \subset \mathbb{R}^n$ is a relation of independent variables $x \in \Omega$, an unknown function u defined in Ω, and a finite number of its partial derivatives. To solve a PDE is to find this unknown function. The *order* of a PDE is the order of the highest derivative in the relation. Hence for a positive integer m, the general form of an mth-order PDE in a domain $\Omega \subset \mathbb{R}^n$ is given by

$$F\big(x, u, \nabla u(x), \nabla^2 u(x), \cdots, \nabla^m u(x)\big) = 0 \quad \text{for } x \in \Omega.$$

Here F is a function which is continuous in all its arguments, and u is a C^m-function in Ω. A C^m-solution u satisfying the above equation in the pointwise sense in Ω is often called a *classical solution*. Sometimes, we need to relax regularity requirements for solutions when classical solutions are not known to exist. Instead of going into details, we only mention that it is an important method to establish first the existence of *weak solutions*, functions with less regularity than C^m and satisfying the equation in some weak sense, and then to prove that these weak solutions actually possess the required regularity to be classical solutions.

A PDE is *linear* if it is linear in the unknown functions and their derivatives, with coefficients depending on independent variables x. A general mth-order linear PDE in Ω is given by

$$\sum_{|\alpha| \le m} a_\alpha(x) \partial^\alpha u = f(x) \quad \text{for } x \in \Omega.$$

Here a_α is the coefficient of $\partial^\alpha u$ and f is the *nonhomogeneous term* of the equation. A PDE of order m is *quasilinear* if it is linear in the derivatives of solutions of order m, with coefficients depending on independent variables x and the derivatives of solutions of order $< m$. In general, an mth-order quasilinear PDE in Ω is given by

$$\sum_{|\alpha| = m} a_\alpha(x, u, \cdots, \nabla^{m-1} u) \partial^\alpha u = f(x, u, \cdots, \nabla^{m-1} u) \quad \text{for } x \in \Omega.$$

Several PDEs involving one or more unknown functions and their derivatives form a *partial differential system*. We define linear and quasilinear partial differential systems accordingly.

In this book, we will focus on first-order and second-order linear PDEs and first-order linear differential systems. On a few occasions, we will diverge to nonlinear PDEs.

1.2. Well-Posed Problems

What is the meaning of *solving* partial differential equations? Ideally, we obtain explicit solutions in terms of elementary functions. In practice this is only possible for very simple PDEs or very simple solutions of more general

PDEs. In general, it is impossible to find explicit expressions of all solutions of all PDEs. In the absence of explicit solutions, we need to seek methods to prove existence of solutions of PDEs and discuss properties of these solutions. In many PDE problems, this is all we need to do.

A given PDE may not have solutions at all or may have many solutions. When it has many solutions, we intend to assign *extra conditions* to pick up the most relevant solutions. Those extra conditions usually are in the form of boundary values or initial values. For example, when we consider a PDE in a domain, we can require that solutions, when restricted to the boundary, have prescribed values. This is the so-called *boundary-value problems*. When one variable is identified as the time and a part of the boundary is identified as an initial hypersurface, values prescribed there are called *initial values*. We use *data* to refer to boundary values or initial values and certain known functions in the equation, such as the nonhomogeneous term if the equation is linear.

Hadamard introduced the notion of *well-posed problems*. A given problem for a partial differential equation is *well-posed* if

(i) there is a solution;

(ii) this solution is unique;

(iii) the solution depends continuously in some suitable sense on the data given in the problem, i.e., the solution changes by a small amount if the data change by a small amount.

We usually refer to (i), (ii) and (iii) as the *existence*, *uniqueness* and *continuous dependence*, respectively. We need to emphasize that the well-posedness goes beyond the existence and uniqueness of solutions. The continuous dependence is particularly important when PDEs are used to model phenomena in the natural world. This is because measurements are always associated with errors. The model can make useful predictions only if solutions depend on data in a controllable way.

In practice, both the uniqueness and the continuous dependence are proved by *a priori estimates*. Namely, we assume solutions already exist and then derive certain norms of solutions in terms of data in the problem. It is important to note that establishing *a priori* estimates is in fact the first step in proving the existence of solutions. A closely related issue here is the regularity of solutions such as continuity and differentiability. Solutions of a particular PDE can only be obtained if the right kind of regularity, or the right kind of norms, are employed. Two classes of norms are used often, sup-norms and L^2-norms.

Let Ω be a domain in $\mathbb{R}^n$. For any bounded function u in Ω, we define the sup-norm of u in Ω by

$$|u|_{L^\infty(\Omega)} = \sup_{\Omega} |u|.$$

For a bounded continuous function u in Ω, we may also write $|u|_{C(\Omega)}$ instead of $|u|_{L^\infty(\Omega)}$. Let m be a positive integer. For any function u in Ω with bounded derivatives up to order m, we define the C^m-norm of u in Ω by

$$|u|_{C^m(\Omega)} = \sum_{|\alpha| \leq m} |\partial^\alpha u|_{L^\infty(\Omega)}.$$

If Ω is a bounded C^m-domain in $\mathbb{R}^n$, then $C^m(\bar{\Omega})$, the collection of functions which are C^m in $\bar{\Omega}$, is a Banach space equipped with the C^m-norm.

Next, for any Lebesgue measurable function u in Ω, we define the L^2-norm of u in Ω by

$$\|u\|_{L^2(\Omega)} = \left(\int_\Omega u^2 \, dx \right)^{\frac{1}{2}},$$

where integration is in the Lebesgue sense. The L^2-space in Ω is the collection of all Lebesgue measurable functions in Ω with finite L^2-norms and is denoted by $L^2(\Omega)$. We learned from real analysis that $L^2(\Omega)$ is a Banach space equipped with the L^2-norm.

Other norms will also be used. We will introduce them as needed.

The basic formula for integration is the formula of *integration by parts*. Let Ω be a piecewise C^1-domain in $\mathbb{R}^n$ and $\nu = (\nu_1, \cdots, \nu_n)$ be the unit exterior normal vector to $\partial\Omega$. Then for any $u, v \in C^1(\Omega) \cap C(\bar{\Omega})$,

$$\int_\Omega u_{x_i} v \, dx = -\int_\Omega u v_{x_i} \, dx + \int_{\partial\Omega} u v \nu_i \, dS,$$

for $i = 1, \cdots, n$. Such a formula is the basis for L^2-estimates.

In deriving a priori estimates, we follow a common practice and use the "variable constant" convention. The same letter C is used to denote constants which may change from line to line, as long as it is clear from the context on what quantities the constants depend. In most cases, we are not interested in the value of the constant, but only in its existence.

1.3. Overview

There are eight chapters in this book.

The main topic in Chapter 2 is first-order PDEs. In Section 2.1, we introduce the basic notion of noncharacteristic hypersurfaces for initial-value problems for first-order PDEs. We discuss first-order linear PDEs, quasilinear PDEs and general nonlinear PDEs. In Section 2.2, we solve initial-value problems by the method of characteristics if initial values are prescribed on noncharacteristic hypersurfaces. We demonstrate that solutions of a system of ordinary differential equations (ODEs) yield solutions of the initial-value problems for first-order PDEs. In Section 2.3, we derive estimates of solutions of initial-value problems for first-order linear PDEs. The L^∞-norms and the L^2-norms of solutions are estimated in terms of those of initial values and nonhomogeneous terms. In doing so, we only assume the existence of solutions and do not use any explicit expressions of solutions. These estimates provide quantitative properties of solutions.

Chapter 3 should be considered as an introduction to the theory of second-order linear PDEs. In Section 3.1, we introduce the Laplace equation, the heat equation and the wave equation. We also introduce their general forms, elliptic equations, parabolic equations and hyperbolic equations, which will be studied in detail in subsequent chapters. In Section 3.2, we derive energy estimates of solutions of certain boundary-value problems. Consequences of such energy estimates are the uniqueness of solutions and the continuous dependence of solutions on boundary values and nonhomogeneous terms. In Section 3.3, we solve these boundary-value problems in the plane by separation of variables. Our main focus is to demonstrate different regularity patterns for solutions of different differential equations, the Laplace equation, the heat equation and the wave equation.

In Chapter 4, we discuss the Laplace equation and the Poisson equation. The Laplace equation is probably the most important PDE with the widest range of applications. In the first three sections, we study harmonic functions (i.e., solutions of the Laplace equation), by three different methods: the fundamental solution, the mean-value property and the maximum principle. These three sections are relatively independent of each other. In Section 4.1, we solve the Dirichlet problem for the Laplace equation in balls and derive Poisson integral formula. Then we discuss regularity of harmonic functions using the fundamental solution. In Section 4.2, we study the mean-value property of harmonic functions and its consequences. In Section 4.3, we discuss the maximum principle for harmonic functions and its applications. In particular, we use the maximum principle to derive interior gradient estimates for harmonic functions and the Harnack inequality for positive harmonic functions. We also solve the Dirichlet problem for the Laplace equation in a large class of bounded domains by Perron's method. Last in Section 4.4, we briefly discuss classical solutions and weak solutions of the Poisson equation.

In Chapter 5, we study the heat equation, which describes the temperature of a body conducting heat, when the density is constant. In Section 5.1, we introduce Fourier transforms briefly and derive formally an explicit expression for solutions of the initial-value problem for the heat equation. In Section 5.2, we prove that such an expression indeed yields a classical solution under appropriate assumptions on initial values. We also discuss regularity of arbitrary solutions of the heat equation by the fundamental solution. In Section 5.3, we discuss the maximum principle for the heat equation and its applications. In particular, we use the maximum principle to derive interior gradient estimates for solutions of the heat equation and the Harnack inequality for positive solutions of the heat equation.

In Chapter 6, we study the n-dimensional wave equation, which represents vibrations of strings or propagation of sound waves in tubes for $n = 1$, waves on the surface of shallow water for $n = 2$, and acoustic or light waves for $n = 3$. In Section 6.1, we discuss initial-value problems and various initial/boundary-value problems for the one-dimensional wave equation. In Section 6.2, we study initial-value problems for the wave equation in higher-dimensional spaces. We derive explicit expressions of solutions in odd dimensions by the method of spherical average and in even dimensions by the method of descent. We also discuss global behaviors of solutions. Then in Section 6.3, we derive energy estimates for solutions of initial-value problems. Chapter 6 is relatively independent of Chapter 4 and Chapter 5 and can be taught after Chapter 3.

In Chapter 7, we discuss partial differential systems of first order and focus on existence of local solutions. In Section 7.1, we introduce noncharacteristic hypersurfaces for partial differential equations and systems of arbitrary order. We demonstrate that partial differential systems of arbitrary order can always be changed to those of first order. In Section 7.2, we discuss the Cauchy-Kovalevskaya theorem, which asserts the existence of analytic solutions of noncharacteristic initial-value problems for differential systems if all data are analytic. In Section 7.3, we construct a first order linear differential system in $\mathbb{R}^3$ which does not admit smooth solutions in any subsets of $\mathbb{R}^3$. In this system, coefficient matrices are analytic and the nonhomogeneous term is a suitably chosen smooth function.

In Chapter 8, we discuss several differential equations we expect to study in more advanced PDE courses. Discussions in this chapter will be brief. In Section 8.1, we discuss basic second-order linear differential equations, including elliptic, parabolic and hyperbolic equations, and first-order linear symmetric hyperbolic differential systems. We will introduce appropriate boundary-value problems and initial-value problems and introduce appropriate function spaces to study these problems. In Section 8.2, we introduce several important nonlinear equations and focus on their background. This chapter is designed to be introductory.

Each chapter, except this introduction and the final chapter, ends with exercises. Level of difficulty varies considerably. Some exercises, at the most difficult level, may require long lasting efforts.

Chapter 2

First-Order Differential Equations

In this chapter, we discuss initial-value problems for first-order PDEs. Main topics include noncharacteristic conditions, methods of characteristics and a priori estimates in L^∞-norms and in L^2-norms.

In Section 2.1, we introduce the basic notion of noncharacteristic hypersurfaces for initial-value problems. In an attempt to solve initial-value problems, we illustrate that we are able to compute all derivatives of solutions on initial hypersurfaces if initial values are prescribed on noncharacteristic initial hypersurfaces. For first-order linear PDEs, the noncharacteristic condition is determined by equations and initial hypersurfaces, independent of initial values. However, for first-order nonlinear equations, initial values also play a role. Noncharacteristic conditions will also be introduced for second-order linear PDEs in Section 3.1 and for linear PDEs of arbitrary order in Section 7.1, where multi-indices will be needed.

In Section 2.2, we solve initial-value problems by the method of characteristics if initial values are prescribed on noncharacteristic hypersurfaces. For first-order homogeneous linear PDEs, special curves are introduced along which solutions are constant. These curves are given by solutions of a system of ordinary differential equations (ODEs), the so-called characteristic ODEs. For nonlinear PDEs, characteristic ODEs also include additional equations for solutions of PDEs and their derivatives. Solutions of the characteristic ODEs yield solutions of the initial-value problems for first-order PDEs.

In Section 2.3, we derive estimates of solutions of initial-value problems for first-order linear PDEs. The L^∞-norms and the L^2-norms of solutions are estimated in terms of those of initial values and nonhomogeneous terms. In doing so, we only assume the existence of solutions and do not use any explicit expressions of solutions. These estimates provide quantitative properties of solutions. In the final part of this section, we discuss briefly the existence of weak solutions as a consequence of the L^2-estimates. The method is from functional analysis and the Riesz representation theorem plays an essential role.

2.1. Noncharacteristic Hypersurfaces

Let Ω be a domain in $\mathbb{R}^n$ and $F = F(x, u, p)$ be a smooth function of $(x, u, p) \in \Omega \times \mathbb{R} \times \mathbb{R}^n$. A first-order PDE in Ω is given by

$$F\big(x, u, \nabla u\big) = 0 \quad \text{for } x \in \Omega. \tag{2.1.1}$$

Solving (2.1.1) in the classical sense means finding a smooth function u satisfying (2.1.1) in Ω. We first examine a simple example.

Example 2.1.1. We consider in $\mathbb{R}^2 = \{(x, t)\}$ the equation

$$u_x + u_t = 0.$$

This is probably the simplest first-order PDE. Obviously, $u(x, t) = x - t$ is a solution. In general, $u(x, t) = u_0(x - t)$ is also a solution for any C^1-function u_0. Such a solution has a physical interpretation. We note that $u(x, t) = u_0(x-t)$ is constant along straight lines $x-t = x_0$. By interpreting x as location and t as time, we can visualize such a solution as a wave propagating to the right with velocity 1 without changing shape. When interpreted in this way, the solution u at later time $(t > 0)$ is determined uniquely by its value at the initial time $(t = 0)$, which is given by $u_0(x)$. The function u_0 is called an *initial value*.

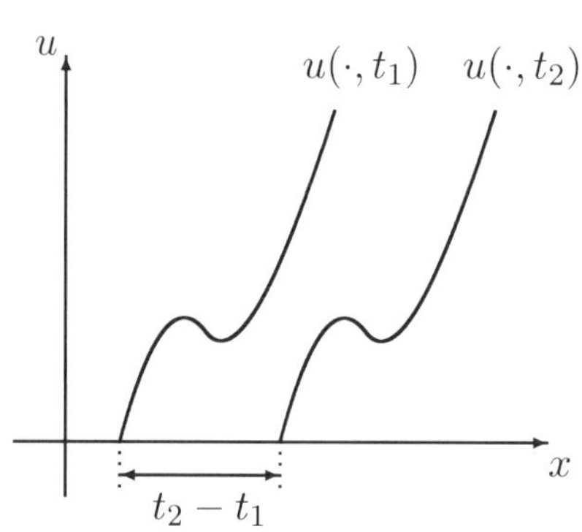

Figure 2.1.1. Graphs of u at different times $t_2 > t_1$.

In light of Example 2.1.1, we will introduce initial values for (2.1.1) and discuss whether initial values determine solutions.

Let Σ be a smooth hypersurface in $\mathbb{R}^n$ with $\Omega \cap \Sigma \neq \emptyset$. We intend to prescribe u on Σ to find a solution of (2.1.1). To be specific, let u_0 be a given smooth function on Σ. We will find a solution u of (2.1.1) also satisfying

$$u = u_0 \quad \text{on } \Sigma. \tag{2.1.2}$$

We usually call Σ the *initial hypersurface* and u_0 the *initial value* or *Cauchy value*. The problem of solving (2.1.1) together with (2.1.2) is called the *initial-value problem* or *Cauchy problem*. Our main focus is to solve such an initial-value problem under appropriate conditions.

We start with the following question. *Given an initial value (2.1.2) for equation (2.1.1), can we compute all derivatives of u at each point of the initial hypersurface Σ?* This should be easier than solving the initial-value problem (2.1.1)–(2.1.2).

To illustrate the main ideas, we first consider linear PDEs. Let Ω be a domain in $\mathbb{R}^n$ containing the origin and a_i, b and f be smooth functions in Ω, for any $i = 1, \cdots, n$. We consider

$$\sum_{i=1}^{n} a_i(x)u_{x_i} + b(x)u = f(x) \quad \text{in } \Omega. \tag{2.1.3}$$

Here, a_i and b are coefficients of u_{x_i} and u, respectively. The function f is called the *nonhomogeneous term*. If $f \equiv 0$, (2.1.3) is called a *homogeneous equation*.

We first consider a special case where the initial hypersurface Σ is given by the hyperplane $\{x_n = 0\}$. For $x \in \mathbb{R}^n$, we write $x = (x', x_n)$ for $x' = (x_1, \cdots, x_{n-1}) \in \mathbb{R}^{n-1}$. Let u_0 be a given smooth function in a neighborhood of the origin in $\mathbb{R}^{n-1}$. The initial condition (2.1.2) has the form

$$u(x', 0) = u_0(x'), \tag{2.1.4}$$

for any $x' \in \mathbb{R}^{n-1}$ sufficiently small.

Let u be a smooth solution of (2.1.3) and (2.1.4). In the following, we will investigate whether we can compute all derivatives of u at the origin in terms of the equation and the initial value. It is obvious that we can find all x'-derivatives of u at the origin in terms of those of u_0. In particular, we have, for $i = 1, \cdots, n-1$,

$$u_{x_i}(0) = u_{0,x_i}(0).$$

To find $u_{x_n}(0)$, we need to use the equation. We note that a_n is the coefficient of u_{x_n} in (2.1.3). If we assume

$$a_n(0) \neq 0, \tag{2.1.5}$$

then by (2.1.3)

$$u_{x_n}(0) = -\frac{1}{a_n(0)}\left(\sum_{i=1}^{n-1} a_i(0)u_{x_i}(0) + b(0)u(0) - f(0)\right).$$

Hence, we can compute all first-order derivatives of u at 0 in terms of the coefficients and the nonhomogeneous term in (2.1.3) and the initial value u_0 in (2.1.4). In fact, we can compute all derivatives of u of any order at the origin by using u_0 and differentiating (2.1.3). We illustrate this by finding all the second-order derivatives. We first note that

$$u_{x_ix_j}(0) = u_{0,x_ix_j}(0),$$

for $i, j = 1, \cdots, n-1$. To find $u_{x_kx_n}$ for $k = 1, \cdots, n$, we differentiate (2.1.3) with respect to x_k to get

$$\sum_{i=1}^{n} a_i u_{x_ix_k} + \sum_{i=1}^{n} a_{i,x_k} u_{x_i} + bu_{x_k} + b_{x_k} u = f_{x_k}.$$

For $k = 1, \cdots, n-1$, the only unknown expression at the origin is $u_{x_kx_n}$, whose coefficient is a_n. If (2.1.5) holds, we can find $u_{x_kx_n}(0)$ for $k = 1, \cdots, n-1$. For $k = n$, with $u_{x_ix_n}(0)$ already determined for $i = 1, \cdots, n-1$, we can find $u_{x_nx_n}(0)$ similarly. This process can be repeated for derivatives of arbitrary order. In summary, we can find all derivatives of u of any order at the origin under the condition (2.1.5), which will be defined as the noncharacteristic condition later on.

More generally, consider a hypersurface Σ given by $\{\varphi = 0\}$ for a smooth function φ in a neighborhood of the origin with $\nabla\varphi \neq 0$. Assume that Σ passes through the origin, i.e., $\varphi(0) = 0$. We note that $\nabla\varphi$ is normal to Σ at each point of Σ. Without loss of generality, we assume $\varphi_{x_n}(0) \neq 0$. Then by the implicit function theorem, we can solve $\varphi = 0$ around $x = 0$ for $x_n = \psi(x_1, \cdots, x_{n-1})$. Consider a change of variables

$$x \mapsto y = \big(x_1, \cdots, x_{n-1}, \varphi(x)\big).$$

This is a well-defined transform in a neighborhood of the origin. Its Jacobian matrix J is given by

$$J = \frac{\partial(y_1, \cdots, y_n)}{\partial(x_1, \cdots, x_n)} = \begin{pmatrix} & & & 0 \\ & Id & & \vdots \\ & & & 0 \\ \varphi_{x_1} & \cdots & \varphi_{x_{n-1}} & \varphi_{x_n} \end{pmatrix}.$$

Hence $\det J(0) = \varphi_{x_n}(0) \neq 0$.

In the following, we denote by L the first-order linear differential operator defined by the left-hand side of (2.1.3), i.e.,

$$Lu = \sum_{i=1}^{n} a_i(x)u_{x_i} + b(x)u. \tag{2.1.6}$$

By the chain rule,

$$u_{x_i} = \sum_{k=1}^{n} y_{k,x_i} u_{y_k}.$$

We write the operator L in the y-coordinates as

$$Lu = \sum_{k=1}^{n} \left(\sum_{i=1}^{n} a_i(x(y)) y_{k,x_i} \right) u_{y_k} + b(x(y))u.$$

In the y-coordinates, the initial hypersurface Σ is given by $\{y_n = 0\}$. With $y_n = \varphi$, the coefficient of u_{y_n} is given by

$$\sum_{i=1}^{n} a_i(x)\varphi_{x_i}.$$

Hence, for the initial-value problem (2.1.3) and (2.1.2), we can find all derivatives of u at $0 \in \Sigma$ if

$$\sum_{i=1}^{n} a_i(0)\varphi_{x_i}(0) \neq 0.$$

We recall that $\nabla\varphi = (\varphi_{x_1}, \cdots, \varphi_{x_n})$ is normal to $\Sigma = \{\varphi = 0\}$. When $\Sigma = \{x_n = 0\}$ or $\varphi(x) = x_n$, then $\nabla\varphi = (0, \cdots, 0, 1)$ and

$$\sum_{i=1}^{n} a_i(x)\varphi_{x_i} = a_n(x).$$

This reduces to the special case we discussed earlier.

Definition 2.1.2. Let L be a first-order linear differential operator as in (2.1.6) in a neighborhood of $x_0 \in \mathbb{R}^n$ and Σ be a smooth hypersurface containing x_0. Then Σ is *noncharacteristic* at x_0 if

$$\sum_{i=1}^{n} a_i(x_0)\nu_i \neq 0, \tag{2.1.7}$$

where $\nu = (\nu_1, \cdots, \nu_n)$ is normal to Σ at x_0. Otherwise, Σ is *characteristic* at x_0.

A hypersurface is noncharacteristic if it is noncharacteristic at every point. Strictly speaking, a hypersurface is characteristic if it is not noncharacteristic, i.e., if it is characteristic at some point. In this book, we will abuse this terminology. When we say a hypersurface is characteristic, we mean it is characteristic *everywhere*. This should cause few confusions. In

$\mathbb{R}^2$, hypersurfaces are curves, so we shall speak of characteristic curves and noncharacteristic curves.

The noncharacteristic condition has a simple geometric interpretation. If we view $a = (a_1, \cdots, a_n)$ as a vector in $\mathbb{R}^n$, then condition (2.1.7) holds if and only if $a(x_0)$ is not a tangent vector to Σ at x_0. This condition assures that we can compute all derivatives of solutions at x_0.

It is straightforward to check that (2.1.7) is maintained under C^1-changes of coordinates.

The discussion leading to Definition 2.1.2 can be easily generalized to first-order quasilinear equations. Let Ω be a domain in $\mathbb{R}^n$ containing the origin as before and a_i and f be smooth functions in $\Omega \times \mathbb{R}$, for any $i = 1, \cdots, n$. We consider

$$\sum_{i=1}^{n} a_i(x,u)u_{x_i} = f(x,u) \quad \text{in } \Omega. \tag{2.1.8}$$

Again, we first consider a special case where the initial hypersurface Σ is given by the hyperplane $\{x_n = 0\}$ and an initial value is given by (2.1.4) for a given smooth function u_0 in a neighborhood of the origin in $\mathbb{R}^{n-1}$. Let u be a smooth solution of (2.1.8) and (2.1.4). Then

$$u_{x_i}(0) = u_{0,x_i}(0),$$

for $i = 1, \cdots, n-1$, and

$$u_{x_n}(0) = -\frac{1}{a_n\big(0, u_0(0)\big)}\left(\sum_{i=1}^{n-1} a_i\big(0, u_0(0)\big)u_{x_i}(0) - f\big(0, u_0(0)\big)\right)$$

if

$$a_n\big(0, u_0(0)\big) \neq 0.$$

Similar to (2.1.5), this is the noncharacteristic condition for (2.1.8) at the origin if the initial hypersurface Σ is given by $\{x_n = 0\}$.

In general, let x_0 be a point in $\mathbb{R}^n$ and Σ be a smooth hypersurface containing x_0. Let u_0 be a prescribed smooth function on Σ and a_i and f be smooth functions in a neighborhood of $(x_0, u(x_0)) \in \mathbb{R}^n \times \mathbb{R}$, for $i = 1, \cdots, n$. Then for quasilinear PDE (2.1.8), Σ is *noncharacteristic* at x_0 with respect to u_0 if

$$\sum_{i=1}^{n} a_i\big(x_0, u_0(x_0)\big)\nu_i \neq 0, \tag{2.1.9}$$

where $\nu = (\nu_1, \cdots, \nu_n)$ is normal to Σ at x_0.

There is a significant difference between (2.1.7) for linear PDEs and (2.1.9) for quasilinear PDEs. For linear PDEs, the noncharacteristic condition depends on initial hypersurfaces and equations, specifically, the coefficients of first-order derivatives. For quasilinear PDEs, it also depends on initial values.

Next, we turn to general nonlinear partial differential equations as in (2.1.1). Let Ω be a domain in $\mathbb{R}^n$ containing the origin as before and let F be a smooth function in $\Omega \times \mathbb{R} \times \mathbb{R}^n$. Consider

$$F(x, u, \nabla u) = 0 \quad \text{in } \Omega.$$

We ask the same question as for linear equations. Given an initial hypersurface Σ containing the origin and an initial value u_0 on Σ, can we compute all derivatives of solutions at the origin? Again, we first consider a special case where the initial hypersurface Σ is given by the hyperplane $\{x_n = 0\}$ and an initial value is given by (2.1.4) for a given smooth function u_0 in a neighborhood of the origin in $\mathbb{R}^{n-1}$.

Example 2.1.3. Consider

$$\sum_{i=1}^{n} u_{x_i}^2 = 1$$

and

$$u(x', 0) = u_0(x').$$

It is obvious that $u = x_i$ is a solution for $u_0(x') = x_i$, $i = 1, \cdots, n-1$. However, if $|\nabla_{x'} u_0(x')|^2 > 1$, there are no solutions for such an initial value.

In light of Example 2.1.3, we first assume that there exists a smooth function v in a neighborhood of the origin having the given initial value u_0 and satisfying $F = 0$ at the origin, i.e.,

$$F\big(0, v(0), \nabla v(0)\big) = 0.$$

Now we can proceed as in the discussion of linear PDEs and ask whether we can find u_{x_n} at the origin. By the implicit function theorem, this is possible if

$$F_{u_{x_n}}\big(0, v(0), \nabla v(0)\big) \neq 0.$$

This is the noncharacteristic condition for $F = 0$ at the origin.

Now we return to Example 2.1.3. We set

$$F(x, u, p) = |p|^2 - 1 \quad \text{for any } p \in \mathbb{R}^n.$$

We claim that the noncharacteristic condition holds at 0 with respect to u_0 if

$$|\nabla_{x'} u_0(0)| < 1.$$

In fact, let $v = u_0 + cx_n$, for a constant c to be determined. Then

$$|\nabla v(0)|^2 = |\nabla_{x'} u_0(0)|^2 + c^2.$$

By choosing

$$c = \pm\sqrt{1 - |\nabla_{x'} u_0(0)|^2} \neq 0,$$

v satisfies the equation at $x = 0$. For such two choices of v, we have

$$F_{u_{x_n}}(0, v(0), \nabla v(0)) = 2v_{x_n}(0) = 2c \neq 0.$$

This proves the claim.

In general, let $F = 0$ be a first-order nonlinear PDE as in (2.1.1) in a neighborhood of $x_0 \in \mathbb{R}^n$. Let Σ be a hypersurface containing x_0 and u_0 be a prescribed function on Σ. Then Σ is *noncharacteristic* at $x_0 \in \Sigma$ with respect to u_0 if there exists a function v such that $v = u_0$ on Σ, $F(x_0, v(x_0), \nabla v(x_0)) = 0$ and

$$\sum_{i=1}^{n} F_{u_{x_i}}\big(x_0, v(x_0), \nabla v(x_0)\big)\nu_i \neq 0,$$

where $\nu = (\nu_1, \cdots, \nu_n)$ is normal to Σ at x_0.

2.2. The Method of Characteristics

In this section, we solve initial-value problems for first-order PDEs by the method of characteristics. We demonstrate that solutions of any first-order PDEs with initial values prescribed on noncharacteristic hypersurfaces can be obtained by solving systems of ordinary differential equations (ODEs).

Let $\Omega \subset \mathbb{R}^n$ be a domain and F a smooth function in $\Omega \times \mathbb{R} \times \mathbb{R}^n$. The general form of first-order PDEs in Ω is given by

$$F(x, u, \nabla u) = 0 \quad \text{for any } x \in \Omega.$$

Let Σ be a smooth hypersurface in $\mathbb{R}^n$ with $\Sigma \cap \Omega \neq \emptyset$ and u_0 be a smooth function on Σ. Then we prescribe an initial value on Σ by

$$u = u_0 \quad \text{on } \Sigma \cap \Omega.$$

If Ω is a domain containing the origin and Σ is noncharacteristic at the origin with respect to u_0, then we are able to compute derivatives of u of arbitrary order at the origin by discussions in the previous section. Next, we investigate whether we can solve the initial-value problem at least in a neighborhood of the origin.

Throughout this section, we always assume that Ω is a domain containing the origin and that the initial hypersurface Σ is given by the hyperplane $\{x_n = 0\}$. Obviously, $\{x_n = 0\}$ has $(0, \cdots, 0, 1)$ as a normal vector field. If

$x \in \mathbb{R}^n$, we write $x = (x', x_n)$, where $x' \in \mathbb{R}^{n-1}$. Our goal is to solve the initial-value problem

$$\begin{aligned} F(x, u, \nabla u) &= 0, \\ u(x', 0) &= u_0(x'). \end{aligned}$$

2.2.1. Linear Homogeneous Equations. We start with first-order linear homogeneous equations. Let a_i be smooth in a neighborhood of $0 \in \mathbb{R}^n$, $i = 1, \cdots, n$, and u_0 be smooth in a neighborhood of $0 \in \mathbb{R}^{n-1}$. Consider

$$\begin{aligned} \sum_{i=1}^{n} a_i(x) u_{x_i} &= 0, \\ u(x', 0) &= u_0(x'). \end{aligned} \tag{2.2.1}$$

By introducing $a = (a_1, \cdots, a_n)$, we simply write the equation in (2.2.1) as

$$a(x) \cdot \nabla u = 0.$$

Here $a(x)$ is regarded as a vector field in a neighborhood of $0 \in \mathbb{R}^n$. Then $a(x) \cdot \nabla$ is a directional derivative along $a(x)$ at x. In the following, we assume that the hyperplane $\{x_n = 0\}$ is noncharacteristic at the origin, i.e.,

$$a_n(0) \neq 0.$$

Here we assume that a solution u of (2.2.1) exists. Our strategy is as follows. For any $\bar{x} \in \mathbb{R}^n$ close to the origin, we construct a special curve along which u is constant. If such a curve starts from $\bar{x}$ and intersects $\mathbb{R}^{n-1} \times \{0\}$ at $(\bar{y}, 0)$ for a small $\bar{y} \in \mathbb{R}^{n-1}$, then $u(\bar{x}) = u_0(\bar{y})$. To find such a curve $x = x(s)$, we consider the restriction of u to it and obtain a one-variable function $u(x(s))$. Now we calculate the s-derivative of this function and obtain

$$\frac{d}{ds}\big(u(x(s))\big) = \sum_{i=1}^{n} u_{x_i} \frac{dx_i}{ds}.$$

In order to have a constant value of u along this curve, we require

$$\frac{d}{ds}\big(u(x(s))\big) = 0.$$

A simple comparison with the equation in (2.2.1) yields

$$\frac{dx_i}{ds} = a_i(x) \quad \text{for } i = 1, \cdots, n.$$

This naturally leads to the following definition.

Definition 2.2.1. Let $a = a(x) : \Omega \to \mathbb{R}^n$ be a smooth vector field in Ω and $x = x(s)$ be a smooth curve in Ω. Then $x = x(s)$ is an *integral curve* of a if

$$\frac{dx}{ds} = a(x). \tag{2.2.2}$$

The calculation preceding Definition 2.2.1 shows that the solution u of (2.2.1) is constant along integral curves of the coefficient vector field. This yields the following method of solving (2.2.1). For any $\bar{x} \in \mathbb{R}^n$ near the origin, we find an integral curve of the coefficient vector field through $\bar{x}$ by solving

$$\begin{aligned} \frac{dx}{ds} &= a(x), \\ x(0) &= \bar{x}. \end{aligned} \tag{2.2.3}$$

If it intersects the hyperplane $\{x_n = 0\}$ at $(\bar{y}, 0)$ for some $\bar{y}$ sufficiently small, then we let $u(\bar{x}) = u_0(\bar{y})$.

Since (2.2.3) is an autonomous system (i.e., the independent variable s does not appear explicitly), we may start integral curves from initial hyperplanes. Instead of (2.2.3), we consider the system

$$\begin{aligned} \frac{dx}{ds} &= a(x), \\ x(0) &= (y, 0). \end{aligned} \tag{2.2.4}$$

In (2.2.4), integral curves start from $(y, 0)$. By allowing $y \in \mathbb{R}^{n-1}$ to vary in a neighborhood of the origin, we expect integral curves $x(y, s)$ to reach any $x \in \mathbb{R}^n$ in a neighborhood of the origin for small s. This is confirmed by the following result.

Lemma 2.2.2. *Let a be a smooth vector field in a neighborhood of the origin with $a_n(0) \neq 0$. Then for any sufficiently small $y \in \mathbb{R}^{n-1}$ and any sufficiently small s, the solution $x = x(y, s)$ of (2.2.4) defines a diffeomorphism in a neighborhood of the origin in $\mathbb{R}^n$.*

Proof. This follows easily from the implicit function theorem. By standard results in ordinary differential equations, (2.2.4) admits a smooth solution $x = x(y, s)$ for any sufficiently small $(y, s) \in \mathbb{R}^{n-1} \times \mathbb{R}$. We treat it as a map $(y, s) \mapsto x$ and calculate its Jacobian matrix J at $(y, s) = (0, 0)$. By $x(y, 0) = (y, 0)$, we have

$$J(0) = \frac{\partial x}{\partial(y, s)}\bigg|_{(y,s)=(0,0)} = \begin{pmatrix} & & & a_1(0) \\ & Id & & \vdots \\ & & & a_{n-1}(0) \\ 0 & \cdots & 0 & a_n(0) \end{pmatrix}.$$

Hence $\det J(0) = a_n(0) \neq 0$. □

Therefore, for any sufficiently small $\bar{x}$, we can solve

$$x(\bar{y}, \bar{s}) = \bar{x}$$

uniquely for small $\bar{y}$ and $\bar{s}$. Then $u(\bar{x}) = u_0(\bar{y})$ yields a solution of (2.2.1). Note that $\bar{s}$ is not present in the expression of solutions. Hence the value

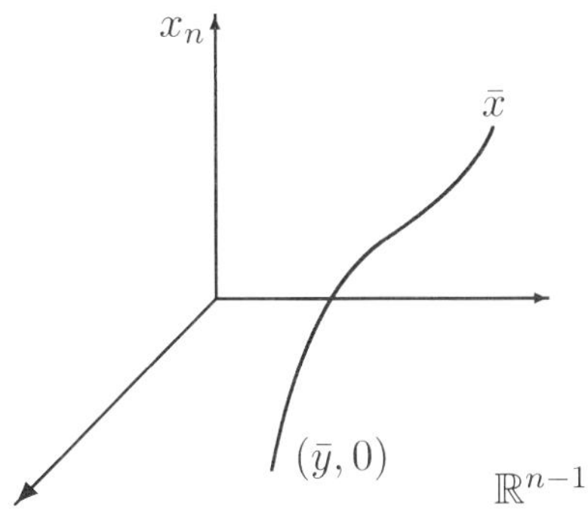

Figure 2.2.1. Solutions by integral curves.

of the solution $u(\bar{x})$ depends only on the initial value u_0 at $(\bar{y}, 0)$ and, meanwhile, the initial value u_0 at $(\bar{y}, 0)$ influences the solution u along the integral curve starting from $(\bar{y}, 0)$. Therefore, we say the *domain of dependence* of the solution $u(\bar{x})$ on the initial value is represented by the single point $(\bar{y}, 0)$ and the *range of influence* of the initial value at a point $(\bar{y}, 0)$ on solutions consists of the integral curve starting from $(\bar{y}, 0)$.

For $n = 2$, integral curves are exactly characteristic curves. This can be seen easily by (2.2.2) and Definition 2.1.2. Hence the ODE (2.2.2) is often referred to as the *characteristic ODE*. This term is adopted for arbitrary dimensions. We have demonstrated how to solve homogeneous first-order linear PDEs by using characteristic ODEs. Such a method is called the *method of characteristics*. Later on, we will develop a similar method to solve general first-order PDEs.

We need to emphasize that solutions constructed by the method of characteristics are only local. In other words, they exist only in a neighborhood of the origin. A natural question here is whether there exists a global solution for globally defined a and u_0. There are several reasons that local solutions cannot be extended globally. First, $u(\bar{x})$ cannot be evaluated at $\bar{x} \in \mathbb{R}^n$ if $\bar{x}$ is not on an integral curve from the initial hypersurface, or equivalently, the integral curve from $\bar{x}$ does not intersect the initial hypersurface. Second, $u(\bar{x})$ cannot be evaluated at $\bar{x} \in \mathbb{R}^n$ if the integral curve starting from $\bar{x}$ intersects the initial hypersurface more than once. In this case, we cannot prescribe initial values arbitrarily. They must satisfy a compatibility condition.

Example 2.2.3. We discuss the initial-value problem for the equation in Example 2.1.1. We denote by (x, t) points in $\mathbb{R}^2$ and let u_0 be a smooth function in $\mathbb{R}$. We consider

$$\begin{aligned} u_t + u_x &= 0 \quad \text{in } \mathbb{R} \times (0, \infty), \\ u(\cdot, 0) &= u_0 \quad \text{on } \mathbb{R}. \end{aligned}$$

It is easy to verify that $\{t = 0\}$ is noncharacteristic. The characteristic ODE and corresponding initial values are given by

$$\frac{dx}{ds} = 1, \quad \frac{dt}{ds} = 1,$$

and

$$x(0) = x_0, \quad t(0) = 0.$$

Here, both x and t are treated as functions of s. Hence

$$x = s + x_0, \quad t = s.$$

By eliminating s, we have

$$x - t = x_0.$$

This is a straight line containing $(x_0, 0)$ and with a slope 1. Along this straight line, u is constant. Hence

$$u(x, t) = u_0(x - t).$$

We interpreted the fact that u is constant along the straight line $x - t = x_0$ in Example 2.1.1. With t as time, the graph of the solution represents a wave propagating to the right with velocity 1 without changing shape. It is clear that u exists globally in $\mathbb{R}^2$.

2.2.2. Quasilinear Equations. Next, we discuss initial-value problems for first-order quasilinear PDEs. Let $\Omega \subset \mathbb{R}^n$ be a domain containing the origin and a_i and f be smooth functions in $\Omega \times \mathbb{R}$. For a given smooth function u_0 in a neighborhood of $0 \in \mathbb{R}^{n-1}$, we consider

$$\begin{aligned} \sum_{i=1}^{n} a_i(x, u) u_{x_i} &= f(x, u), \\ u(x', 0) &= u_0(x'). \end{aligned} \tag{2.2.5}$$

Assume the hyperplane $\{x_n = 0\}$ is noncharacteristic at the origin with respect to u_0, i.e.,

$$a_n\big(0, u_0(0)\big) \neq 0.$$

Suppose (2.2.5) admits a smooth solution u. We first examine integral curves

$$\begin{aligned} \frac{dx}{ds} &= a\big(x, u\big), \\ x(0) &= (y, 0), \end{aligned}$$

where $y \in \mathbb{R}^{n-1}$. Contrary to the case of homogenous linear equations we studied earlier, we are unable to solve this ODE since u, the unknown function we intend to find, is present. However, viewing u as a function of s along these curves, we can calculate how u changes. A similar calculation as before yields

$$\frac{d}{ds}\big(u(x(s))\big) = \sum_{i=1}^{n} u_{x_i}\frac{dx_i}{ds} = \sum_{i=1}^{n} a_i(x,u)u_{x_i} = f(x,u).$$

Then

$$\begin{aligned}\frac{du}{ds} &= f(x,u),\\ u(0) &= u_0(y).\end{aligned}$$

Hence we have an ordinary differential system for x and u. This leads to the following method for quasilinear PDEs.

Consider the ordinary differential system

$$\begin{aligned}\frac{dx}{ds} &= a(x,u),\\ \frac{du}{ds} &= f(x,u),\end{aligned}$$

with initial values

$$\begin{aligned}x(0) &= (y,0),\\ u(0) &= u_0(y),\end{aligned}$$

where $y \in \mathbb{R}^{n-1}$. In formulating this system, we treat both x and u as functions of s. This system consists of $n+1$ equations for $n+1$ functions and is the *characteristic ODE* of the first-order quasilinear PDE (2.2.5). By solving the characteristic ODE, we have a solution given by

$$x = x(y,s), \quad u = \varphi(y,s).$$

As in the proof of Lemma 2.2.2, we can prove that the map $(y,s) \mapsto x$ is a diffeomorphism. Hence, for any $\bar{x} \in \mathbb{R}^n$ sufficiently small, there exist unique $\bar{y} \in \mathbb{R}^{n-1}$ and $\bar{s} \in \mathbb{R}$ sufficiently small such that

$$\bar{x} = x(\bar{y}, \bar{s}).$$

Then the solution u at $\bar{x}$ is given by

$$u(\bar{x}) = \varphi(\bar{y}, \bar{s}).$$

We now consider an initial-value problem for a nonhomogeneous linear equation.

Example 2.2.4. We denote by (x,t) points in $\mathbb{R}^2$ and let f be a smooth function in $\mathbb{R}\times(0,\infty)$ and u_0 be a smooth function in $\mathbb{R}$. We consider

$$\begin{aligned}u_t - u_x &= f \quad \text{in } \mathbb{R}\times(0,\infty),\\ u(\cdot,0) &= u_0 \quad \text{on } \mathbb{R}.\end{aligned}$$

It is easy to verify that $\{t = 0\}$ is noncharacteristic. The characteristic ODE and corresponding initial values are given by

$$-\frac{dx}{ds} = 1, \quad \frac{dt}{ds} = 1, \quad \frac{du}{ds} = f,$$

and

$$x(0) = x_0, \quad t(0) = 0, \quad u(0) = u_0(x_0).$$

Here, x, t and u are all treated as functions of s. By solving for x and t first, we have

$$x = x_0 - s, \quad t = s.$$

Then the equation for u can be written as

$$\frac{du}{ds} = f(x_0 - s, s).$$

A simple integration yields

$$u = u_0(x_0) + \int_0^s f(x_0 - \tau, \tau)\, d\tau.$$

By substituting x_0 and s by x and t, we obtain

$$u(x,t) = u_0(x+t) + \int_0^t f(x + t - \tau, \tau)\, d\tau.$$

Next, we consider an initial-value problem for a quasilinear equation.

Example 2.2.5. We denote by (x, t) points in $\mathbb{R}^2$ and let u_0 be a smooth function in $\mathbb{R}$. Consider the initial-value problem for *Burgers' equation*

$$\begin{aligned} u_t + uu_x &= 0 \quad \text{in } \mathbb{R} \times (0, \infty), \\ u(\cdot, 0) &= u_0 \quad \text{on } \mathbb{R}. \end{aligned}$$

It is easy to check that $\{t = 0\}$ is noncharacteristic with respect to any u_0. The characteristic ODE and corresponding initial values are given by

$$\frac{dx}{ds} = u, \quad \frac{dt}{ds} = 1, \quad \frac{du}{ds} = 0,$$

and

$$x(0) = x_0, \quad t(0) = 0, \quad u(0) = u_0(x_0).$$

Here, x, t and u are all treated as functions of s. By solving for t and u first and then for x, we obtain

$$\begin{aligned} x &= u_0(x_0)s + x_0, \\ t &= s, \\ u &= u_0(x_0). \end{aligned}$$

By eliminating s from the expressions of x and t, we have

$$x = u_0(x_0)t + x_0. \tag{2.2.6}$$

By the implicit function theorem, we can solve for x_0 in terms of (x,t) in a neighborhood of the origin in $\mathbb{R}^2$. If we denote such a function by

$$x_0 = x_0(x,t),$$

then we have a solution

$$u = u_0\big(x_0(x,t)\big),$$

for any (x,t) sufficiently small. By eliminating x_0 and s from the expressions of x, t and u, we may also write the solution u implicitly by

$$u = u_0(x - ut).$$

It is interesting to ask whether such a solution can be extended to $\mathbb{R}^2$. Let c_{x_0} be the characteristic curve given by (2.2.6). It is a straight line in $\mathbb{R}^2$ with a slope $1/u_0(x_0)$, along which u is the constant $u_0(x_0)$. For $x_0 < x_1$ with $u_0(x_0) > u_0(x_1)$, two characteristic curves c_{x_0} and c_{x_1} intersect at (X,T) with

$$T = -\frac{x_0 - x_1}{u_0(x_0) - u_0(x_1)}.$$

Hence, u cannot be extended as a smooth solution up to (X,T), even as a continuous function. Such a positive T always exists unless u_0 is nondecreasing. In a special case where u_0 is strictly decreasing, any two characteristic curves intersect.

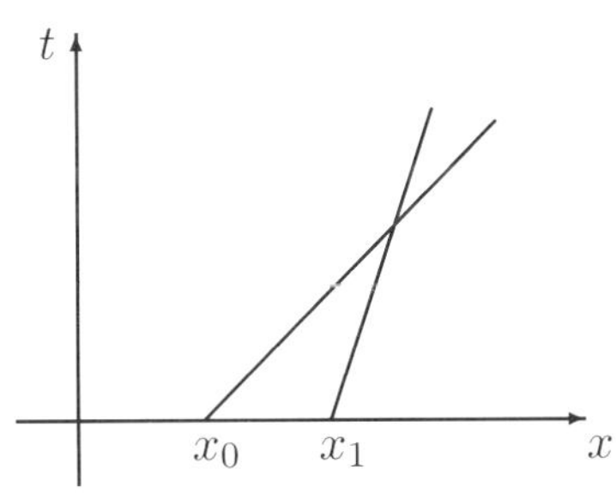

Figure 2.2.2. Intersecting characteristic curves.

Now we examine a simple case. Let

$$u_0(x) = -x.$$

Obviously, this is strictly decreasing. In this case, c_{x_0} in (2.2.6) is given by

$$x = x_0 - x_0 t,$$

and the solution on this line is given by $u = -x_0$. We note that each c_{x_0} contains the point $(x,t) = (0,1)$ and hence any two characteristic curves intersect at $(0,1)$. Then, u cannot be extended up to $(0,1)$ as a smooth solution. In fact, we can solve for x_0 easily to get

$$x_0 = \frac{x}{1-t}.$$

Therefore, u is given by

$$u(x,t) = \frac{x}{t-1} \quad \text{for any } (x,t) \in \mathbb{R} \times (0,1).$$

Clearly, u is not defined at $t=1$.

In general, smooth solutions of first-order nonlinear PDEs may not exist globally. When two characteristic curves intersect at a positive time T, solutions develop a singularity and the method of characteristics breaks down. A natural question arises whether we can define solutions beyond the time T. We expect that less regular functions, if interpreted appropriately, may serve as solutions.

For an illustration, we return to Burgers' equation and employ its *divergence* structure. We note that Burgers' equation can be written as

$$u_t + \left(\frac{u^2}{2}\right)_x = 0.$$

This is an example of a *scalar conservation law*, that is, it is a first-order quasilinear PDE of the form

$$u_t + F(u)_x = 0 \quad \text{in } \mathbb{R} \times (0,\infty), \tag{2.2.7}$$

where $F : \mathbb{R} \to \mathbb{R}$ is a given smooth function. By taking a C^1-function φ of compact support in $\mathbb{R} \times (0,\infty)$ and integrating by parts the product of φ and the equation in (2.2.7), we obtain

$$\int_{\mathbb{R}\times(0,\infty)} \big(u\varphi_t + F(u)\varphi_x\big)\, dxdt = 0. \tag{2.2.8}$$

The integration by parts is justified since φ is zero outside a compact set in $\mathbb{R} \times (0,\infty)$. By comparing (2.2.7) and (2.2.8), we note that derivatives are transferred from u in (2.2.7) to φ in (2.2.8). Hence, functions u with no derivatives are allowed in (2.2.8). A locally bounded function u is called an *integral solution* of (2.2.7) if it satisfies (2.2.8) for any C^1-function φ of compact support in $\mathbb{R} \times (0,\infty)$. The function φ in (2.2.8) is often referred to as a *test function*. In this formulation, discontinuous functions are admitted to be integral solutions. Even for continuous initial values, a discontinuity along a curve, called a *shock*, may develop for integral solutions. Conservation laws and shock waves are an important subject in PDEs. The brief discussion here serves only as an introduction to this field. It is beyond the scope of this book to give a presentation of conservation laws and shock waves.

Now we return to our study of initial-value problems of general first-order PDEs. So far in our discussion, initial values are prescribed on noncharacteristic hypersurfaces. In general, solutions are not expected to exist if initial values are prescribed on characteristic hypersurfaces. We illustrate this by the initial-value problem (2.2.5) for quasilinear equations. Suppose the initial hyperplane $\{x_n = 0\}$ is characteristic at the origin with respect to the initial value u_0. Then

$$a_n\big(0, u_0(0)\big) = 0.$$

Hence $u_{x_n}(0)$ is absent from the equation in (2.2.5) when evaluated at $x = 0$. Therefore, (2.2.5) implies

$$\sum_{i=1}^{n-1} a_i\big(0, u_0(0)\big) u_{0,x_i}(0) = f\big(0, u_0(0)\big). \tag{2.2.9}$$

This is the *compatibility condition* for the initial value u_0. Even if the origin is the only point where $\{x_n = 0\}$ is characteristic, solutions may not exist in any neighborhood of the origin for initial values satisfying the compatibility condition (2.2.9). Refer to Exercise 2.5.

2.2.3. General Nonlinear Equations. Next, we discuss general first-order nonlinear PDEs. Let $\Omega \subset \mathbb{R}^n$ be a domain containing the origin and $F = F(x, u, p)$ be a smooth function in $(x, u, p) \in \Omega \times \mathbb{R} \times \mathbb{R}^n$. Consider

$$F(x, u, \nabla u) = 0, \tag{2.2.10}$$

for any $x \in \Omega$, and prescribe an initial value on $\{x_n = 0\}$ by

$$u(x', 0) = u_0(x'), \tag{2.2.11}$$

for any x' with $(x', 0) \in \Omega$. Assume there is a scalar a_0 such that

$$F\big(0, u_0(0), \nabla_{x'} u_0(0), a_0\big) = 0.$$

The noncharacteristic condition with respect to u_0 and a_0 is given by

$$F_{p_n}\big(0, u_0(0), \nabla_{x'} u_0(0), a_0\big) \neq 0. \tag{2.2.12}$$

By (2.2.12) and the implicit function theorem, there exists a smooth function $a(x')$ in a neighborhood of the origin in $\mathbb{R}^{n-1}$ such that $a(0) = a_0$ and

$$F\big(x', 0, u_0(x'), \nabla_{x'} u_0(x'), a(x')\big) = 0, \tag{2.2.13}$$

for any $x' \in \mathbb{R}^{n-1}$ sufficiently small. In the following, we will seek a solution of (2.2.10)–(2.2.11) and

$$u_{x_n}(x', 0) = a(x'),$$

for any x' small.

We start with a formal consideration. Suppose we have a smooth solution u. Set

$$p_i = u_{x_i} \quad \text{for } i = 1, \cdots, n. \tag{2.2.14}$$

Then

$$F(x_1, \cdots, x_n, u, p_1, \cdots, p_n) = 0. \tag{2.2.15}$$

Differentiating (2.2.15) with respect to x_i, we have

$$\sum_{j=1}^{n} F_{p_j} p_{j,x_i} + F_{x_i} + F_u u_{x_i} = 0 \quad \text{for } i = 1, \cdots, n.$$

By $p_{j,x_i} = p_{i,x_j}$, we obtain

$$\sum_{j=1}^{n} F_{p_j} p_{i,x_j} = -F_{x_i} - F_u u_{x_i} \quad \text{for } i = 1, \cdots, n. \tag{2.2.16}$$

We view (2.2.16) as a first-order quasilinear equation for p_i, for each fixed $i = 1, \cdots, n$. An important feature here is that the coefficient for p_{i,x_j} is F_{p_j}, which is independent of i. For each fixed $i = 1, \cdots, n$, the characteristic ODE associated with (2.2.16) is given by

$$\begin{aligned} \frac{dx_j}{ds} &= F_{p_j} \quad \text{for } j = 1, \cdots, n, \\ \frac{dp_i}{ds} &= -F_u p_i - F_{x_i}. \end{aligned}$$

We also have

$$\frac{du}{ds} = \sum_{j=1}^{n} u_{x_j} \frac{dx_j}{ds} = \sum_{j=1}^{n} p_j F_{p_j}.$$

Now we collect ordinary differential equations for x_j, u and p_i.

The *characteristic ODE* for the first-order nonlinear PDE (2.2.10) is the ordinary differential system

$$\begin{aligned} \frac{dx_j}{ds} &= F_{p_j}(x, u, p) \quad \text{for } j = 1, \cdots, n, \\ \frac{dp_i}{ds} &= -F_u(x, u, p)p_i - F_{x_i}(x, u, p) \quad \text{for } i = 1, \cdots, n, \\ \frac{du}{ds} &= \sum_{j=1}^{n} p_j F_{p_j}(x, u, p), \end{aligned} \tag{2.2.17}$$

with initial values at $s = 0$,

$$
\begin{aligned}
x(0) &= (y, 0), \\
u(0) &= u_0(y), \\
p_i(0) &= u_{0,x_i}(y) \quad \text{for } i = 1, \cdots, n-1, \\
p_n(0) &= a(y),
\end{aligned} \tag{2.2.18}
$$

where $y \in \mathbb{R}^{n-1}$, u_0 is the initial value as in (2.2.11) and a is the function chosen to satisfy (2.2.13). This is an ordinary differential system of $2n+1$ equations for the $2n+1$ functions x, u and p. Here we view x, u and p as functions of s. Compare this with a similar ordinary differential system of $n+1$ equations for $n+1$ functions x and u for first-order quasilinear PDEs. Solving (2.2.17) with (2.2.18) near $(y, s) = (0, 0)$, we have

$$x = x(y, s), \quad u = \varphi(y, s), \quad p = p(y, s),$$

for any y and s sufficiently small. We will prove that the map $(y, s) \mapsto x$ is a diffeomorphism near the origin in $\mathbb{R}^n$. Hence for any given $\bar{x}$ near the origin, there exist unique $\bar{y} \in \mathbb{R}^{n-1}$ and $\bar{s} \in \mathbb{R}$ such that

$$\bar{x} = x(\bar{y}, \bar{s}).$$

Then we define u by

$$u(\bar{x}) = \varphi(\bar{y}, \bar{s}).$$

Theorem 2.2.6. *The function u defined above is a solution of* (2.2.10)–(2.2.11).

We should note that this solution u depends on the choice of the scalar a_0 and the function $a(x')$.

Proof. The proof consists of several steps.

Step 1. The map $(y, s) \mapsto x$ is a diffeomorphism near $(0, 0)$. This is proved as in the proof of Lemma 2.2.2. In fact, the Jacobian matrix of the map $(y, s) \mapsto x$ at $(0, 0)$ is given by

$$J(0) = \left.\frac{\partial x}{\partial(y, s)}\right|_{y=0, s=0} = \begin{pmatrix} & & & * \\ & Id & & \vdots \\ & & & * \\ 0 & \cdots & 0 & \frac{dx_n}{ds}(0,0) \end{pmatrix},$$

where

$$\frac{dx_n}{ds}(0, 0) = F_{p_n}(0, u_0(0), \nabla_{x'} u_0(0), a_0).$$

Hence $\det J(0) \neq 0$ by the noncharacteristic condition (2.2.12). By the implicit function theorem, for any $x \in \mathbb{R}^n$ sufficiently small, we can solve

$x = x(y, s)$ uniquely for $y \in \mathbb{R}^{n-1}$ and $s \in \mathbb{R}$ sufficiently small. Then define

$$u(x) = \varphi(y, s).$$

We will prove that this is the desired solution and

$$p_i(y, s) = u_{x_i}(x(y, s)) \quad \text{for } i = 1, \cdots, n.$$

Step 2. We claim that

$$F\big(x(y, s), \varphi(y, s), p(y, s)\big) \equiv 0,$$

for any y and s sufficiently small. Denote by $f(s)$ the function in the left-hand side. Then by (2.2.18)

$$f(0) = F\big(y, 0, u_0(y), \nabla_{x'} u_0(y), a(y)\big) = 0.$$

Next, we have by (2.2.17)

$$\begin{aligned}
\frac{df(s)}{ds} &= \frac{d}{ds} F\big(x(y, s), \varphi(y, s), p(y, s)\big) \\
&= \sum_{i=1}^{n} F_{x_i} \frac{dx_i}{ds} + F_u \frac{du}{ds} + \sum_{j=1}^{n} F_{p_j} \frac{dp_j}{ds} \\
&= \sum_{i=1}^{n} F_{x_i} F_{p_i} + F_u \sum_{j=1}^{n} p_j F_{p_j} + \sum_{j=1}^{n} F_{p_j}(-F_u p_j - F_{x_j}) = 0.
\end{aligned}$$

Hence $f(s) \equiv 0$.

Step 3. We claim that

$$p_i(y, s) = u_{x_i}(x(y, s)) \quad \textit{for } i = 1, \cdots, n,$$

for any y and s sufficiently small. Let

$$w_i(s) = u_{x_i}(x(y, s)) - p_i(y, s) \quad \text{for } i = 1, \cdots, n.$$

We will prove that

$$w_i(s) = 0 \quad \text{for any } s \text{ small and } i = 1, \cdots, n.$$

We first evaluate w_i at $s = 0$. By initial values (2.2.18), we have $w_i(0) = 0$ for $i = 1, \cdots, n-1$. Next, we note that, by (2.2.17),

$$(2.2.19)\quad 0 = \frac{du}{ds} - \sum_{j=1}^{n} p_j F_{p_j} = \sum_{j=1}^{n} \left(u_{x_j} \frac{dx_j}{ds} - p_j F_{p_j} \right) = \sum_{j=1}^{n} F_{p_j}(u_{x_j} - p_j),$$

or

$$\sum_{j=1}^{n} F_{p_j} w_j = 0.$$

This implies $w_n(0) = 0$ since $w_i(0) = 0$ for $i = 1, \cdots, n-1$, and $F_{p_n}|_{s=0} \neq 0$ by the noncharacteristic condition (2.2.12).

Next, we claim that $\frac{dw_i}{ds}$ is a linear combination of w_j, $j = 1, \cdots, n$, i.e.,

$$\frac{dw_i}{ds} = \sum_{j=1}^{n} a_{ij} w_j \quad \text{for } i = 1, \cdots, n,$$

for some functions a_{ij}, $i, j = 1, \cdots, n$. Then basic ODE theory implies $w_i \equiv 0$ for $i = 1, \cdots, n$. To prove the claim, we first note that, by (2.2.17),

$$\frac{dw_i}{ds} = \sum_{j=1}^{n} u_{x_i x_j} \frac{dx_j}{ds} - \frac{dp_i}{ds} = \sum_{j=1}^{n} u_{x_i x_j} F_{p_j} + F_u p_i + F_{x_i}.$$

To eliminate the second-order derivatives of u, we differentiate (2.2.19) with respect to x_i and get

$$\sum_{j=1}^{n} F_{p_j} \cdot (u_{x_i x_j} - p_{j,x_i}) + \sum_{j=1}^{n} (F_{p_j})_{x_i} w_j = 0.$$

A simple substitution implies

$$\frac{dw_i}{ds} = \sum_{j=1}^{n} F_{p_j} p_{j,x_i} - \sum_{j=1}^{n} (F_{p_j})_{x_i} w_j + F_u p_i + F_{x_i}.$$

By Step 2,

$$F\big(x, u(x), p_1(x), \cdots, p_n(x)\big) = 0.$$

Differentiating with respect to x_i, we have

$$F_{x_i} + F_u u_{x_i} + \sum_{j=1}^{n} F_{p_j} p_{j,x_i} = 0.$$

Hence

$$\begin{aligned} \frac{dw_i}{ds} &= - F_{x_i} - F_u u_{x_i} - \sum_{j=1}^{n} (F_{p_j})_{x_i} w_j + F_u p_i + F_{x_i} \\ &= - F_u w_i - \sum_{j=1}^{n} (F_{p_j})_{x_i} w_j, \end{aligned}$$

or

$$\frac{dw_i}{ds} = - \sum_{j=1}^{n} \big(F_u \delta_{ij} + (F_{p_j})_{x_i}\big) w_j.$$

This ends the proof of Step 3.

Step 2 and Step 3 imply that u is the desired solution. □

To end this section, we briefly compare methods we used to solve first-order linear or quasi-linear PDEs and general first-order nonlinear PDEs. In solving a first-order quasi-linear PDE, we formulate an ordinary differential system of $n+1$ equations for $n+1$ functions x and u. For a general first-order nonlinear PDE, the corresponding ordinary differential system consists of $2n+1$ equations for $2n+1$ functions x, u and ∇u. Here, we need to take into account the gradient of u by adding n more equations for ∇u. In other words, we regard our first-order nonlinear PDE as a relation for (u, p) with a constraint $p = \nabla u$. We should emphasize that this is a unique feature for single first-order PDEs. For PDEs of higher order or for first-order partial differential systems, nonlinear equations are dramatically different from linear equations. In the rest of the book, we concentrate only on linear equations.

2.3. A Priori Estimates

A priori estimates play a fundamental role in PDEs. Usually, they are the starting point for establishing existence and regularity of solutions. To derive a priori estimates, we first assume that solutions already exist and then estimate certain norms of solutions by those of known functions in equations, for example, nonhomogeneous terms, coefficients and initial values. Two frequently used norms are L^∞-norms and L^2-norms. The importance of L^2-norm estimates lies in the Hilbert space structure of the L^2-space. Once L^2-estimates of solutions and their derivatives have been derived, we can employ standard results about Hilbert spaces, for example, the Riesz representation theorem, to establish the existence of solutions.

In this section, we will use first-order linear PDEs to demonstrate how to derive a priori estimates in L^∞-norms and L^2-norms. We first examine briefly first-order linear ordinary differential equations. Let β be a constant and $f = f(t)$ be a continuous function. Consider

$$\frac{du}{dt} - \beta u = f(t).$$

A simple calculation shows that

$$u(t) = e^{\beta t} u(0) + \int_0^t e^{\beta(t-s)} f(s)\, ds.$$

For any fixed $T > 0$, we have

$$|u(t)| \le e^{\beta t} \left(|u(0)| + T \sup_{[0,T]} |f| \right) \quad \text{for any } t \in (0, T).$$

Here, we estimate the sup-norm of u in $[0, T]$ by the initial value $u(0)$ and the sup-norm of the nonhomogeneous term f in $[0, T]$.

Now we turn to PDEs. For convenience, we work in $\mathbb{R}^n \times (0, \infty)$ and denote points by (x, t), with $x \in \mathbb{R}^n$ and $t \in (0, \infty)$. In many applications, we interpret x as the space variable and t the time variable.

2.3.1. L^∞-Estimates. Let a_i, b and f be continuous functions in $\mathbb{R}^n \times [0,\infty)$ and u_0 be a continuous function in $\mathbb{R}^n$. We assume that $a = (a_1, \cdots, a_n)$ satisfies

$$|a| \le \frac{1}{\kappa} \quad \text{in } \mathbb{R}^n \times [0,\infty), \tag{2.3.1}$$

for a positive constant κ. Consider

$$\begin{aligned} u_t + \sum_{i=1}^n a_i(x,t)u_{x_i} + b(x,t)u &= f(x,t) \quad \text{in } \mathbb{R}^n \times (0,\infty), \\ u(x,0) &= u_0(x) \quad \text{in } \mathbb{R}^n. \end{aligned} \tag{2.3.2}$$

It is obvious that the initial hypersurface $\{t = 0\}$ is noncharacteristic. We may write the equation in (2.3.2) as

$$u_t + a(x,t) \cdot \nabla_x u + b(x,t)u = f(x,t).$$

We note that $a(x,t) \cdot \nabla_x + \partial_t$ is a directional derivative along the direction $(a(x,t), 1)$. With (2.3.1), it is easy to see that the vector $(a(x,t),1)$ (starting from the origin) is in fact in the cone given by

$$\{(y,s) : \ \kappa|y| \le s\} \subset \mathbb{R}^n \times \mathbb{R}.$$

This is a cone opening upward and with vertex at the origin.

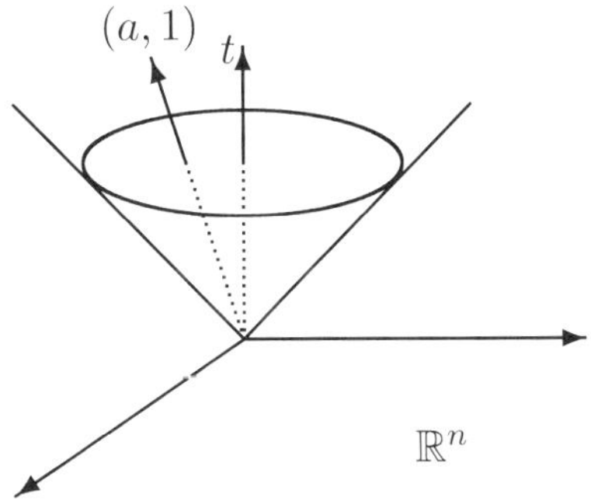

Figure 2.3.1. The cone with the vertex at the origin.

For any point $P = (X,T) \in \mathbb{R}^n \times (0,\infty)$, consider the cone $C_\kappa(P)$ (opening downward) with vertex at P defined by

$$C_\kappa(P) = \{(x,t) : \ 0 < t < T, \ \kappa|x - X| < T - t\}.$$

We denote by $\partial_s C_\kappa(P)$ and $\partial_- C_\kappa(P)$ the side and bottom of the boundary, respectively, i.e.,

$$\begin{aligned} \partial_s C_\kappa(P) &= \{(x,t) : \ 0 < t \le T, \ \kappa|x - X| = T - t\}, \\ \partial_- C_\kappa(P) &= \{(x,0) : \ \kappa|x - X| \le T\}. \end{aligned}$$

We note that $\partial_- C_\kappa(P)$ is simply the closed ball in $\mathbb{R}^n \times \{0\}$ centered at $(X, 0)$ with radius T/κ. For any $(x,t) \in \partial_s C_\kappa(P)$, let $a(x,t)$ be a vector in $\mathbb{R}^n$ satisfying (2.3.1). Then the vector $(a(x,t),1)$, if positioned at (x,t), points outward from the cone $C_\kappa(P)$ or along the boundary $\partial_s C_\kappa(P)$. Hence for a function defined only in $\bar{C}_\kappa(P)$, it makes sense to calculate $u_t + a \cdot \nabla_x u$ at (x,t), which is viewed as a directional derivative of u along $(a(x,t),1)$ at (x,t). This holds in particular when (x,t) is the vertex P.

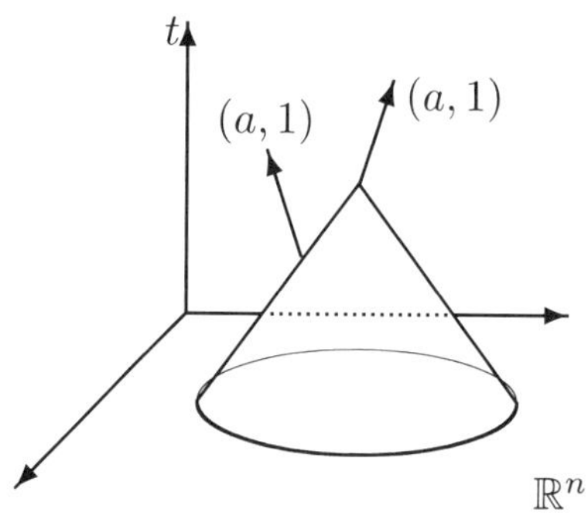

Figure 2.3.2. The cone $C_\kappa(P)$ and positions of vectors.

Now we calculate the unit outward normal vector of $\partial_s C_\kappa(P) \setminus \{P\}$. Set

$$\varphi(x,t) = \kappa|x - X| - (T - t).$$

Obviously, $\partial_s C_\kappa(P) \setminus \{P\}$ is a part of $\{\varphi = 0\}$. Then for any $(x,t) \in \partial_s C_\kappa(P) \setminus \{P\}$,

$$\nabla \varphi = (\nabla_x \varphi, \varphi_t) = \left(\kappa \frac{x - X}{|x - X|}, 1\right).$$

Therefore, the unit outward normal vector ν of $\partial_s C_\kappa(P) \setminus \{P\}$ at (x,t) is given by

$$\nu = \frac{1}{\sqrt{\kappa^2 + 1}} \left(\kappa \frac{x - X}{|x - X|}, 1\right).$$

For $n = 1$, the cone $C_\kappa(P)$ is a triangle bounded by the straight lines $\pm\kappa(x - X) = T - t$ and $t = 0$. The side of the cone consists of two line segments,

$$\text{the left segment:} \quad -\kappa(x - X) = T - t, \ 0 < t \le T, \quad \text{with a normal vector } (-\kappa, 1),$$

and

$$\text{the right segment:} \quad \kappa(x - X) = T - t, \ 0 < t \le T, \quad \text{with a normal vector } (\kappa, 1).$$

It is easy to see that the integral curve associated with (2.3.2) starting from P and going to the initial hypersurface $\mathbb{R}^n \times \{0\}$ stays in $C_\kappa(P)$. In fact, this is true for any point $(x,t) \in C_\kappa(P)$. This suggests that solutions in $C_\kappa(P)$ should depend only on f in $C_\kappa(P)$ and the initial value u_0 on $\partial_- C_\kappa(P)$. The following result, proved by a maximum principle type argument, confirms this.

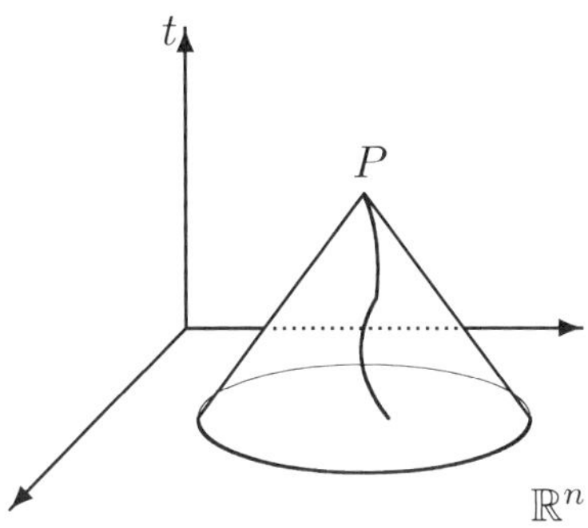

Figure 2.3.3. The domain of dependence.

Theorem 2.3.1. *Let a_i, b and f be continuous functions in $\mathbb{R}^n \times [0,\infty)$ satisfying (2.3.1) and u_0 be a continuous function in $\mathbb{R}^n$. Suppose $u \in C^1(\mathbb{R}^n \times (0,\infty)) \cap C(\mathbb{R}^n \times [0,\infty))$ is a solution of (2.3.2). Then for any $P = (X,T) \in \mathbb{R}^n \times (0,\infty)$,*

$$\sup_{C_\kappa(P)} |e^{-\beta t} u| \leq \sup_{\partial_- C_\kappa(P)} |u_0| + T \sup_{C_\kappa(P)} |e^{-\beta t} f|,$$

where β is a nonnegative constant such that

$$b \geq -\beta \quad \text{in } C_\kappa(P).$$

If $b \geq 0$, we take $\beta = 0$ and have

$$\sup_{C_\kappa(P)} |u| \leq \sup_{\partial_- C_\kappa(P)} |u_0| + T \sup_{C_\kappa(P)} |f|.$$

Proof. Take any positive number $\beta' > \beta$ and set

$$M = \sup_{\partial_- C_\kappa(P)} |u_0|, \quad F = \sup_{C_\kappa(P)} |e^{-\beta' t} f|.$$

We will prove

$$|e^{-\beta' t} u(x,t)| \leq M + tF \quad \text{for any } (x,t) \in C_\kappa(P).$$

For the upper bound, we consider

$$w(x,t) = e^{-\beta' t} u(x,t) - M - tF.$$

A simple calculation shows that

$$w_t + \sum_{i=1}^{n} a_i w_{x_i} + (b+\beta')w = -(b+\beta')(M+tF) + e^{-\beta' t} f - F.$$

Since $b+\beta' > 0$, the right-hand side is nonpositive by the definition of M and F. Hence

$$w_t + a \cdot \nabla_x w + \big(b+\beta'\big)w \le 0 \quad \text{in } C_\kappa(P).$$

Let w attain its maximum in $\overline{C_\kappa(P)}$ at $(x_0, t_0) \in \overline{C_\kappa(P)}$. We prove

$$w(x_0, t_0) \le 0.$$

First, it is obvious if $(x_0, t_0) \in \partial_- C_\kappa(P)$, since $w(x_0, t_0) = u_0(x_0) - M \le 0$ by the definition of M. If $(x_0, t_0) \in C_\kappa(P)$, i.e., (x_0, t_0) is an interior maximum point, then

$$(w_t + a \cdot \nabla_x w)|_{(x_0,t_0)} = 0.$$

If $(x_0, t_0) \in \partial_s C_\kappa(P)$, by the position of the vector $(a(x_0, t_0), 1)$ relative to the cone $C_\kappa(P)$, we can take the directional derivative along $(a(x_0, t_0), 1)$, obtaining

$$(w_t + a \cdot \nabla_x w)|_{(x_0,t_0)} \ge 0.$$

Hence, in both cases, we obtain

$$(b+\beta')w|_{(x_0,t_0)} \le 0.$$

Since $b+\beta' > 0$, this implies $w(x_0, t_0) \le 0$. (We need the positivity of $b+\beta'$ here!) Hence $w(x_0, t_0) \le 0$ in all three cases. Therefore, $w \le 0$ in $C_\kappa(P)$, or

$$u(x,t) \le e^{\beta' t}(M + tF) \quad \text{for any } (x,t) \in C_\kappa(P).$$

We simply let $\beta' \to \beta$ to get the desired upper bound. For the lower bound, we consider

$$v(x,t) = e^{-\beta' t} u(x,t) + M + tF.$$

The argument is similar and hence omitted. □

For $n = 1$, (2.3.2) has the form

$$u_t + a(x,t)u_x + b(x,t)u = f(x,t).$$

In this case, it is straightforward to see that

$$(w_t + a w_x)|_{(x_0,t_0)} \ge 0,$$

if w assumes its maximum at $(x_0, t_0) \in \partial_s C_\kappa(P)$. To prove this, we first note that $\partial_t + \frac{1}{\kappa}\partial_x$ and $\partial_t - \frac{1}{\kappa}\partial_x$ are directional derivatives along the straight lines $t - t_0 = \kappa(x - x_0)$ and $t - t_0 = -\kappa(x - x_0)$, respectively. Since w assumes its maximum at (x_0, t_0), we have

$$\left(w_t + \frac{1}{\kappa} w_x\right)\bigg|_{(x_0,t_0)} \ge 0, \quad \left(w_t - \frac{1}{\kappa} w_x\right)\bigg|_{(x_0,t_0)} \ge 0.$$

In fact, one of them is zero if $(x_0, t_0) \in \partial_s C_\kappa(P) \setminus \{P\}$. Then we obtain

$$w_t(x_0, t_0) \geq \frac{1}{\kappa}|w_x|(x_0, t_0) \geq |aw_x|(x_0, t_0).$$

One consequence of Theorem 2.3.1 is the uniqueness of solutions of (2.3.2).

Corollary 2.3.2. *Let a_i, b and f be continuous functions in $\mathbb{R}^n \times [0, \infty)$ satisfying* (2.3.1) *and u_0 be a continuous function in $\mathbb{R}^n$. Then there exists at most one $C^1(\mathbb{R}^n \times (0, \infty)) \cap C(\mathbb{R}^n \times [0, \infty))$-solution of* (2.3.2).

Proof. Let u_1 and u_2 be two solutions of (2.3.2). Then $u_1 - u_2$ satisfies (2.3.2) with $f = 0$ in $C_\kappa(P)$ and $u_0 = 0$ on $\partial_- C_\kappa(P)$. Hence $u_1 - u_2 = 0$ in $C_\kappa(P)$ by Theorem 2.3.1. □

Another consequence of Theorem 2.3.1 is the continuous dependence of solutions on initial values and nonhomogeneous terms.

Corollary 2.3.3. *Let a_i, b, f_1, f_2 be continuous functions in $\mathbb{R}^n \times [0, \infty)$ satisfying* (2.3.1) *and u_{01}, u_{02} be continuous functions in $\mathbb{R}^n$. Suppose $u_1, u_2 \in C^1(\mathbb{R}^n \times (0, \infty)) \cap C(\mathbb{R}^n \times [0, \infty))$ are solutions of* (2.3.2), *with f_1, f_2 replacing f and u_{01}, u_{02} replacing u_0, respectively. Then for any $P = (X, T) \in \mathbb{R}^n \times (0, \infty)$,*

$$\sup_{C_\kappa(P)} |e^{-\beta t}(u_1 - u_2)| \leq \sup_{\partial_- C_\kappa(P)} |u_{01} - u_{02}| + T \sup_{C_\kappa(P)} |e^{-\beta t}(f_1 - f_2)|,$$

where β is a nonnegative constant such that

$$b \geq -\beta \quad \text{in } C_\kappa(P).$$

The proof is similar to that of Corollary 2.3.2 and is omitted.

Theorem 2.3.1 also shows that the value $u(P)$ depends only on f in $C_\kappa(P)$ and u_0 on $\partial_- C_\kappa(P)$. Hence $C_\kappa(P)$ contains the domain of dependence of $u(P)$ on f, and $\partial_- C_\kappa(P)$ contains the domain of dependence of $u(P)$ on u_0. In fact, the domain of dependence of $u(P)$ on f is the integral curve through P in $C_\kappa(P)$, and the domain of dependence of $u(P)$ on u_0 is the intercept of this integral curve with the initial hyperplane $\{t = 0\}$. We now consider this from another point of view. For simplicity, we assume that f is identically zero in $\mathbb{R}^n \times (0, \infty)$ and the initial value u_0 at $t = 0$ is zero outside a bounded domain $D_0 \subset \mathbb{R}^n$. Then for any $t > 0$, $u(\cdot, t) = 0$ outside

$$D_t = \{(x, t) : \ \kappa|x - x_0| < t \text{ for some } x_0 \in D_0\}.$$

In other words, u_0 influences u only in $\bigcup_{\{t>0\}} D_t$. This is the *finite-speed propagation.*

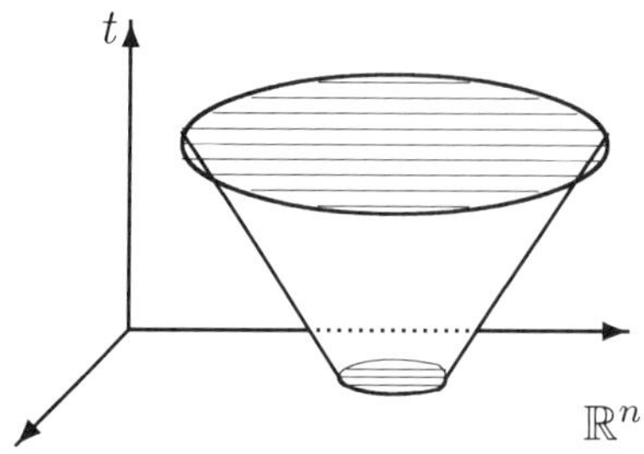

Figure 2.3.4. The range of influence.

2.3.2. L^2-Estimates. Next, we derive an estimate of the L^2-norm of u in terms of the L^2-norms of f and u_0.

Theorem 2.3.4. *Let a_i be C^1-functions in $\mathbb{R}^n \times [0,\infty)$ satisfying (2.3.1), b and f be continuous functions in $\mathbb{R}^n \times [0,\infty)$ and u_0 be a continuous function in $\mathbb{R}^n$. Suppose $u \in C^1(\mathbb{R}^n \times (0,\infty)) \cap C(\mathbb{R}^n \times [0,\infty))$ is a solution of (2.3.2). Then for any $P = (X,T) \in \mathbb{R}^n \times (0,\infty)$,*

$$\int_{C_\kappa(P)} e^{-\alpha t} u^2\, dxdt \le \int_{\partial_- C_\kappa(P)} u_0^2\, dx + \int_{C_\kappa(P)} e^{-\alpha t} f^2\, dxdt,$$

where α is a positive constant depending only on the C^1-norms of a_i and the sup-norm of b in $C_\kappa(P)$.

Proof. For a nonnegative constant α to be determined, we multiply the equation in (2.3.2) by $2e^{-\alpha t}u$. In view of

$$\begin{aligned} 2e^{-\alpha t}uu_t &= (e^{-\alpha t}u^2)_t + \alpha e^{-\alpha t}u^2, \\ 2a_i e^{-\alpha t}uu_{x_i} &= (e^{-\alpha t}a_i u^2)_{x_i} - e^{-\alpha t}a_{i,x_i}u^2, \end{aligned}$$

we have

$$(e^{-\alpha t}u^2)_t + \sum_{i=1}^n (e^{-\alpha t}a_i u^2)_{x_i} + e^{-\alpha t}\left(\alpha + 2b - \sum_{i=1}^n a_{i,x_i}\right)u^2 = 2e^{-\alpha t}uf.$$

An integration in $C_\kappa(P)$ yields

$$\begin{aligned} &\int_{\partial_s C_\kappa(P)} e^{-\alpha t}\left(\nu_t + \sum_{i=1}^n a_i\nu_i\right)u^2\, dS \\ &\qquad + \int_{C_\kappa(P)} e^{-\alpha t}\left(\alpha + 2b - \sum_{i=1}^n a_{i,x_i}\right)u^2\, dxdt \\ &\quad = \int_{\partial_- C_\kappa(P)} u_0^2\, dx + \int_{C_\kappa(P)} 2e^{-\alpha t}uf\, dxdt, \end{aligned}$$

where the unit exterior normal vector on $\partial_s C_\kappa(P)$ is given by

$$(\nu_x, \nu_t) = (\nu_1, \cdots, \nu_n, \nu_t) = \frac{1}{\sqrt{1+\kappa^2}}\left(\kappa\frac{x-X}{|x-X|}, 1\right).$$

By (2.3.1) and the Cauchy inequality, we have

$$\left|\sum_{i=1}^n a_i\nu_i\right| \leq |a||\nu_x| = \frac{\kappa|a|}{\sqrt{1+\kappa^2}} \leq \nu_t,$$

and hence

$$\nu_t + \sum_{i=1}^n a_i\nu_i \geq 0 \quad \text{on } \partial_s C_\kappa(P).$$

Next, we choose α such that

$$\alpha + 2b - \sum_{i=1}^n a_{i,x_i} \geq 2 \quad \text{in } C_\kappa(P).$$

Then

$$2\int_{C_\kappa(P)} e^{-\alpha t}u^2\,dxdt \leq \int_{\partial_- C_\kappa(P)} u_0^2\,dx + \int_{C_\kappa(P)} 2e^{-\alpha t}uf\,dxdt.$$

Here we simply dropped the integral over $\partial_s C_\kappa(P)$ since it is nonnegative. The Cauchy inequality implies

$$\int_{C_\kappa(P)} 2e^{-\alpha t}uf\,dxdt \leq \int_{C_\kappa(P)} e^{-\alpha t}u^2\,dxdt + \int_{C_\kappa(P)} e^{-\alpha t}f^2\,dxdt.$$

We then have the desired result. □

The proof illustrates a typical method of deriving L^2-estimates. We multiply the equation by its solution u and rewrite the product as a linear combination of u^2 and its derivatives. Upon integrating by parts, domain integrals of derivatives are reduced to boundary integrals. Hence, the resulting integral identity consists of domain integrals and boundary integrals of u^2 itself. Derivatives of u are eliminated.

We note that the estimate in Theorem 2.3.4 is similar to that in Theorem 2.3.1, with the L^2-norms replacing the L^∞-norms. As consequences of Theorem 2.3.4, we have the uniqueness of solutions of (2.3.2) and the continuous dependence of solutions on initial values and nonhomogeneous terms in L^2-norms. We can also discuss domains of dependence and ranges of influence using Theorem 2.3.4.

We now derive an L^2-estimate of solutions in the entire space.

Theorem 2.3.5. *Let a_i be bounded C^1-functions, b and f be continuous functions in $\mathbb{R}^n \times [0,\infty)$ and u_0 be a continuous function in $\mathbb{R}^n$. Suppose*

$u \in C^1(\mathbb{R}^n \times (0,\infty)) \cap C(\mathbb{R}^n \times [0,\infty))$ is a solution of (2.3.2). For any $T > 0$, if $f \in L^2(\mathbb{R}^n \times (0,T))$ and $u_0 \in L^2(\mathbb{R}^n)$, then

$$\int_{\mathbb{R}^n \times \{T\}} e^{-\alpha t} u^2\, dx + \int_{\mathbb{R}^n \times (0,T)} e^{-\alpha t} u^2\, dxdt \\ \leq \int_{\mathbb{R}^n} u_0^2\, dx + \int_{\mathbb{R}^n \times (0,T)} e^{-\alpha t} f^2\, dxdt,$$

where α is a positive constant depending only on the C^1-norms of a_i and the sup-norm of b in $\mathbb{R}^n \times (0,T)$.

Proof. We first take $\kappa > 0$ such that (2.3.1) holds. Take any $\bar{t} > T$ and

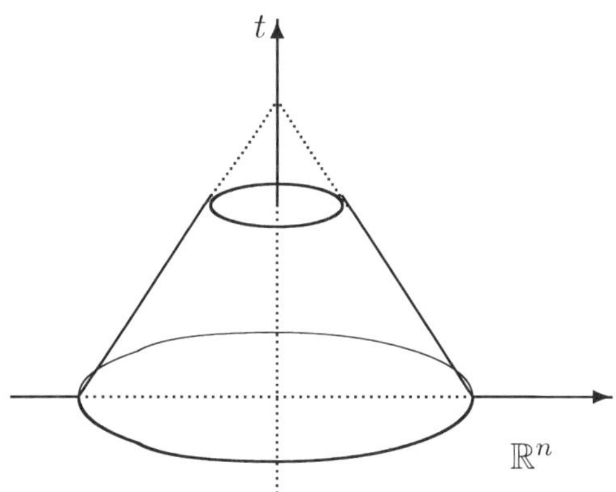

Figure 2.3.5. A domain of integration.

consider

$$D(\bar{t}) = \{(x,t) : \; \kappa|x| < \bar{t} - t, \; 0 < t < T\}.$$

In other words,

$$D(\bar{t}) = C_\kappa(0,\bar{t}) \cap \{0 < t < T\}.$$

We denote by $\partial_- D(\bar{t})$, $\partial_s D(\bar{t})$ and $\partial_+ D(\bar{t})$ the bottom, the side and the top of the boundary, i.e.,

$$\begin{aligned} \partial_- D(\bar{t}) &= \{(x,0) : \; \kappa|x| < \bar{t}\}, \\ \partial_s D(\bar{t}) &= \{(x,t) : \; \kappa|x| = \bar{t} - t, \; 0 < t < T\}, \\ \partial_+ D(\bar{t}) &= \{(x,T) : \; \kappa|x| < \bar{t} - T\}. \end{aligned}$$

We now proceed as in the proof of Theorem 2.3.4, with $D(\bar{t})$ replacing $C_\kappa(P)$. We note that there is an extra portion $\partial_+ D(\bar{t})$ in the boundary $\partial D(\bar{t})$. A similar integration over $D(\bar{t})$ yields

$$\int_{\partial_+ D(\bar{t})} e^{-\alpha t} u^2 \, dx + \int_{D(\bar{t})} e^{-\alpha t} u^2 \, dxdt \\ \leq \int_{\partial_- D(\bar{t})} u_0^2 \, dx + \int_{D(\bar{t})} e^{-\alpha t} f^2 \, dxdt.$$

We note that $\bar{t}$ enters this estimate only through the domain $D(\bar{t})$. Hence, we may let $\bar{t} \to \infty$ to get the desired result. □

We point out that there are no decay assumptions on u as $x \to \infty$ in Theorem 2.3.5.

2.3.3. Weak Solutions. Anyone beginning to study PDEs might well ask, what a priori estimates are good for. One consequence is of course the uniqueness of solutions, as shown in Corollary 2.3.2. In fact, one of the most important applications of Theorem 2.3.5 is to prove the existence of a *weak solution* of the initial-value problem (2.3.2). We illustrate this with the homogeneous initial value, i.e., $u_0 = 0$.

To introduce the notion of a weak solution, we fix a $T > 0$ and consider functions in $\mathbb{R}^n \times (0, T)$. Set

$$Lu = u_t + \sum_{i=1}^n a_i u_{x_i} + bu \quad \text{in } \mathbb{R}^n \times (0, T). \tag{2.3.3}$$

Obviously, L is a linear differential operator defined in $C^1(\mathbb{R}^n \times (0, T))$. For any $u, v \in C^1(\mathbb{R}^n \times (0, T)) \cap C(\mathbb{R}^n \times [0, T])$, we integrate vLu in $\mathbb{R}^n \times (0, T)$. To this end, we write

$$vLu = u\left(-v_t - \sum_{i=1}^n (a_i v)_{x_i} + bv\right) + (uv)_t + \sum_{i=1}^n (a_i uv)_{x_i}.$$

This identity naturally leads to an introduction of the *adjoint differential operator* L^* of L defined by

$$L^* v = -v_t - \sum_{i=1}^n (a_i v)_{x_i} + bv = -v_t - \sum_{i=1}^n a_i v_{x_i} + \left(b - \sum_{i=1}^n a_{i,x_i}\right) v.$$

Then

$$vLu = uL^* v + (uv)_t + \sum_{i=1}^n (a_i uv)_{x_i}.$$

We now require that u and v vanish for large x. Then by a simple integration in $\mathbb{R}^n \times (0,T)$, we obtain

$$\int_{\mathbb{R}^n\times(0,T)} vLu\,dxdt = \int_{\mathbb{R}^n\times(0,T)} uL^*v\,dxdt + \int_{\mathbb{R}^n\times\{t=T\}} uv\,dx - \int_{\mathbb{R}^n\times\{t=0\}} uv\,dx.$$

We note that derivatives are transferred from u in the left-hand side to v in the right-hand side. Integrals over $\{t=0\}$ and $\{t=T\}$ will disappear if we require, in addition, that $uv=0$ on $\{t=0\}$ and $\{t=T\}$.

Definition 2.3.6. Let f and u be functions in $L^2(\mathbb{R}^n \times (0,T))$. Then u is a *weak solution* of $Lu = f$ in $\mathbb{R}^n \times (0,T)$ if

$$\int_{\mathbb{R}^n\times(0,T)} uL^*v\,dxdt = \int_{\mathbb{R}^n\times(0,T)} fv\,dxdt, \tag{2.3.4}$$

for any C^1-functions v of compact support in $\mathbb{R}^n \times (0,T)$.

The functions v in (2.3.4) are called *test functions*. It is worth restating that no derivatives of u are involved.

Now we are ready to prove the existence of weak solutions of (2.3.3) with homogeneous initial values. The proof requires the Hahn-Banach theorem and the Riesz representation theorem in functional analysis.

Theorem 2.3.7. *Let a_i be bounded C^1-functions in $\mathbb{R}^n\times(0,T)$, $i=1,\cdots,n$, and b a bounded continuous function in $\mathbb{R}^n \times (0,T)$. Then for any $f \in L^2(\mathbb{R}^n \times (0,T))$, there exists a $u \in L^2(\mathbb{R}^n \times (0,T))$ such that*

$$\int_{\mathbb{R}^n\times(0,T)} uL^*v\,dxdt = \int_{\mathbb{R}^n\times(0,T)} fv\,dxdt,$$

for any $v \in C^1(\mathbb{R}^n \times (0,T)) \cap C(\mathbb{R}^n \times [0,T])$ with $v(x,t)=0$ for any (x,t) with large x and any $(x,t)=(x,T)$.

The function u in Theorem 2.3.7 is called a *weak solution* of the initial-value problem

$$\begin{aligned} Lu &= f \quad \text{in } \mathbb{R}^n \times (0,T), \\ u &= 0 \quad \text{on } \mathbb{R}^n. \end{aligned} \tag{2.3.5}$$

We note that test functions v in Theorem 2.3.7 are not required to vanish on $\{t=0\}$.

To prove Theorem 2.3.7, we first introduce some notation. We denote by $C_0^1(\mathbb{R}^n \times (0,T))$ the collection of C^1-functions in $\mathbb{R}^n \times (0,T)$ with compact support, and we denote by $\widetilde{C}_0^1(\mathbb{R}^n \times (0,T))$ the collection of C^1-functions in $\mathbb{R}^n \times (0,T)$ with compact support in x-directions. In other words, functions in $C_0^1(\mathbb{R}^n\times(0,T))$ vanish for large x and for t close to 0 and T, and functions in $\widetilde{C}_0^1(\mathbb{R}^n\times(0,T))$ vanish only for large x.

We note that, with L in (2.3.3), we can rewrite the estimate in Theorem 2.3.5 as

$$\|u\|_{L^2(\mathbb{R}^n\times(0,T))}\le C\big(\|u(\cdot,0)\|_{L^2(\mathbb{R}^n)}+\|Lu\|_{L^2(\mathbb{R}^n\times(0,T))}\big),$$

where C is a positive constant depending only on T, the C^1-norms of a_i and the sup-norm of b in $\mathbb{R}^n\times(0,T)$. This holds for any $u\in C^1(\mathbb{R}^n\times(0,T))\cap C(\mathbb{R}^n\times[0,T))$ with $Lu\in L^2(\mathbb{R}^n\times(0,T))$ and $u(\cdot,0)\in L^2(\mathbb{R}^n)$. In particular, we have

$$\|u\|_{L^2(\mathbb{R}^n\times(0,T))}\le C\|Lu\|_{L^2(\mathbb{R}^n\times(0,T))}, \tag{2.3.6}$$

for any $u\in\widetilde{C}_0^1(\mathbb{R}^n\times(0,T))\cap C(\mathbb{R}^n\times[0,T))$ with $u=0$ on $\{t=0\}$.

Proof of Theorem 2.3.7. In the following, we denote by $(\cdot,\cdot)_{L^2(\mathbb{R}^n\times(0,T))}$ the L^2-inner product in $\mathbb{R}^n\times(0,T)$.

Now L^* is like L, but the terms involving derivatives have opposite signs. When we consider an initial-value problem for L^* in $\mathbb{R}^n\times(0,T)$, we view $\{t=T\}$ as the initial hyperplane for the domain $\mathbb{R}^n\times(0,T)$. Thus (2.3.6) also holds for L^*, and we obtain

$$\|v\|_{L^2(\mathbb{R}^n\times(0,T))}\le C\|L^*v\|_{L^2(\mathbb{R}^n\times(0,T))}, \tag{2.3.7}$$

for any $v\in\widetilde{C}_0^1(\mathbb{R}^n\times(0,T))\cap C(\mathbb{R}^n\times(0,T])$ with $v=0$ on $\{t=T\}$, where C is a positive constant depending only on T, the C^1-norms of a_i and the sup-norm of b in $\mathbb{R}^n\times(0,T)$. We denote by $\widehat{C}^1(\mathbb{R}^n\times(0,T))$ the collection of functions $v\in\widetilde{C}_0^1(\mathbb{R}^n\times(0,T))\cap C(\mathbb{R}^n\times(0,T])$ with $v=0$ on $\{t=T\}$.

Consider the linear functional $F:L^*\widehat{C}^1(\mathbb{R}^n\times(0,T))\to\mathbb{R}$ given by

$$F(L^*v)=(f,v)_{L^2(\mathbb{R}^n\times(0,T))},$$

for any $v\in\widehat{C}^1(\mathbb{R}^n\times(0,T))$. We note that F acting on L^*v in the left-hand side is defined in terms of v itself in the right-hand side. Hence we need to verify that such a definition makes sense. In other words, we need to prove that $L^*v_1=L^*v_2$ implies

$$(f,v_1)_{L^2(\mathbb{R}^n\times(0,T))}=(f,v_2)_{L^2(\mathbb{R}^n\times(0,T))},$$

for any $v_1,v_2\in\widehat{C}^1(\mathbb{R}^n\times(0,T))$. By linearity, it suffices to prove that $L^*v=0$ implies $v=0$ for any $v\in\widehat{C}^1(\mathbb{R}^n\times(0,T))$. We note that it is a consequence of (2.3.7). Hence, F is a well-defined linear functional on $L^*\widehat{C}^1(\mathbb{R}^n\times(0,T))$. Moreover, by the Cauchy inequality and (2.3.7) again, we have

$$\begin{aligned}|F(L^*v)|&\le\|f\|_{L^2(\mathbb{R}^n\times(0,T))}\|v\|_{L^2(\mathbb{R}^n\times(0,T))}\\&\le C\|f\|_{L^2(\mathbb{R}^n\times(0,T))}\|L^*v\|_{L^2(\mathbb{R}^n\times(0,T))},\end{aligned}$$

for any $v \in \widehat{C}^1(\mathbb{R}^n \times (0,T))$. Therefore, F is a well-defined bounded linear functional on the subspace $L^*\widehat{C}^1(\mathbb{R}^n \times (0,T))$ of $L^2(\mathbb{R}^n \times (0,T))$. Thus we apply the Hahn-Banach theorem to obtain a bounded linear extension of F (also denoted by F) defined on $L^2(\mathbb{R}^n \times (0,T))$ such that

$$\|F\| \leq C\|f\|_{L^2(\mathbb{R}^n \times (0,T))}.$$

Here, $\|F\|$ is the norm of the linear functional F on $L^2(\mathbb{R}^n \times (0,T))$. By the Riesz representation theorem, there exists a $u \in L^2(\mathbb{R}^n \times (0,T))$ such that

$$F(w) = (u, w)_{L^2(\mathbb{R}^n \times (0,T))} \quad \text{for any } w \in L^2(\mathbb{R}^n \times (0,T)),$$

and

$$\|u\|_{L^2(\mathbb{R}^n \times (0,T))} = \|F\|.$$

In particular, we have

$$F(L^*v) = (u, L^*v)_{L^2(\mathbb{R}^n \times (0,T))},$$

for any $v \in \widehat{C}^1(\mathbb{R}^n \times (0,T))$, and hence by the definition of F,

$$(u, L^*v)_{L^2(\mathbb{R}^n \times (0,T))} = (f, v)_{L^2(\mathbb{R}^n \times (0,T))}.$$

Then u is the desired function. □

Theorem 2.3.7 asserts the existence of a weak solution of (2.3.5). Now we show that the weak solution u is a classical solution if u is C^1 in $\mathbb{R}^n \times (0,T)$ and continuous up to $\{t = 0\}$. Under these extra assumptions on u, we integrate uL^*v by parts to get

$$\int_{\mathbb{R}^n \times (0,T)} vLu \, dxdt + \int_{\mathbb{R}^n \times \{t=0\}} uv \, dx = \int_{\mathbb{R}^n \times (0,T)} fv \, dxdt,$$

for any $v \in \widetilde{C}_0^1(\mathbb{R}^n \times (0,T)) \cap C(\mathbb{R}^n \times [0,T])$ with $v = 0$ on $\{t = T\}$. There are no boundary integrals on the vertical sides and on the upper side since v vanishes there. In particular,

$$\int_{\mathbb{R}^n \times (0,T)} vLu \, dxdt = \int_{\mathbb{R}^n \times (0,T)} fv \, dxdt,$$

for any $v \in C_0^1(\mathbb{R}^n \times (0,T))$. Since $C_0^1(\mathbb{R}^n \times (0,T))$ is dense in $L^2(\mathbb{R}^n \times (0,T))$, we conclude that

$$Lu = f \quad \text{in } \mathbb{R}^n \times (0,T).$$

Therefore,

$$\int_{\mathbb{R}^n \times \{t=0\}} uv \, dx = 0,$$

for any $v \in \widetilde{C}_0^1(\mathbb{R}^n \times (0,T)) \cap C(\mathbb{R}^n \times [0,T])$ with $v = 0$ on $\{t = T\}$. This implies

$$\int_{\mathbb{R}^n} u(\cdot, 0)\varphi \, dx = 0 \quad \text{for any } \varphi \in C_0^1(\mathbb{R}^n).$$

Again by the density of $C_0^1(\mathbb{R}^n)$ in $L^2(\mathbb{R}^n)$, we conclude that

$$u(\cdot, 0) = 0 \quad \text{on } \mathbb{R}^n.$$

We note that a crucial step in passing from weak solutions to classical solutions is to improve the regularity of weak solutions.

Now we summarize the process of establishing solutions by using a priori estimates in the following four steps:

Step 1. Prove a priori estimates.

Step 2. Prove the existence of a weak solution by methods of functional analysis.

Step 3. Improve the regularity of a weak solution.

Step 4. Prove that a weak solution with sufficient regularity is a classical solution.

In discussions above, we carried out Steps 1, 2 and 4. Now we make several remarks on Steps 3 and 4. We recall that in Step 4 we proved that weak solutions with continuous derivatives are classical solutions. The requirement of continuity of derivatives can be weakened. It suffices to assume that u has derivatives in the L^2-sense and to verify that the integration by parts can be performed. Then we can conclude that $Lu = f$ almost everywhere. Because of this relaxed regularity requirement, we need only prove that weak solutions possess derivatives in the L^2-sense in Step 3. The proof is closely related to a priori estimates of derivatives of solutions. The brief discussion here suggests the necessity of introducing new spaces of functions, functions with derivatives in L^2. These are the *Sobolev spaces*, which play a fundamental role in PDEs. In subsequent chapters, Sobolve spaces will come up for different classes of equations. We should point out that Sobolev spaces and weak solutions are not among the main topics in this book. Their appearance in this book serves only as an illustration of their importance.

2.4. Exercises

Exercise 2.1. Find solutions of the following initial-value problems in $\mathbb{R}^2$:

(1) $2u_y - u_x + xu = 0$ with $u(x, 0) = 2xe^{x^2/2}$;

(2) $u_y + (1 + x^2)u_x - u = 0$ with $u(x, 0) = \arctan x$.

Exercise 2.2. Solve the following initial-value problems:

(1) $u_y + u_x = u^2$ with $u(x, 0) = h(x)$;

(2) $u_z + xu_x + yu_y = u$ with $u(x, y, 0) = h(x, y)$.

Exercise 2.3. Let B_1 be the unit disc in $\mathbb{R}^2$ and a and b be continuous functions in $\bar{B}_1$ with $a(x,y)x + b(x,y)y > 0$ on ∂B_1. Assume u is a C^1-solution of

$$a(x,y)u_x + b(x,y)u_y = -u \quad \text{in } \bar{B}_1.$$

Prove that u vanishes identically.

Exercise 2.4. Find a smooth function $a = a(x,y)$ in $\mathbb{R}^2$ such that, for the equation of the form

$$u_y + a(x,y)u_x = 0,$$

there does not exist any solution in the entire $\mathbb{R}^2$ for any nonconstant initial value prescribed on $\{y = 0\}$.

Exercise 2.5. Let α be a number and $h = h(x)$ be a continuous function in $\mathbb{R}$. Consider

$$\begin{aligned} yu_x + xu_y &= \alpha u, \\ u(x,0) &= h(x). \end{aligned}$$

(1) Find all points on $\{y = 0\}$ where $\{y = 0\}$ is characteristic. What is the compatibility condition on h at these points?

(2) Away from the points in (1), find the solution of the initial-value problem. What is the domain of this solution in general?

(3) For the cases $h(x) = x, \alpha = 1$ and $h(x) = x, \alpha = 3$, check whether this solution can be extended over the points in (1).

(4) For each point in (1), find all characteristic curves containing it. What is the relation of these curves and the domain in (2)?

Exercise 2.6. Let $\alpha \in \mathbb{R}$ be a real number and $h = h(x)$ be continuous in $\mathbb{R}$ and C^1 in $\mathbb{R} \setminus \{0\}$. Consider

$$\begin{aligned} xu_x + yu_y &= \alpha u, \\ u(x,0) &= h(x). \end{aligned}$$

(1) Check that the straight line $\{y = 0\}$ is characteristic at each point.

(2) Find all h satisfying the compatibility condition on $\{y = 0\}$. (Consider three cases, $\alpha > 0$, $\alpha = 0$ and $\alpha < 0$.)

(3) For $\alpha > 0$, find two solutions with the given initial value on $\{y = 0\}$. (This is easy to do simply by inspecting the equation, especially for $\alpha = 2$.)

Exercise 2.7. In the plane, solve $u_y = 4u_x^2$ near the origin with $u(x,0) = x^2$ on $\{y = 0\}$.

Exercise 2.8. In the plane, find two solutions of the initial-value problem

$$xu_x + yu_y + \frac{1}{2}(u_x^2 + u_y^2) = u,$$
$$u(x,0) = \frac{1}{2}(1 - x^2).$$

Exercise 2.9. In the plane, find two solutions of the initial-value problem

$$\frac{1}{4}u_x^2 + uu_y = u,$$
$$u\left(x, \frac{1}{2}x^2\right) = -\frac{1}{2}x^2.$$

Exercise 2.10. Let a_i, b and f be continuous functions satisfying (2.3.1) and u be a C^1-solution of (2.3.2) in $\mathbb{R}^n \times [0, \infty)$. Prove that, for any $P = (X, T) \in \mathbb{R}^n \times (0, \infty)$,

$$\sup_{C_\kappa(P)} |e^{-\alpha t} u| \le \sup_{\partial_- C_\kappa(P)} |u_0| + \frac{1}{\alpha + \inf_{C_\kappa(P)} b} \sup_{C_\kappa(P)} |e^{-\alpha t} f|,$$

where α is a constant such that

$$\alpha + \inf_{C_\kappa(P)} b > 0.$$

Exercise 2.11. Let a_i, b and f be C^1-functions in $\mathbb{R}^n \times [0, \infty)$ satisfying (2.3.1) and u_0 be a C^1-function in $\mathbb{R}^n$. Suppose u is a C^2-solution of (2.3.2) in $\mathbb{R}^n \times [0, \infty)$. Prove that, for any $P = (X, T) \in \mathbb{R}^n \times (0, \infty)$,

$$|u|_{C^1(C_\kappa(P))} \le C\big(|u_0|_{C^1(\partial_- C_\kappa(P))} + |f|_{C^1(C_\kappa(P))}\big),$$

where C is a positive constant depending only on T and the C^1-norms of a_i and b in $C_\kappa(P)$.

Exercise 2.12. Let a be a C^1-function in $\mathbb{R} \times [0, \infty)$ satisfying

$$|a(x,t)| \le \frac{1}{\kappa},$$

and let b_{ij} be continuous in $\mathbb{R} \times [0, \infty)$, for $i, j = 1, 2$. Suppose (u, v) is a C^1-solution in $\mathbb{R} \times (0, \infty)$ of the first-order differential system

$$u_t - a(x,t)v_x + b_{11}(x,t)u + b_{12}(x,t)v = f_1(x,t),$$
$$v_t - a(x,t)u_x + b_{21}(x,t)u + b_{22}(x,t)v = f_2(x,t),$$

with

$$u(x,0) = u_0(x), \quad v(x,0) = v_0(x).$$

Derive an L^2-estimate of (u, v) in appropriate cones.

Chapter 3

An Overview of Second-Order PDEs

This chapter should be considered as an introduction to second-order linear PDEs.

In Section 3.1, we introduce the notion of noncharacteristics for initial-value problems. We proceed here for second-order linear PDEs as we did for first-order linear PDEs in Section 2.1. We show that we can compute all derivatives of solutions on initial hypersurfaces if initial values are prescribed on noncharacteristic initial hypersurfaces. We also introduce the Laplace equation, the heat equation and the wave equation, as well as their general forms, elliptic equations, parabolic equations and hyperbolic equations.

In Section 3.2, we discuss boundary-value problems for the Laplace equation and initial/boundary-value problems for the heat equation and the wave equation. Our main tool is a priori estimates. For homogeneous boundary values, we derive estimates of L^2-norms of solutions in terms of those of nonhomogeneous terms and initial values. These estimates yield uniqueness of solutions and continuous dependence of solutions on nonhomogeneous terms and initial values.

In Section 3.3, we use separation of variables to solve Dirichlet problems for the Laplace equation in the unit disc in $\mathbb{R}^2$ and initial/boundary-value problems for the 1-dimensional heat equation and the 1-dimensional wave equation. We derive explicit expressions of solutions in Fourier series and discuss the regularity of these solutions. Our main focus in this section is to demonstrate different regularity patterns for solutions. Indeed, a solution of the heat equation is smooth for all $t > 0$ regardless of the regularity of its initial values, while the regularity of a solution of the wave equation is similar to the regularity of its initial values. Such a difference in regularity suggests that different methods are needed to study these two equations.

3.1. Classifications

The main focus in this section is the second-order linear PDEs. We proceed as in Section 2.1.

Let Ω be a domain in $\mathbb{R}^n$ containing the origin and a_{ij}, b_i and c be continuous functions in Ω, for $i, j = 1, \cdots, n$. Consider a second-order linear differential operator L defined by

$$Lu = \sum_{i,j=1}^{n} a_{ij}(x)u_{x_i x_j} + \sum_{i=1}^{n} b_i(x)u_{x_i} + c(x)u \quad \text{in } \Omega. \tag{3.1.1}$$

Here a_{ij}, b_i, c are called *coefficients* of $u_{x_i x_j}$, u_{x_i}, u, respectively. We usually assume $a_{ij} = a_{ji}$, for any $i, j = 1, \cdots, n$. Hence, (a_{ij}) is a symmetric matrix in Ω. For the operator L, we define its *principal symbol* by

$$p(x; \xi) = \sum_{i,j=1}^{n} a_{ij}(x)\xi_i \xi_j,$$

for any $x \in \Omega$ and $\xi \in \mathbb{R}^n$.

Let f be a continuous function in Ω. We consider the equation

$$Lu = f(x) \quad \text{in } \Omega. \tag{3.1.2}$$

The function f is called the *nonhomogeneous term* of the equation. Let Σ be the hyperplane $\{x_n = 0\}$. We now prescribe values of u and its normal derivative on Σ so that we can at least find all derivatives of u at the origin. Let u_0, u_1 be functions defined in a neighborhood of the origin in $\mathbb{R}^{n-1}$. Now we prescribe

$$u(x', 0) = u_0(x'), \ u_{x_n}(x', 0) = u_1(x'), \tag{3.1.3}$$

for any $x' \in \mathbb{R}^{n-1}$ small. We call Σ the *initial hypersurface* and u_0, u_1 the *initial values* or *Cauchy values*. The problem of solving (3.1.2) together with (3.1.3) is called the *initial-value problem* or *the Cauchy problem.*

Let u be a C^2-solution of (3.1.2) and (3.1.3) in a neighborhood of the origin. In the following, we will investigate whether we can compute all derivatives of u at the origin in terms of the equation and initial values. It is obvious that we can find all x'-derivatives of u and u_{x_n} at the origin in terms of those of u_0 and u_1. In particular, we can find all first derivatives and all second derivatives, except $u_{x_n x_n}$, at the origin in terms of appropriate derivatives of u_0 and u_1. In fact,

$$\begin{aligned} u_{x_i}(0) &= u_{0,x_i}(0) \quad \text{for } i = 1, \cdots, n-1, \\ u_{x_n}(0) &= u_1(0), \end{aligned}$$

and

$$\begin{aligned} u_{x_i x_j}(0) &= u_{0,x_i x_j}(0) \quad \text{for } i, j = 1, \cdots, n-1, \\ u_{x_i x_n}(0) &= u_{1,x_i}(0) \quad \text{for } i = 1, \cdots, n-1. \end{aligned}$$

To compute $u_{x_n x_n}(0)$, we need to use the equation. We note that a_{nn} is the coefficient of $u_{x_n x_n}$ in (3.1.2). If we assume

$$a_{nn}(0) \neq 0, \tag{3.1.4}$$

then by (3.1.2)

$$\begin{aligned} u_{x_n x_n}(0) = -\frac{1}{a_{nn}(0)} \Bigg(&\sum_{(i,j)\neq(n,n)} a_{ij}(0) u_{x_i x_j}(0) \\ &+ \sum_{i=1}^{n} b_i(0) u_{x_i}(0) + c(0)u(0) - f(0) \Bigg). \end{aligned}$$

Hence, we can compute all first-order and second-order derivatives at 0 in terms of the coefficients and nonhomogeneous term in (3.1.2) and the initial values u_0 and u_1 in (3.1.3). In fact, if all functions involved are smooth, we can compute all derivatives of u of any order at the origin by using u_0, u_1 and their derivatives and differentiating (3.1.2). In summary, we can find all derivatives of u of any order at the origin under the condition (3.1.4), which will be defined as the noncharacteristic condition later on.

Comparing the initial-value problem (3.1.2) and (3.1.3) here with the initial-value problem (2.1.3) and (2.1.4) for first-order PDEs, we note that there is an extra condition in (3.1.3). This reflects the general fact that two conditions are needed for initial-value problems for second-order PDEs.

More generally, consider the hypersurface Σ given by $\{\varphi = 0\}$ for a smooth function φ in a neighborhood of the origin with $\nabla\varphi \neq 0$. We note that the vector field $\nabla\varphi$ is normal to the hypersurface Σ at each point of Σ. We take a point on Σ, say the origin. Then $\varphi(0) = 0$. Without loss of generality, we assume $\varphi_{x_n}(0) \neq 0$. Then by the implicit function theorem, we can solve $\varphi = 0$ for $x_n = \psi(x_1, \cdots, x_{n-1})$ in a neighborhood of the origin. Consider the change of variables

$$x \mapsto y = (x_1, \cdots, x_{n-1}, \varphi(x)).$$

This is a well-defined transformation with a nonsingular Jacobian in a neighborhood of the origin. With

$$u_{x_i} = \sum_{k=1}^{n} y_{k,x_i} u_{y_k}$$

and

$$u_{x_i x_j} = \sum_{k,l=1}^{n} y_{k,x_i} y_{l,x_j} u_{y_k y_l} + \sum_{k=1}^{n} y_{k,x_i x_j} u_{y_k},$$

we can write the operator L in the y-coordinates as

$$\begin{aligned} Lu = &\sum_{k,l=1}^{n} \left(\sum_{i,j=1}^{n} a_{ij} y_{k,x_i} y_{l,x_j} \right) u_{y_k y_l} \\ &+ \sum_{k=1}^{n} \left(\sum_{i=1}^{n} b_i y_{k,x_i} + \sum_{i,j=1}^{n} a_{ij} y_{k,x_i x_j} \right) u_{y_k} + cu. \end{aligned}$$

The initial hypersurface Σ is given by $\{y_n = 0\}$ in the y-coordinates. With $y_n = \varphi$, the coefficient of $u_{y_n y_n}$ is given by

$$\sum_{i,j=1}^{n} a_{ij}(x) \varphi_{x_i} \varphi_{x_j}.$$

This is the principal symbol $p(x; \xi)$ evaluated at $\xi = \nabla\varphi(x)$.

Definition 3.1.1. Let L be a second-order linear differential operator as in (3.1.1) in a neighborhood of $x_0 \in \mathbb{R}^n$ and Σ be a smooth hypersurface containing x_0. Then Σ is *noncharacteristic* at x_0 if

$$\sum_{i,j=1}^{n} a_{ij}(x_0) \nu_i \nu_j \neq 0, \tag{3.1.5}$$

where $\nu = (\nu_1, \cdots, \nu_n)$ is normal to Σ at x_0. Otherwise, it is *characteristic* at x_0.

A hypersurface is noncharacteristic if it is noncharacteristic at every point. Strictly speaking, a hypersurface is characteristic if it is not noncharacteristic, i.e., if it is characteristic at some point. In this book, we will abuse this terminology. When we say a hypersurface is characteristic, we mean it is characteristic *everywhere*. This should cause few confusions. In $\mathbb{R}^2$, hypersurfaces are curves, so we shall speak of characteristic curves and noncharacteristic curves.

When the hypersurface Σ is given by $\{\varphi = 0\}$ with $\nabla\varphi \neq 0$, its normal vector field is given by $\nabla\varphi = (\varphi_{x_1}, \cdots, \varphi_{x_n})$. Hence we may take $\nu = \nabla\varphi(x_0)$ in (3.1.5). We note that the condition (3.1.5) is preserved under

C^2-changes of coordinates. Using this condition, we can find successively the values of all derivatives of u at x_0, as far as they exist. Then, we could write *formal* power series at x_0 for solutions of initial-value problems. If the initial hypersurface is analytic and the coefficients, nonhomogeneous terms and initial values are analytic, then this formal power series converges to an analytic solution. This is the content of the Cauchy-Kovalevskaya theorem, which we will discuss in Section 7.2.

Now we introduce a special class of linear differential operators.

Definition 3.1.2. Let L be a second-order linear differential operator as in (3.1.1) defined in a neighborhood of $x_0 \in \mathbb{R}^n$. Then L is *elliptic* at x_0 if

$$\sum_{i,j=1}^{n} a_{ij}(x_0)\xi_i\xi_j \neq 0,$$

for any $\xi \in \mathbb{R}^n \setminus \{0\}$.

An operator L defined in Ω is called elliptic in Ω if it is elliptic at every point in Ω.

According to Definition 3.1.2, linear differential operators are elliptic if every hypersurface is noncharacteristic. We already assumed that (a_{ij}) is an $n \times n$ symmetric matrix. Then L is elliptic at x_0 if $(a_{ij}(x_0))$ is a definite matrix—positive definite or negative definite.

We now turn our attention to second-order linear differential equations in $\mathbb{R}^2$, where complete classifications are available. Let Ω be a domain in $\mathbb{R}^2$ and consider

$$Lu = \sum_{i,j=1}^{2} a_{ij}(x)u_{x_ix_j} + \sum_{i=1}^{2} b_i(x)u_{x_i} + c(x)u = f(x) \quad \text{in } \Omega. \tag{3.1.6}$$

Here we assume (a_{ij}) is a 2×2 symmetric matrix.

Definition 3.1.3. Let L be a differential operator defined in a neighborhood of $x_0 \in \mathbb{R}^2$ as in (3.1.6). Then

(1) L is *elliptic* at $x_0 \in \Omega$ if $\det(a_{ij}(x_0)) > 0$;

(2) L is *hyperbolic* at $x_0 \in \Omega$ if $\det(a_{ij}(x_0)) < 0$;

(3) L is *degenerate* at $x_0 \in \Omega$ if $\det(a_{ij}(x_0)) = 0$.

The operator L defined in $\Omega \subset \mathbb{R}^2$ is called elliptic (or hyperbolic) in Ω if it is elliptic (or hyperbolic) at every point in Ω.

It is obvious that the ellipticity defined in Definition 3.1.3 coincides with that in Definition 3.1.2 for $n = 2$.

For the operator L in (3.1.6), the symmetric matrix (a_{ij}) always has two (real) eigenvalues. Then

L is elliptic if the two eigenvalues have the same sign;

L is hyperbolic if the two eigenvalues have different signs;

L is degenerate if at least one of the eigenvalues vanishes.

The number of characteristic curves is determined by the type of the operator. For the operator L in (3.1.6),

there are two characteristic curves if L is hyperbolic;

there are no characteristic curves if L is elliptic.

We shall study several important linear differential operators in $\mathbb{R}^2$. The first of these is the Laplacian. In $\mathbb{R}^2$, the *Laplace operator* Δ is defined by

$$\Delta u = u_{x_1x_1} + u_{x_2x_2}.$$

It is easy to see that the Laplace operator is elliptic. In polar coordinates

$$x_1 = r\cos\theta, \quad x_2 = r\sin\theta,$$

the Laplace operator Δ can be expressed by

$$\Delta u = u_{rr} + \frac{1}{r}u_r + \frac{1}{r^2}u_{\theta\theta}.$$

The equation

$$\Delta u = 0$$

is called the *Laplace equation* and its solutions are called *harmonic functions*. By writing $x = x_1$ and $y = x_2$, we can associate with a harmonic function $u(x, y)$ a conjugate harmonic function $v(x, y)$ such that u and v satisfy the first-order system of Cauchy-Riemann equations

$$u_x = v_y, \quad u_y = -v_x.$$

Any such a pair gives an analytic function

$$f(z) = u(x, y) + iv(x, y)$$

of the complex argument $z = x + iy$, if we identify $\mathbb{C}$ with $\mathbb{R}^2$. Physically, $(u, -v)$ is the velocity field of an irrotational, incompressible flow. Conversely, for any analytic function f, functions $u = \operatorname{Re} f$ and $v = \operatorname{Im} f$ are harmonic. In this way, we can find many nontrivial harmonic functions in the plane. For example, for any positive integer k, $\operatorname{Re}(x + iy)^k$ and $\operatorname{Im}(x+iy)^k$ are homogeneous harmonic polynomials of degree k. Next, with $e^z = e^{x+iy} = e^x\cos y + ie^x \sin y$, we know $e^x \cos y$ and $e^x \sin y$ are harmonic functions.

Although there are no characteristic curves for the Laplace operator, initial-value problems are not well-posed.

Example 3.1.4. Consider the Laplace equation in $\mathbb{R}^2$

$$u_{xx} + u_{yy} = 0,$$

with initial values prescribed on $\{y = 0\}$. For any positive integer k, set

$$u_k(x, y) = \frac{1}{k} \sin(kx) e^{ky}.$$

Then u_k is harmonic. Moreover,

$$u_{k,x}(x, y) = \cos(kx) e^{ky}, \quad u_{k,y}(x, y) = \sin(kx) e^{ky},$$

and hence

$$|\nabla u_k(x, y)|^2 = u_{k,x}^2(x, y) + u_{k,y}^2(x, y) = e^{2ky}.$$

Therefore,

$$|\nabla u_k(x, 0)| = 1 \quad \text{for any } x \in \mathbb{R} \text{ and any } k,$$

and

$$|\nabla u_k(x, y)| \to \infty \quad \text{as } k \to \infty, \text{ for any } x \in \mathbb{R} \text{ and } y > 0.$$

There is no continuous dependence on initial values in C^1-norms.

In $\mathbb{R}^2$, the *wave operator* $\Box$ is given by

$$\Box u = u_{x_2 x_2} - u_{x_1 x_1}.$$

It is easy to see that the wave operator is hyperbolic. It is actually called the one-dimensional wave operator. This is because the wave equation $\Box u = 0$ in $\mathbb{R}^2$ represents vibrations of strings or propagation of sound waves in tubes. Because of its physical interpretation, we write u as a function of two independent variables x and t. The variable x is commonly identified with position and t with time. Then the wave equation in $\mathbb{R}^2$ has the form

$$u_{tt} - u_{xx} = 0.$$

The two families of straight lines $t = \pm x + c$, where c is a constant, are characteristic.

The *heat operator* in $\mathbb{R}^2$ is given by

$$Lu = u_{x_2} - u_{x_1 x_1}.$$

This is a degenerate operator. The heat equation $u_{x_2} - u_{x_1 x_1} = 0$ is satisfied by the temperature distribution in a heat-conducting insulated wire. As with the wave operator, we refer to the one-dimensional heat operator and we write u as a function of the independent variables x and t. Then the heat equation in $\mathbb{R}^2$ has the form

$$u_t - u_{xx} = 0.$$

It is easy to see that $\{t = 0\}$ is characteristic. If we prescribe $u(x, 0) = u_0(x)$ in an interval of $\{t = 0\}$, then using the equation we can compute all derivatives there. However, u_0 does not determine a unique solution even in a neighborhood of this interval. We will see later on that we need initial values on the entire initial line $\{t = 0\}$ to compute local solutions.

Many important differential operators do not have a definite type. In other words, they are neither elliptic nor hyperbolic in the domain where they are defined. We usually say a differential operator is of *mixed type* if it is elliptic in a subdomain and hyperbolic in another subdomain. In general, it is more difficult to study equations of mixed type.

Example 3.1.5. Consider the *Tricomi equation*

$$u_{x_2x_2} + x_2u_{x_1x_1} = f \quad \text{in } \mathbb{R}^2.$$

It is elliptic if $x_2 > 0$, hyperbolic if $x_2 < 0$ and degenerate if $x_2 = 0$.

Characteristic curves also arise naturally in connection with the propagation of singularities. We consider a simple case.

Let Ω be a domain in $\mathbb{R}^2$, Γ be a continuous curve in Ω and w be a continuous function in $\Omega \setminus \Gamma$. For simplicity, we assume Γ divides Ω into two parts, Ω_+ and Ω_-. Take a point $x_0 \in \Gamma$. Then w is said to have a *jump* at x_0 across Γ if the two limits

$$w_-(x_0) = \lim_{x \to x_0,\ x \in \Omega_-} w(x), \quad w_+(x_0) = \lim_{x \to x_0,\ x \in \Omega_+} w(x)$$

exist. The difference

$$[w](x_0) = w_+(x_0) - w_-(x_0)$$

is called the *jump* of w at x_0 across Γ. The function w has a jump across Γ if it has a jump at every point of Γ across Γ. If w has a jump across Γ, then $[w]$ is a well-defined function on Γ. It is easy to see that $[w] = 0$ on Γ if w is continuous in Ω.

Proposition 3.1.6. *Let Ω be a domain in $\mathbb{R}^2$ and Γ be a C^1-curve in Ω dividing Ω into two parts. Suppose a_{ij}, b_i, c, f are continuous functions in Ω and $u \in C^1(\Omega) \cap C^2(\Omega \setminus \Gamma)$ satisfies*

$$\sum_{i,j=1}^{2} a_{ij}u_{x_ix_j} + \sum_{i=1}^{2} b_iu_{x_i} + cu = f \quad \text{in } \Omega \setminus \Gamma.$$

If $\nabla^2 u$ has a jump across Γ, then Γ is a characteristic curve.

Proof. Since u is C^1 in Ω, we have

$$[u] = [u_{x_1}] = [u_{x_2}] = 0 \quad \text{on } \Gamma.$$

Let the vector field (ν_1, ν_2) be normal to Γ. Then $\nu_2\partial_{x_1} - \nu_1\partial_{x_2}$ is a directional derivative along Γ. Hence on Γ

$$(\nu_2\partial_{x_1} - \nu_1\partial_{x_2})[u_{x_1}] = 0,$$

and

$$(\nu_2\partial_{x_1} - \nu_1\partial_{x_2})[u_{x_2}] = 0.$$

Then we conclude

$$\nu_2[u_{x_1x_1}] - \nu_1[u_{x_1x_2}] = 0,$$

and

$$\nu_2[u_{x_1x_2}] - \nu_1[u_{x_2x_2}] = 0.$$

By the continuity of a_{ij}, b_i, c and f in Ω, we have

$$a_{11}[u_{x_1x_1}] + 2a_{12}[u_{x_1x_2}] + a_{22}[u_{x_2x_2}] = 0 \quad \text{on } \Gamma.$$

Thus, the nontrivial vector $([u_{x_1x_1}], [u_{x_1x_2}], [u_{x_2x_2}])$ satisfies a 3×3 homogeneous linear system on Γ. Hence the coefficient matrix is singular. That is, on Γ

$$\det \begin{pmatrix} \nu_2 & -\nu_1 & 0 \\ 0 & \nu_2 & -\nu_1 \\ a_{11} & 2a_{12} & a_{22} \end{pmatrix} = 0,$$

or

$$a_{11}\nu_1^2 + 2a_{12}\nu_1\nu_2 + a_{22}\nu_2^2 = 0.$$

This yields the desired result. □

The Laplace operator, the wave operator and the heat operator can be generalized to higher dimensions.

Example 3.1.7. The n-dimensional *Laplace operator* in $\mathbb{R}^n$ is defined by

$$\Delta u = \sum_{i=1}^{n} u_{x_ix_i},$$

and the *Laplace equation* is given by $\Delta u = 0$. Solutions are called *harmonic functions*. The principal symbol of the Laplace operator Δ is given by

$$p(x;\xi) = |\xi|^2,$$

for any $\xi \in \mathbb{R}^n$. Obviously, Δ is elliptic. Note that Δ is invariant under rotations. If $x = Ay$ for an orthogonal matrix A, then

$$\sum_{i=1}^{n} u_{x_ix_i} = \sum_{i=1}^{n} u_{y_iy_i}.$$

For a nonzero function f, we call the equation $\Delta u = f$ the *Poisson equation*.

The Laplace equation has a wide variety of physical backgrounds. For example, let u denote a temperature in equilibrium in a domain $\Omega \subset \mathbb{R}^n$ with the flux density $\mathbf{F}$. Then for any smooth subdomain $\Omega' \subset \Omega$, the net flux of u through $\partial\Omega'$ is zero, i.e.,

$$\int_{\partial\Omega'} \mathbf{F} \cdot \nu \, dS = 0,$$

where ν is the unit exterior normal vector to Ω'. Upon integration by parts, we obtain

$$\int_{\Omega'} \operatorname{div} \mathbf{F} \, dx = \int_{\partial\Omega'} \mathbf{F} \cdot \nu \, dS = 0.$$

This implies

$$\operatorname{div} \mathbf{F} = 0 \quad \text{in } \Omega,$$

since Ω' is arbitrary. In a special case where the flux $\mathbf{F}$ is proportional to the gradient ∇u, we have

$$\mathbf{F} = -a\nabla u,$$

for a positive constant a. Here the negative sign indicates that the flow is from regions of higher temperature to those of lower one. Now a simple substitution yields the Laplace equation

$$\Delta u = \operatorname{div}(\nabla u) = 0.$$

Example 3.1.8. We denote points in $\mathbb{R}^{n+1}$ by $(x_1, \cdots, x_n, t)$. The *heat operator* in $\mathbb{R}^{n+1}$ is given by

$$Lu = u_t - \Delta_x u.$$

It is often called the n-dimensional heat operator. Its principal symbol is given by

$$p(x, t; \xi, \tau) = -|\xi|^2,$$

for any $\xi \in \mathbb{R}^n$ and $\tau \in \mathbb{R}$. A hypersurface $\{\varphi(x_1, \cdots, x_n, t) = 0\}$ is noncharacteristic for the heat operator if, at each of its points,

$$-|\nabla_x \varphi|^2 \neq 0.$$

Likewise, a hypersurface $\{\varphi(x, t) = 0\}$ is characteristic if $\nabla_x \varphi = 0$ and $\varphi_t \neq 0$ at each of its points. For example, any horizontal hyperplane $\{t = t_0\}$, for a fixed $t_0 \in \mathbb{R}$, is characteristic.

The heat equation describes the evolution of heat. Let u denote a temperature in a domain $\Omega \subset \mathbb{R}^n$ with the flux density $\mathbf{F}$. Then for any smooth subdomain $\Omega' \subset \Omega$, the rate of change of the total quantity in Ω' equals the negative of the net flux of u through $\partial\Omega'$, i.e.,

$$\frac{d}{dt} \int_{\Omega'} u \, dx = -\int_{\partial\Omega'} \mathbf{F} \cdot \nu \, dS,$$

where ν is the unit exterior normal vector to Ω'. Upon integration by parts, we obtain

$$\frac{d}{dt} \int_{\Omega'} u \, dx = -\int_{\Omega'} \operatorname{div} \mathbf{F} \, dx.$$

This implies

$$u_t = -\operatorname{div} \mathbf{F} \quad \text{in } \Omega,$$

since Ω' is arbitrary. In a special case where the flux $\mathbf{F}$ is proportional to the gradient ∇u, we have

$$\mathbf{F} = -a\nabla u,$$

for a positive constant a. Now a simple substitution yields

$$u_t = a \operatorname{div}(\nabla u) = a\Delta u.$$

This is the heat equation if $a = 1$.

Example 3.1.9. We denote points in $\mathbb{R}^{n+1}$ by $(x_1, \cdots, x_n, t)$. The *wave operator* $\Box$ in $\mathbb{R}^{n+1}$ is given by

$$\Box u = u_{tt} - \Delta_x u.$$

It is often called the n-dimensional wave operator. Its principal symbol is given by

$$p(x, t; \xi, \tau) = \tau^2 - |\xi|^2,$$

for any $\xi \in \mathbb{R}^n$ and $\tau \in \mathbb{R}$. A hypersurface $\{\varphi(x_1, \cdots, x_n, t) = 0\}$ is noncharacteristic for the wave operator if, at each of its points,

$$\varphi_t^2 - |\nabla_x \varphi|^2 \neq 0.$$

For any $(x_0, t_0) \in \mathbb{R}^n \times \mathbb{R}$, the surface

$$|x - x_0|^2 = (t - t_0)^2$$

is characteristic except at (x_0, t_0). We note that this surface, smooth except at (x_0, t_0), is the union of two cones. It is usually called the *characteristic cone*.

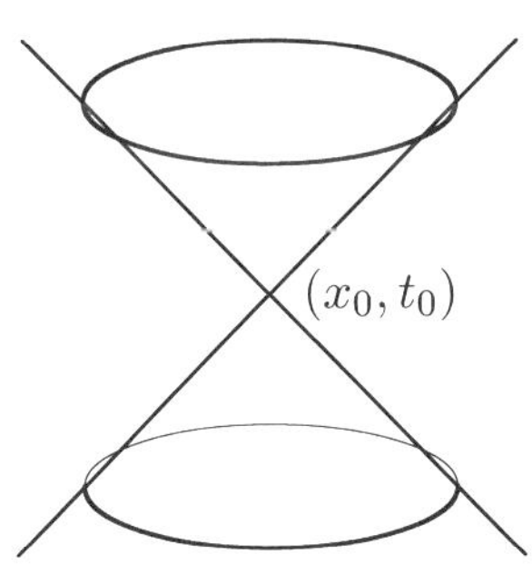

Figure 3.1.1. The characteristic cone.

To interpret the wave equation, we let $u(x, t)$ denote the displacement in some direction of a point $x \in \Omega \subset \mathbb{R}^n$ at time $t \geq 0$. For any smooth subdomain $\Omega' \subset \Omega$, Newton's law asserts that the product of mass and the acceleration equals the net force, i.e.,

$$\frac{d^2}{dt^2}\int_{\Omega'} u\,dx = -\int_{\partial\Omega'} \mathbf{F} \cdot \nu\,dS,$$

where $\mathbf{F}$ is the force acting on Ω' through $\partial\Omega'$ and the mass density is taken to be 1. Upon integration by parts, we obtain

$$\frac{d^2}{dt^2}\int_{\Omega'} u\,dx = -\int_{\Omega'} \operatorname{div}\mathbf{F}\,dx.$$

This implies

$$u_{tt} = -\operatorname{div}\mathbf{F} \quad \text{in } \Omega,$$

since Ω' is arbitrary. In a special case $\mathbf{F} = -a\nabla u$ for a positive constant a, we have

$$u_{tt} = a\ \operatorname{div}(\nabla u) = a\Delta u.$$

This is the wave equation if $a = 1$.

The heat equation and the wave equation can be generalized to parabolic equations and hyperbolic equations in arbitrary dimensions. Again, we denote points in $\mathbb{R}^{n+1}$ by $(x_1, \cdots, x_n, t)$. Let a_{ij}, b_i, c and f be functions defined in a domain in $\mathbb{R}^{n+1}$, $i, j = 1, \cdots, n$. We assume (a_{ij}) is an $n \times n$ positive definite matrix in this domain. An equation of the form

$$u_t - \sum_{i=1}^{n} a_{ij}(x,t)u_{x_ix_j} + \sum_{i=1}^{n} b_i(x,t)u_{x_i} + c(x,t)u = f(x,t)$$

is *parabolic*, and an equation of the form

$$u_{tt} - \sum_{i=1}^{n} a_{ij}(x,t)u_{x_ix_j} + \sum_{i=1}^{n} b_i(x,t)u_{x_i} + c(x,t)u = f(x,t)$$

is *hyperbolic*.

3.2. Energy Estimates

In this section, we discuss the uniqueness of solutions of boundary-value problems for the Laplace equation and initial/boundary-value problems for the heat equation and the wave equation. Our main tool is the energy estimates. Specifically, we derive estimates of L^2-norms of solutions in terms of those of boundary values and/or initial values.

We start with the Laplace equation. Let $\Omega \subset \mathbb{R}^n$ be a bounded C^1-domain and φ be a continuous function on $\partial\Omega$. Consider the *Dirichlet boundary-value problem* for the Laplace equation:

$$\begin{aligned} \Delta u &= 0 \quad \text{in } \Omega, \\ u &= \varphi \quad \text{on } \partial\Omega. \end{aligned}$$

We now prove that a C^2-solution, if it exists, is unique. To see this, let u_1 and u_2 be solutions in $C^2(\Omega) \cap C^1(\bar{\Omega})$. Then, the difference $w = u_1 - u_2$

satisfies

$$\begin{aligned} \Delta w &= 0 \quad \text{in } \Omega, \\ w &= 0 \quad \text{on } \partial\Omega. \end{aligned}$$

We multiply the Laplace equation by w and write the resulting product as

$$0 = w\Delta w = \sum_{i=1}^{n} (ww_{x_i})_{x_i} - |\nabla w|^2.$$

An integration by parts in Ω yields

$$0 = -\int_{\Omega} |\nabla w|^2 \, dx + \int_{\partial\Omega} w \frac{\partial w}{\partial \nu} \, dS.$$

With the homogeneous boundary value $w = 0$ on $\partial\Omega$, we have

$$\int_{\Omega} |\nabla w|^2 \, dx = 0,$$

and then $\nabla w = 0$ in Ω. Hence w is constant and this constant is zero since w is zero on the boundary. Obviously, the argument above applies to Dirichlet problems for the Poisson equation. In general, we have the following result.

Lemma 3.2.1. *Let $\Omega \subset \mathbb{R}^n$ be a bounded C^1-domain, f be a continuous function in $\bar{\Omega}$ and φ be a continuous function on $\partial\Omega$. Then there exists at most one solution in $C^2(\Omega) \cap C^1(\bar{\Omega})$ of the Dirichlet problem*

$$\begin{aligned} \Delta u &= f \quad \text{in } \Omega, \\ u &= \varphi \quad \text{on } \partial\Omega. \end{aligned}$$

By the maximum principle, the solution is in fact unique in $C^2(\Omega) \cap C(\bar{\Omega})$, as we will see in Chapter 4.

Now we discuss briefly the *Neumann boundary-value problem*, where we prescribe normal derivatives on the boundary. Let ψ be a continuous function on $\partial\Omega$. Consider

$$\begin{aligned} \Delta u &= 0 \quad \text{in } \Omega, \\ \frac{\partial u}{\partial \nu} &= \psi \quad \text{on } \partial\Omega. \end{aligned}$$

We can prove similarly that solutions are unique up to additive constants if Ω is connected. We note that if there exists a solution of the Neumann problem, then ψ necessarily satisfies

$$\int_{\partial\Omega} \psi \, dS = 0.$$

This can be seen easily upon integration by parts.

Next, we derive an estimate of a solution of the Dirichlet boundary-value problem for the Poisson equation. We need the following result, which is referred to as the *Poincaré lemma.*

Lemma 3.2.2. *Let Ω be a bounded C^1-domain in $\mathbb{R}^n$ and u be a C^1-function in $\bar{\Omega}$ with $u = 0$ on $\partial\Omega$. Then*

$$\|u\|_{L^2(\Omega)} \leq \text{diam}(\Omega)\|\nabla u\|_{L^2(\Omega)}.$$

Here $\text{diam}(\Omega)$ denotes the diameter of Ω and is defined by

$$\text{diam}(\Omega) = \sup_{x,y\in\Omega} |x-y|.$$

Proof. We write $\mathbb{R}^n = \mathbb{R}^{n-1} \times \mathbb{R}$. For any $x_0' \in \mathbb{R}^{n-1}$, let $l_{x_0'}$ be the straight line containing x_0' and normal to $\mathbb{R}^{n-1} \times \{0\}$. Consider those x_0' such that $l_{x_0'} \cap \Omega \neq \emptyset$. Now $l_{x_0'} \cap \Omega$ is the union of a countable collection of pairwise disjoint open intervals. Let I be such an interval. Then $I \subset \Omega$ and I has the form

$$I = \{(x_0', x_n) : \ a < x_n < b\},$$

where $(x_0', a), (x_0', b) \in \partial\Omega$. Since $u(x_0', a) = 0$, then

$$u(x_0', x_n) = \int_a^{x_n} u_{x_n}(x_0', s)ds \quad \text{for any } x_n \in (a, b).$$

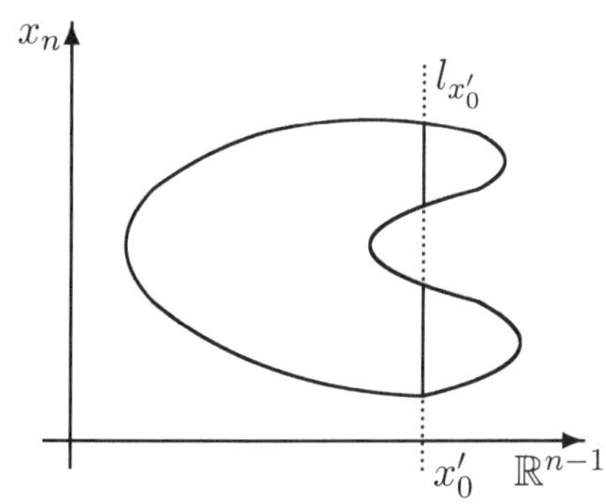

Figure 3.2.1. An integration along $l_{x_0'}$.

The Cauchy inequality yields

$$u^2(x_0', x_n) \leq (x_n - a) \int_a^{x_n} u_{x_n}^2(x_0', s)\, ds \quad \text{for any } x_n \in (a, b).$$

By a simple integration along I, we have

$$\int_a^b u^2(x_0', x_n)\, dx_n \leq (b-a)^2 \int_a^b u_{x_n}^2(x_0', x_n)\, dx_n.$$

By adding the integrals over all such intervals, we obtain

$$\int_{l_{x_0'}\cap\Omega} u^2(x_0', x_n)\, dx_n \le C_{x_0'}^2 \int_{l_{x_0'}\cap\Omega} u_{x_n}^2(x_0', x_n)\, dx_n,$$

where $C_{x_0'}$ is the length of $l_{x_0'}$ in Ω. Now a simple integration over x_0' yields the desired result. □

Now consider

$$\begin{aligned} \Delta u &= f \quad \text{in } \Omega, \\ u &= 0 \quad \text{on } \partial\Omega. \end{aligned} \tag{3.2.1}$$

We note that u has a homogeneous Dirichlet boundary value on $\partial\Omega$.

Theorem 3.2.3. *Let $\Omega \subset \mathbb{R}^n$ be a bounded C^1-domain and f be a continuous function in $\bar{\Omega}$. Suppose $u \in C^2(\Omega) \cap C^1(\bar{\Omega})$ is a solution of* (3.2.1). *Then*

$$\|u\|_{L^2(\Omega)} + \|\nabla u\|_{L^2(\Omega)} \le C\|f\|_{L^2(\Omega)},$$

where C is a positive constant depending only on Ω.

Proof. Multiply the equation in (3.2.1) by u and write the resulting product in the left-hand side as

$$u\Delta u = \sum_{i=1}^n (uu_{x_i})_{x_i} - |\nabla u|^2.$$

Upon integrating by parts in Ω, we obtain

$$\int_{\partial\Omega} u\frac{\partial u}{\partial \nu}\, dS - \int_\Omega |\nabla u|^2\, dx = \int_\Omega uf\, dx.$$

With $u = 0$ on $\partial\Omega$, we have

$$\int_\Omega |\nabla u|^2\, dx = -\int_\Omega uf\, dx.$$

The Cauchy inequality yields

$$\left(\int_\Omega |\nabla u|^2\, dx\right)^2 = \left(\int_\Omega uf\, dx\right)^2 \le \int_\Omega u^2\, dx \cdot \int_\Omega f^2\, dx.$$

By Lemma 3.2.2, we get

$$\int_\Omega |\nabla u|^2\, dx \le \big(\mathrm{diam}(\Omega)\big)^2 \int_\Omega f^2\, dx.$$

Using Lemma 3.2.2 again, we then have

$$\int_\Omega u^2\, dx \le \big(\mathrm{diam}(\Omega)\big)^4 \int_\Omega f^2\, dx.$$

This yields the desired estimate. □

Now we study initial/boundary-value problems for the heat equation. Suppose Ω is a bounded C^1-domain in $\mathbb{R}^n$, f is continuous in $\bar{\Omega}\times[0,\infty)$ and u_0 is continuous in $\bar{\Omega}$. Consider

$$\begin{aligned} u_t - \Delta u &= f \quad \text{in } \Omega\times(0,\infty),\\ u(\cdot,0) &= u_0 \quad \text{in } \Omega,\\ u &= 0 \quad \text{on } \partial\Omega\times(0,\infty). \end{aligned} \tag{3.2.2}$$

The geometric boundary of $\Omega\times(0,\infty)$ consists of three parts, $\Omega\times\{0\}$, $\partial\Omega\times(0,\infty)$ and $\partial\Omega\times\{0\}$. We treat $\Omega\times\{0\}$ and $\partial\Omega\times(0,\infty)$ differently and refer to values prescribed on $\Omega\times\{0\}$ and $\partial\Omega\times(0,\infty)$ as *initial values* and *boundary values*, respectively. Problems of this type are usually called *initial/boundary-value problems* or *mixed problems.* We note that u has a homogeneous Dirichlet boundary value on $\partial\Omega\times(0,\infty)$. We now derive an estimate of the L^2-norm of a solution. For each $t\geq 0$, we denote by $u(t)$ the function defined on Ω by $u(t)=u(\cdot,t)$.

Theorem 3.2.4. *Let Ω be a bounded C^1-domain in $\mathbb{R}^n$, f be continuous in $\bar{\Omega}\times[0,\infty)$ and u_0 be continuous in $\bar{\Omega}$. Suppose $u\in C^2(\Omega\times(0,\infty))\cap C^1(\bar{\Omega}\times[0,\infty))$ is a solution of* (3.2.2). *Then*

$$\|u(t)\|_{L^2(\Omega)} \leq \|u_0\|_{L^2(\Omega)} + \int_0^t \|f(s)\|_{L^2(\Omega)}\,ds \quad \textit{for any } t>0.$$

Theorem 3.2.4 yields the uniqueness of solutions of (3.2.2). In fact, if $f\equiv 0$ and $u_0\equiv 0$, then $u\equiv 0$. We also have the continuous dependence on initial values in L^2-norms. Let f_1, f_2 be continuous in $\bar{\Omega}\times[0,\infty)$ and u_{01}, u_{02} be continuous in $\bar{\Omega}$. Suppose $u_1,u_2\in C^2(\Omega\times(0,\infty))\cap C^1(\bar{\Omega}\times[0,\infty))$ are solutions of (3.2.2) with f_1,u_{01} and f_2,u_{02} replacing f,u_0, respectively. Then for any $t>0$,

$$\|u_1(t)-u_2(t)\|_{L^2(\Omega)} \leq \|u_{01}-u_{02}\|_{L^2(\Omega)} + \int_0^t \|f_1(s)-f_2(s)\|_{L^2(\Omega)}\,ds.$$

Proof. We multiply the equation in (3.2.2) by u and write the product in the left-hand side as

$$uu_t - u\Delta u = \frac{1}{2}(u^2)_t - \sum_{i=1}^n (uu_{x_i})_{x_i} + |\nabla u|^2.$$

Upon integration by parts in Ω for each fixed $t>0$ and $u(t)=0$ on $\partial\Omega$, we have

$$\frac{1}{2}\frac{d}{dt}\int_\Omega u^2(t)\,dx + \int_\Omega |\nabla u(t)|^2\,dx = \int_\Omega f(t)u(t)\,dx.$$

An integration in t yields, for any $t>0$,

$$\int_\Omega u^2(t)\,dx + 2\int_0^t\int_\Omega |\nabla u|^2\,dxds = \int_\Omega u_0^2\,dx + 2\int_0^t\int_\Omega fu\,dxds.$$

Set

$$E(t) = \|u(t)\|_{L^2(\Omega)}.$$

Then

$$\big(E(t)\big)^2 + 2\int_0^t \int_\Omega |\nabla u|^2\, dxds = \big(E(0)\big)^2 + 2\int_0^t \int_\Omega fu\, dxds.$$

Differentiating with respect to t, we have

$$\begin{aligned}2E(t)E'(t) &\le 2E(t)E'(t) + 2\int_\Omega |\nabla u(t)|^2\, dx = 2\int_\Omega f(t)u(t)\, dx\\ &\le 2\|f(t)\|_{L^2(\Omega)} \cdot \|u(t)\|_{L^2(\Omega)} = 2E(t)\|f(t)\|_{L^2(\Omega)}.\end{aligned}$$

Hence

$$E'(t) \le \|f(t)\|_{L^2(\Omega)}.$$

Integrating from 0 to t gives the desired estimate. □

Now we study initial/boundary-value problems for the wave equation. Suppose Ω is a bounded C^1-domain in $\mathbb{R}^n$, f is continuous in $\bar\Omega \times [0,\infty)$, u_0 is C^1 in $\bar\Omega$ and u_1 is continuous in $\bar\Omega$. Consider

$$\begin{aligned}&u_{tt} - \Delta u = f \quad \text{in } \Omega \times (0,\infty),\\ &u(\cdot,0) = u_0,\ u_t(\cdot,0) = u_1 \quad \text{in } \Omega,\\ &u = 0 \quad \text{on } \partial\Omega \times (0,\infty).\end{aligned} \tag{3.2.3}$$

Comparing (3.2.3) with (3.2.2), we note that there is an extra initial condition on u_t in (3.2.3). This relates to the extra order of the t-derivative in the wave equation.

Theorem 3.2.5. *Let Ω be a bounded C^1-domain in $\mathbb{R}^n$, f be continuous in $\bar\Omega \times [0,\infty)$, u_0 be C^1 in $\bar\Omega$ and u_1 be continuous in $\bar\Omega$. Suppose $u \in C^2(\Omega \times (0,\infty)) \cap C^1(\bar\Omega \times [0,\infty))$ is a solution of (3.2.3). Then for any $t > 0$,*

$$\begin{aligned}\big(\|u_t(t)\|^2_{L^2(\Omega)} + \|\nabla_x u(t)\|^2_{L^2(\Omega)}\big)^{\frac12} \le& \big(\|u_1\|^2_{L^2(\Omega)} + \|\nabla_x u_0\|^2_{L^2(\Omega)}\big)^{\frac12}\\ &+ \int_0^t \|f(s)\|_{L^2(\Omega)}\, ds,\end{aligned}$$

and

$$\begin{aligned}\|u(t)\|_{L^2(\Omega)} \le& \|u_0\|_{L^2(\Omega)} + t\big(\|u_1\|^2_{L^2(\Omega)} + \|\nabla_x u_0\|^2_{L^2(\Omega)}\big)^{\frac12}\\ &+ \int_0^t (t-s)\|f(s)\|_{L^2(\Omega)}\, ds.\end{aligned}$$

As a consequence, we also have the uniqueness and continuous dependence on initial values in L^2-norms.

Proof. Multiply the equation in (3.2.3) by u_t and write the resulting product in the left-hand side as

$$u_t u_{tt} - u_t \Delta u = \frac{1}{2}(u_t^2 + |\nabla u|^2)_t - \sum_{i=1}^n (u_t u_{x_i})_{x_i}.$$

Upon integration by parts in Ω for each fixed $t > 0$, we obtain

$$\frac{1}{2}\frac{d}{dt}\int_\Omega \left(u_t^2(t) + |\nabla_x u(t)|^2\right) dx - \int_{\partial\Omega} u_t(t)\frac{\partial u}{\partial \nu}(t)\, dS = \int_\Omega f(t)u_t(t)\, dx.$$

Note that $u_t = 0$ on $\partial\Omega \times (0, \infty)$ since $u = 0$ on $\partial\Omega \times (0, \infty)$. Then

$$\frac{1}{2}\frac{d}{dt}\int_\Omega \left(u_t^2(t) + |\nabla_x u(t)|^2\right) dx = \int_\Omega f(t)u_t(t)\, dx.$$

Define the *energy* by

$$E(t) = \int_\Omega \left(u_t^2(t) + |\nabla_x u(t)|^2\right) dx.$$

If $f \equiv 0$, then

$$\frac{d}{dt}E(t) = 0.$$

Hence for any $t > 0$,

$$E(t) = E(0) = \frac{1}{2}\int_\Omega \left(u_1^2 + |\nabla_x u_0|^2\right) dx.$$

This is the *conservation of energy*. In general,

$$E(t) = E(0) + 2\int_0^t \int_\Omega f u_t \, dx ds.$$

To get an estimate of $E(t)$, set

$$J(t) = \big(E(t)\big)^{\frac{1}{2}}.$$

Then

$$\big(J(t)\big)^2 = \big(J(0)\big)^2 + 2\int_0^t \int_\Omega f u_t \, dx ds.$$

By differentiating with respect to t and applying the Cauchy inequality, we get

$$\begin{aligned} 2J(t)J'(t) &= 2\int_\Omega f(t)u_t(t)\, dx \le 2\|f(t)\|_{L^2(\Omega)}\|u_t(t)\|_{L^2(\Omega)} \\ &\le 2J(t)\|f(t)\|_{L^2(\Omega)}. \end{aligned}$$

Hence for any $t > 0$,

$$J'(t) \le \|f(t)\|_{L^2(\Omega)}.$$

Integrating from 0 to t, we obtain

$$J(t) \le J(0) + \int_0^t \|f(s)\|_{L^2(\Omega)}\, ds.$$

This is the desired estimate for the energy. Next, to estimate the L^2-norm of u, we set

$$F(t) = \|u(t)\|_{L^2(\Omega)},$$

i.e.,

$$\big(F(t)\big)^2 = \int_\Omega u^2(t)\, dx.$$

A simple differentiation yields

$$\begin{aligned} 2F(t)F'(t) &= 2\int_\Omega u(t)u_t(t)\, dx \le 2\|u(t)\|_{L^2(\Omega)}\|u_t(t)\|_{L^2(\Omega)} \\ &= 2F(t)\|u_t(t)\|_{L^2(\Omega)}. \end{aligned}$$

Hence

$$F'(t) \le \|u_t\|_{L^2(\Omega)} \le J(0) + \int_0^t \|f(s)\|_{L^2(\Omega)}\, ds.$$

Integrating from 0 to t, we have

$$\|u(t)\|_{L^2(\Omega)} \le \|u_0\|_{L^2(\Omega)} + tJ(0) + \int_0^t \int_0^{t'} \|f(s)\|_{L^2(\Omega)}\, dsdt'.$$

By interchanging the order of integration in the last term in the right-hand side, we obtain the desired estimate on u. □

There are other forms of estimates on energies. By squaring the first estimate in Theorem 3.2.5 and applying the Cauchy inequality, we obtain

$$\begin{aligned} \int_\Omega \big(u_t^2(t) + |\nabla_x u(t)|^2\big)\, dx \le {} & 2\int_\Omega \big(u_1^2 + |\nabla_x u_0|^2\big)\, dx \\ & + 2t\int_0^t \int_\Omega f^2\, dxds. \end{aligned}$$

Integrating from 0 to t, we get

$$\begin{aligned} \int_0^t \int_\Omega \big(u_t^2 + |\nabla_x u|^2\big)\, dxds \le {} & 2t\int_\Omega \big(u_1^2 + |\nabla_x u_0|^2\big)\, dxds \\ & + t^2\int_0^t \int_\Omega f^2\, dxds. \end{aligned}$$

Next, we briefly review methods used in deriving estimates in Theorems 3.2.3–3.2.5. In the proofs of Theorems 3.2.3–3.2.4, we multiply the Laplace equation and the heat equation by u and integrate the resulting product over Ω, while in the proof of Theorem 3.2.5, we multiply the wave equation by u_t and integrate over Ω. It is important to write the resulting product as a linear combination of u^2, u_t^2, $|\nabla u|^2$ and their derivatives. Upon integrating by parts, domain integrals of derivatives are reduced to boundary integrals. Hence, the resulting integral identity consists of domain integrals and boundary integrals of u^2, u_t^2 and $|\nabla u|^2$. Second-order derivatives of

u are eliminated. These strategies also work for general elliptic equations, parabolic equations and hyperbolic equations. Compare methods in this section with those used to obtain L^2-estimates of solutions of initial-value problems for first-order linear PDEs in Section 2.3.

To end this section, we discuss an elliptic differential equation in the entire space. Let f be a continuous function in $\mathbb{R}^n$. We consider

$$-\Delta u + u = f \quad \text{in } \mathbb{R}^n. \tag{3.2.4}$$

Let u be a C^2-solution in $\mathbb{R}^n$. Next, we demonstrate that we can obtain estimates of L^2-norms of u and its derivatives under the assumption that u and its derivatives decay sufficiently fast at infinity.

To obtain an estimate of u and its first derivatives, we multiply (3.2.4) by u. In view of

$$u\Delta u = \sum_{k=1}^{n} (uu_{x_k})_{x_k} - |\nabla u|^2,$$

we write the resulting product as

$$|\nabla u|^2 + u^2 - \sum_{k=1}^{n} (uu_{x_k})_{x_k} = fu.$$

We now integrate in $\mathbb{R}^n$. Since u and u_{x_k} decay sufficiently fast at infinity, we have

$$\int_{\mathbb{R}^n} (|\nabla u|^2 + u^2)\, dx = \int_{\mathbb{R}^n} fu\, dx.$$

Rigorously, we need to integrate in B_R and let $R \to \infty$ after integrating by parts. By the Cauchy inequality, we get

$$\int_{\mathbb{R}^n} fu\, dx \leq \frac{1}{2} \int_{\mathbb{R}^n} u^2\, dx + \frac{1}{2} \int_{\mathbb{R}^n} f^2\, dx.$$

A simple substitution yields

$$\int_{\mathbb{R}^n} (2|\nabla u|^2 + u^2)\, dx \leq \int_{\mathbb{R}^n} f^2\, dx.$$

Hence, the L^2-norm of f controls the L^2-norms of u and ∇u. In fact, the L^2-norm of f also controls the L^2-norms of the second derivatives of u. To see this, we take square of the equation (3.2.4) to get

$$(\Delta u)^2 - 2u\Delta u + u^2 = f^2.$$

We note that

$$\begin{aligned}(\Delta u)^2 &= \sum_{k,l=1}^{n} u_{x_k x_k} u_{x_l x_l} = \sum_{k,l=1}^{n} (u_{x_k x_k} u_{x_l})_{x_l} - \sum_{k,l=1}^{n} u_{x_k x_k x_l} u_{x_l} \\ &= \sum_{k,l=1}^{n} (u_{x_k x_k} u_{x_l})_{x_l} - \sum_{k,l=1}^{n} (u_{x_k x_l} u_{x_l})_{x_k} + \sum_{k,l=1}^{n} u_{x_k x_l}^2.\end{aligned}$$

Hence

$$\begin{aligned}|\nabla^2 u|^2 + 2|\nabla u|^2 + u^2 + \sum_{k,l=1}^{n} (u_{x_k x_k} u_{x_l})_{x_l} &- \sum_{k,l=1}^{n} (u_{x_k x_l} u_{x_l})_{x_k} \\ &- 2\sum_{k=1}^{n} (u u_{x_k})_{x_k} = f^2.\end{aligned}$$

Integration in $\mathbb{R}^n$ yields

$$\int_{\mathbb{R}^n} (|\nabla^2 u|^2 + 2|\nabla u|^2 + u^2)\, dx = \int_{\mathbb{R}^n} f^2\, dx. \tag{3.2.5}$$

Therefore, the L^2-norm of f controls the L^2-norms of every second derivatives of u, although f is related to u by Δu, which is just one particular combination of second derivatives. As we will see, this is the feature of elliptic differential equations.

We need to point out that it is important to assume that u and its derivatives decay sufficiently fast. Otherwise, the integral identity (3.2.5) does not hold. By taking $f = 0$, we obtain $u = 0$ from (3.2.5) if u and its derivatives decay sufficiently fast. We note that $u(x) = e^{x_1}$ is a nonzero solution of (3.2.4) for $f = 0$.

3.3. Separation of Variables

In this section, we solve boundary-value problems for the Laplace equation and initial/boundary-value problems for the heat equation and the wave equation in the plane by separation of variables.

3.3.1. Dirichlet Problems. In this subsection we use the method of separation of variables to solve the Dirichlet problem for the Laplace equation in the unit disc in $\mathbb{R}^2$. We will use polar coordinates

$$x = r\cos\theta, \quad y = r\sin\theta$$

in $\mathbb{R}^2$, and we will build up solutions from functions that depend only on r and functions that depend only on θ. Our first step is to determine all harmonic functions u in $\mathbb{R}^2$ having the form

$$u(r, \theta) = f(r)g(\theta),$$

where f is defined for $r > 0$ and g is defined on $\mathbb{S}^1$. (Equivalently, we can view g as a 2π-periodic function defined on $\mathbb{R}$.) Then we shall express the solution of a Dirichlet problem as the sum of a suitably convergent infinite series of functions of this form.

In polar coordinates, the Laplace equation is

$$\Delta u \equiv \frac{1}{r}(ru_r)_r + \frac{1}{r^2}u_{\theta\theta} = 0.$$

Thus the function $u(r,\theta) = f(r)g(\theta)$ is harmonic if and only if

$$\frac{1}{r}\big(rf'(r)\big)'g(\theta) + \frac{1}{r^2}f(r)g''(\theta) = 0,$$

that is,

$$\left(f''(r) + \frac{1}{r}f'(r)\right)g(\theta) + \frac{1}{r^2}f(r)g''(\theta) = 0.$$

When $u(r,\theta) \neq 0$, this equation is equivalent to

$$\frac{r^2}{f(r)}\left(f''(r) + \frac{1}{r}f'(r)\right) = -\frac{g''(\theta)}{g(\theta)}.$$

The left-hand side of this equation depends only on r and the right-hand side depends only on θ. Thus there is a constant λ such that

$$\frac{r^2}{f(r)}\left(f''(r) + \frac{1}{r}f'(r)\right) = \lambda = -\frac{g''(\theta)}{g(\theta)}.$$

Hence

$$f''(r) + \frac{1}{r}f'(r) - \frac{\lambda}{r^2}f(r) = 0 \quad \text{for } r > 0,$$

and

$$g''(\theta) + \lambda g(\theta) = 0 \quad \text{for } \theta \in \mathbb{S}^1.$$

Our next step is to analyze the equation for g. Then we shall recall some facts about Fourier series, after which we shall turn to the equation for f. The equation for g describes the eigenvalue problem for $-\frac{d^2}{d\theta^2}$ on $\mathbb{S}^1$. This equation has nontrivial solutions when $\lambda = k^2$, $k = 0, 1, 2, \cdots$. When $\lambda = 0$, the general solution is $g(\theta) = a_0$, where a_0 is a constant. For $\lambda = k^2$, $k = 1, 2, \cdots$, the general solution is

$$g(\theta) = a_k \cos k\theta + b_k \sin k\theta,$$

where a_k and b_k are constants. Moreover, the normalized eigenfunctions

$$\frac{1}{\sqrt{2\pi}}, \ \frac{1}{\sqrt{\pi}}\cos k\theta, \ \frac{1}{\sqrt{\pi}}\sin k\theta, \quad k = 1, 2, \cdots,$$

form an orthonormal basis for $L^2(\mathbb{S}^1)$. In other words, for any $v \in L^2(\mathbb{S}^1)$,

$$v(\theta) = \frac{1}{\sqrt{2\pi}}a_0 + \frac{1}{\sqrt{\pi}}\sum_{k=1}^{\infty}\left(a_k \cos k\theta + b_k \sin k\theta\right),$$

where

$$a_0 = \frac{1}{\sqrt{2\pi}} \int_{\mathbb{S}^1} v(\theta)\, d\theta,$$

and for $k = 1, 2, \cdots$,

$$a_k = \frac{1}{\sqrt{\pi}} \int_{\mathbb{S}^1} v(\theta) \cos k\theta\, d\theta,$$

$$b_k = \frac{1}{\sqrt{\pi}} \int_{\mathbb{S}^1} v(\theta) \sin k\theta\, d\theta.$$

This series for v is its *Fourier series* and a_0, a_k, b_k are its *Fourier coefficients.* The series converges in $L^2(\mathbb{S}^1)$. Moreover,

$$\|v\|_{L^2(\mathbb{S}^1)} = \left(a_0^2 + \sum_{k=1}^{\infty} (a_k^2 + b_k^2) \right)^{\frac{1}{2}}.$$

As for f, when $\lambda = 0$ the general solution is

$$f(r) = c_0 + d_0 \log r,$$

where c_0 and d_0 are constants. Now we want $u(r, \theta) = f(r)g(\theta)$ to be harmonic in $\mathbb{R}^2$, thus f must remain bounded as r tends to 0. Therefore we must have $d_0 = 0$, and so

$$f(r) = c_0$$

is a constant function. For $\lambda = k^2$, $k = 1, 2, \cdots$, the general solution is

$$f(r) = c_k r^k + d_k r^{-k},$$

where c_k and d_k are constants. Again f must remain bounded as r tends to 0, so $d_k = 0$ and

$$f(r) = c_k r^k.$$

In summary, a harmonic function u in $\mathbb{R}^2$ of the form $u(r, \theta) = f(r)g(\theta)$ is given by

$$u(r, \theta) = a_0,$$

or by

$$u(r, \theta) = a_k r^k \cos k\theta + b_k r^k \sin k\theta,$$

for $k = 1, 2, \cdots$, where a_0, a_k, b_k are constants.

Remark 3.3.1. Note that $r^k \cos k\theta$ and $r^k \sin k\theta$ are homogenous harmonic polynomials of degree k in $\mathbb{R}^2$. Taking $z = x + iy$, we see that

$$r^k \cos k\theta + i r^k \sin k\theta = r^k e^{ik\theta} = (x + iy)^k,$$

and hence

$$r^k \cos k\theta = \operatorname{Re}(x + iy)^k, \quad r^k \sin k\theta = \operatorname{Im}(x + iy)^k.$$

Now, we are ready to solve the Dirichlet problem for the Laplace equation in the unit disc $B_1 \subset \mathbb{R}^2$. Let φ be a function on $\partial B_1 = \mathbb{S}^1$ and consider

$$\begin{aligned} \Delta u &= 0 \quad \text{in } B_1, \\ u &= \varphi \quad \text{on } \mathbb{S}^1. \end{aligned} \tag{3.3.1}$$

We first derive an expression for the solution purely formally. We seek a solution of the form

$$u(r,\theta) = \frac{1}{\sqrt{2\pi}} a_0 + \frac{1}{\sqrt{\pi}} \sum_{k=1}^{\infty} \big(a_k r^k \cos k\theta + b_k r^k \sin k\theta \big). \tag{3.3.2}$$

The terms in the series are all harmonic functions of the form $f(r)g(\theta)$ that we discussed above. Thus the sum $u(r,\theta)$ should also be harmonic. Letting $r = 1$ in (3.3.2), we get

$$\varphi(\theta) = u(1,\theta) = \frac{1}{\sqrt{2\pi}} a_0 + \frac{1}{\sqrt{\pi}} \sum_{k=1}^{\infty} \big(a_k \cos k\theta + b_k \sin k\theta \big).$$

Therefore, the constants a_0, a_k and b_k, $k = 1, 2, \cdots$, should be the Fourier coefficients of φ. Hence,

$$a_0 = \frac{1}{\sqrt{2\pi}} \int_{\mathbb{S}^1} \varphi(\theta)\, d\theta, \tag{3.3.3}$$

and for $k = 1, 2, \cdots$,

$$\begin{aligned} a_k &= \frac{1}{\sqrt{\pi}} \int_{\mathbb{S}^1} \varphi(\theta) \cos k\theta \, d\theta, \\ b_k &= \frac{1}{\sqrt{\pi}} \int_{\mathbb{S}^1} \varphi(\theta) \sin k\theta \, d\theta. \end{aligned} \tag{3.3.4}$$

Theorem 3.3.2. *Suppose $\varphi \in L^2(\mathbb{S}^1)$ and u is given by (3.3.2), (3.3.3) and (3.3.4). Then u is smooth in B_1 and satisfies*

$$\Delta u = 0 \quad \textit{in } B_1.$$

Moreover,

$$\lim_{r \to 1} \| u(r,\cdot) - \varphi \|_{L^2(\mathbb{S}^1)} = 0.$$

Proof. Since $\varphi \in L^2(\mathbb{S}^1)$, we have

$$\|\varphi\|^2_{L^2(\mathbb{S}^1)} = a_0^2 + \sum_{k=0}^{\infty} (a_k^2 + b_k^2) < \infty.$$

In the following, we fix an $R \in (0,1)$.

First, we set

$$S_{00}(r,\theta)=\sum_{k=1}^{\infty}|a_kr^k\cos k\theta+b_kr^k\sin k\theta|.$$

By (3.3.2), we have

$$|u(r,\theta)|\le\frac{1}{\sqrt{2\pi}}|a_0|+\frac{1}{\sqrt{\pi}}S_{00}(r,\theta).$$

To estimate S_{00}, we note that, for any $r\in[0,R]$ and any $\theta\in\mathbb{S}^1$,

$$S_{00}(r,\theta)\le\sum_{k=1}^{\infty}\left(a_k^2+b_k^2\right)^{\frac{1}{2}}R^k.$$

By the Cauchy inequality, we get

$$S_{00}(r,\theta)\le\left(\sum_{k=1}^{\infty}\left(a_k^2+b_k^2\right)\right)^{\frac{1}{2}}\left(\sum_{k=1}^{\infty}R^{2k}\right)^{\frac{1}{2}}<\infty.$$

Hence, the series defining u in (3.3.2) is convergent absolutely and uniformly in $\bar{B}_R$. Therefore, u is continuous in $\bar{B}_R$.

Next, we take any positive integer m and any nonnegative integers m_1 and m_2 with $m_1+m_2=m$. For any $r\in[0,R]$ and any $\theta\in\mathbb{S}^1$, we have formally

$$\partial_x^{m_1}\partial_y^{m_2}u(r,\theta)=\frac{1}{\sqrt{\pi}}\sum_{k=1}^{\infty}\partial_x^{m_1}\partial_y^{m_2}\big(a_kr^k\cos k\theta+b_kr^k\sin k\theta\big).$$

In order to justify the interchange of the order of differentiation and summation, we need to prove that the series in the right-hand side is convergent absolutely and uniformly in $\bar{B}_R$. Set

$$S_{m_1m_2}(r,\theta)=\sum_{k=1}^{\infty}\left|\partial_x^{m_1}\partial_y^{m_2}\big(a_kr^k\cos k\theta+b_kr^k\sin k\theta\big)\right|. \tag{3.3.5}$$

(We note that this is S_{00} defined earlier if $m_1=m_2=0$.) By using rectangular coordinates, it is easy to check that, for $k<m$,

$$\partial_x^{m_1}\partial_y^{m_2}\big(a_kr^k\cos k\theta+b_kr^k\sin k\theta\big)=0,$$

and for $k\ge m$,

$$\begin{aligned}&\left|\partial_x^{m_1}\partial_y^{m_2}\big(a_kr^k\cos k\theta+b_kr^k\sin k\theta\big)\right|\\&\qquad\le k(k-1)\cdots(k-m+1)\big(a_k^2+b_k^2\big)^{\frac{1}{2}}r^{k-m}.\end{aligned}$$

Hence, for any $r\in[0,R]$ and any $\theta\in\mathbb{S}^1$,

$$S_{m_1m_2}(r,\theta)\le\sum_{k=m}^{\infty}\big(a_k^2+b_k^2\big)^{\frac{1}{2}}k^mR^{k-m}.$$

By the Cauchy inequality, we have

$$S_{m_1m_2}(r,\theta) \le \left(\sum_{k=m}^{\infty}\left(a_k^2+b_k^2\right)\right)^{\frac{1}{2}}\left(\sum_{k=m}^{\infty}k^{2m}R^{2(k-m)}\right)^{\frac{1}{2}} < \infty.$$

This verifies that the series defining $\partial_x^{m_1}\partial_y^{m_2}u$ is convergent absolutely and uniformly in $\bar{B}_R$, for any m_1 and m_2 with $m_1+m_2\ge 1$. Hence, u is smooth in $\bar{B}_R$ for any $R<1$ and all derivatives of u can be obtained from term-by-term differentiation in (3.3.2). Then it is easy to conclude that $\Delta u = 0$.

We now prove the L^2-convergence. First, by the series expansions of u and φ, we have

$$u(r,\theta)-\varphi(\theta) = \frac{1}{\sqrt{\pi}}\sum_{k=1}^{\infty}\big(a_k\cos k\theta + b_k\sin k\theta\big)\left(r^k-1\right),$$

and then

$$\int_{\mathbb{S}^1}|u(r,\theta)-\varphi(\theta)|^2\,d\theta = \sum_{k=1}^{\infty}\left(a_k^2+b_k^2\right)\left(r^k-1\right)^2.$$

We note that $r^k\to 1$ as $r\to 1$ for each fixed $k\ge 1$. For a positive integer K to be determined, we write

$$\int_{\mathbb{S}^1}|u(r,\theta)-\varphi(\theta)|^2\,d\theta = \sum_{k=1}^{K}\left(a_k^2+b_k^2\right)\left(r^k-1\right)^2 + \sum_{k=K+1}^{\infty}\left(a_k^2+b_k^2\right)\left(r^k-1\right)^2.$$

For any $\varepsilon>0$, there exists a positive integer $K=K(\varepsilon)$ such that

$$\sum_{k=K+1}^{\infty}\left(a_k^2+b_k^2\right) < \varepsilon.$$

Then there exists a $\delta>0$, depending on ε and K, such that

$$\sum_{k=1}^{K}\left(a_k^2+b_k^2\right)\left(r^k-1\right)^2 < \varepsilon \quad \text{for any } r\in(1-\delta,1),$$

since the series in the left-hand side consists of finitely many terms. Therefore, we obtain

$$\int_{\mathbb{S}^1}|u(r,\theta)-\varphi(\theta)|^2\,d\theta < 2\varepsilon \quad \text{for any } r\in(1-\delta,1).$$

This implies the desired L^2-convergence as $r\to 1$. □

We note that u is smooth in B_1 even if the boundary value φ is only L^2. Naturally, we ask whether u in Theorem 3.3.2 is continuous up to ∂B_1, or, more generally, whether u is smooth up to ∂B_1. We note that a function is smooth in $\bar{B}_1$ if all its derivatives are continuous in $\bar{B}$.

Theorem 3.3.3. *Suppose $\varphi \in C^\infty(\mathbb{S}^1)$ and u is given by (3.3.2), (3.3.3) and (3.3.4). Then u is smooth in $\bar{B}_1$ with $u(1,\cdot) = \varphi$.*

Proof. Let m_1 and m_2 be nonnegative integers with $m_1 + m_2 = m$. We need to prove that the series defining $\partial_x^{m_1}\partial_y^{m_2}u(r,\theta)$ converges absolutely and uniformly in $\bar{B}_1$. Let $S_{m_1m_2}$ be the series defined in (3.3.5). Then for any $r \in [0,1]$ and $\theta \in \mathbb{S}^1$,

$$S_{m_1m_2}(r,\theta) \le \sum_{k=1}^{\infty} \left(a_k^2 + b_k^2\right)^{\frac{1}{2}} k^m.$$

To prove that the series in the right-hand side is convergent, we need to improve estimates of a_k and b_k, $k = 1, 2, \cdots$. By definitions of a_k and b_k in (3.3.4) and integrations by parts, we have

$$\begin{aligned} a_k &= \frac{1}{\sqrt{\pi}} \int_{\mathbb{S}^1} \varphi(\theta) \cos k\theta \, d\theta = -\frac{1}{\sqrt{\pi}} \int_{\mathbb{S}^1} \varphi'(\theta) \frac{\sin k\theta}{k} \, d\theta, \\ b_k &= \frac{1}{\sqrt{\pi}} \int_{\mathbb{S}^1} \varphi(\theta) \sin k\theta \, d\theta = \frac{1}{\sqrt{\pi}} \int_{\mathbb{S}^1} \varphi'(\theta) \frac{\cos k\theta}{k} \, d\theta. \end{aligned}$$

Hence, $\{kb_k, -ka_k\}$ are the coefficients of the Fourier series of φ', so

$$\sum_{k=1}^{\infty} k^2(a_k^2 + b_k^2) = \|\varphi'\|_{L^2(\mathbb{S}^1)}^2.$$

By continuing this process, we obtain for any positive integer ℓ,

$$\sum_{k=1}^{\infty} k^{2\ell}(a_k^2 + b_k^2) = \|\varphi^{(\ell)}\|_{L^2(\mathbb{S}^1)}^2 < \infty.$$

Hence by the Cauchy inequality, we have for any $r \in [0,1]$ and $\theta \in \mathbb{S}^1$,

$$\begin{aligned} S_{m_1m_2}(r,\theta) &\le \sum_{k=1}^{\infty} \left(a_k^2 + b_k^2\right)^{\frac{1}{2}} k^m \\ &\le \left(\sum_{k=1}^{\infty} k^{2(m+1)}\left(a_k^2 + b_k^2\right)\right)^{\frac{1}{2}} \left(\sum_{k=1}^{\infty} k^{-2}\right)^{\frac{1}{2}}. \end{aligned}$$

This implies

$$S_{m_1m_2}(r,\theta) \le C_m \|\varphi^{(m+1)}\|_{L^2(\mathbb{S}^1)},$$

where C_m is a positive constant depending only on m. Then the series defining $\partial_x^{m_1}\partial_y^{m_2}u$ converges absolutely and uniformly in $\bar{B}_1$. Therefore, $\partial_x^{m_1}\partial_y^{m_1}u$ is continuous in $\bar{B}_1$. $\square$

By examining the proofs of Theorem 3.3.2 and Theorem 3.3.3, we have the following estimates. For any integer $m \geq 0$ and any $R \in (0, 1)$,

$$\|u\|_{C^m(\bar{B}_R)} \leq C_{m,R} \|\varphi\|_{L^2(\mathbb{S}^1)},$$

where $C_{m,R}$ is a positive constant depending only on m and R. This estimate controls the C^m-norm of u in $\bar{B}_R$ in terms of the L^2-norm of φ on $\mathbb{S}^1$. It is referred to as an *interior estimate*. Moreover, for any integer $m \geq 0$,

$$\|u\|_{C^m(\bar{B}_1)} \leq C_m \sum_{i=0}^{m+1} \|\varphi^{(i)}\|_{L^2(\mathbb{S}^1)},$$

where C_m is a positive constant depending only on m. This is referred to as a *global estimate*.

If we are interested only in the continuity of u up to ∂B_1, we have the following result.

Corollary 3.3.4. *Suppose* $\varphi \in C^1(\mathbb{S}^1)$ *and* u *is given by* (3.3.2), (3.3.3) *and* (3.3.4). *Then* u *is smooth in* B_1, *continuous in* $\bar{B}_1$ *and satisfies* (3.3.1).

Proof. It follows from Theorem 3.3.2 that u is smooth in B_1 and satisfies $\Delta u = 0$ in B_1. The continuity of u up to ∂B_1 follows from the proof of Theorem 3.3.3 with $m_1 = m_2 = 0$. □

The regularity assumption on φ in Corollary 3.3.4 does not seem to be optimal. It is natural to ask whether it suffices to assume that φ is in $C(\mathbb{S}^1)$ instead of $C^1(\mathbb{S}^1)$. To answer this question, we need to analyze pointwise convergence of Fourier series. We will not pursue along this direction in this book. An alternative approach is to rewrite the solution u in (3.3.2). With the explicit expressions of a_0, a_k, b_k in terms of φ as in (3.3.3) and (3.3.4), we can write

$$u(r, \theta) = \int_{\mathbb{S}^1} K(r, \theta, \eta)\varphi(\eta)\, d\eta, \tag{3.3.6}$$

where

$$K(r, \theta, \eta) = \frac{1}{2\pi} + \frac{1}{\pi} \sum_{k=1}^{\infty} r^k \cos k(\theta - \eta).$$

The integral expression (3.3.6) is called the *Poisson integral formula* and the function K is called the *Poisson kernel.* We can verify that

$$K(r, \theta, \eta) = \frac{1}{2\pi} \cdot \frac{1 - r^2}{1 - 2r\cos(\theta - \eta) + r^2}. \tag{3.3.7}$$

We leave this verification as an exercise. In Section 4.1, we will prove that u is continuous up to ∂B_1 if φ is continuous on ∂B_1. In fact, we will derive Poisson integral formulas for arbitrary dimension and prove they provide solutions of Dirichlet problems for the Laplace equation in balls with continuous boundary values.

Next, we compare the regularity results in Theorems 3.3.2–3.3.3. For Dirichlet problems for the Laplace equation in the unit disc, solutions are always smooth in B_1 even with very weak boundary values, for example, with L^2-boundary values. This is the *interior smoothness*, i.e., solutions are always smooth inside the domain regardless of the regularity of boundary values. Moreover, solutions are smooth up to the boundary if boundary values are also smooth. This is the *global smoothness.*

3.3.2. Initial/Boundary-Value Problems. In the following, we solve initial/boundary-value problems for the 1-dimensional heat equation and the 1-dimensional wave equation by separation of variables, and discuss regularity of these solutions. We denote by (x, t) points in $[0, \pi] \times [0, \infty)$, with x identified as the space variable and t as the time variable.

We first discuss the 1-dimensional heat equation. Let u_0 be a continuous function in $[0, \pi]$. Consider the initial/boundary-value problem

$$\begin{aligned} &u_t - u_{xx} = 0 \quad \text{in } (0, \pi) \times (0, \infty), \\ &u(x, 0) = u_0(x) \quad \text{for } x \in (0, \pi), \\ &u(0, t) = u(\pi, t) = 0 \quad \text{for } t \in (0, \infty). \end{aligned} \tag{3.3.8}$$

Physically, u represents the temperature in an insulated rod with ends kept at zero temperature.

We first consider

$$\begin{aligned} &u_t - u_{xx} = 0 \quad \text{in } (0, \pi) \times (0, \infty), \\ &u(0, t) = u(\pi, t) = 0 \quad \text{for } t \in (0, \infty). \end{aligned} \tag{3.3.9}$$

We intend to find its solutions by separation of variables. Set

$$u(x, t) = a(t)w(x) \quad \text{for } (x, t) \in (0, \pi) \times (0, \infty).$$

Then

$$a'(t)w(x) - a(t)w''(x) = 0,$$

and hence

$$\frac{a'(t)}{a(t)} = \frac{w''(x)}{w(x)}.$$

Since the left-hand side is a function of t and the right-hand side is a function of x, there is a constant λ such that each side is $-\lambda$. Then

$$a'(t) + \lambda a(t) = 0 \quad \text{for } t \in (0, \infty),$$

and

$$\begin{aligned} &w''(x) + \lambda w(x) = 0 \quad \text{for } x \in (0, \pi), \\ &w(0) = w(\pi) = 0. \end{aligned} \tag{3.3.10}$$

We note that (3.3.10) describes the homogeneous eigenvalue problem for $-\frac{d^2}{dx^2}$ in $(0, \pi)$. The eigenvalues of this problem are $\lambda_k = k^2$, $k = 1, 2, \cdots$, and the corresponding normalized eigenfunctions

$$w_k(x) = \sqrt{\frac{2}{\pi}} \sin kx$$

form a complete orthonormal set in $L^2(0, \pi)$. For any $v \in L^2(0, \pi)$, the Fourier series of v with respect to $\{\sqrt{\frac{2}{\pi}} \sin kx\}$ is given by

$$v(x) = \sqrt{\frac{2}{\pi}} \sum_{k=1}^{\infty} v_k \sin kx,$$

where

$$v_k = \sqrt{\frac{2}{\pi}} \int_0^{\pi} v(x) \sin kx \, dx.$$

The Fourier series converges to v in $L^2(0, \pi)$, and

$$\|v\|_{L^2(0,\pi)} = \left(\sum_{k=1}^{\infty} v_k^2 \right)^{\frac{1}{2}}.$$

For $k = 1, 2, \cdots$, let

$$u_k(x, t) = a_k(t) w_k(x)$$

be a solution of (3.3.9). Then $a_k(t)$ satisfies the ordinary differential equation

$$a_k'(t) + k^2 a_k(t) = 0.$$

Thus, $a_k(t)$ has the form

$$a_k(t) = a_k e^{-k^2 t},$$

where a_k is constant. Therefore, for $k = 1, 2, \cdots$, we have

$$u_k(x, t) = \sqrt{\frac{2}{\pi}} a_k e^{-k^2 t} \sin kx \quad \text{for } (x, t) \in (0, \pi) \times (0, \infty).$$

We note that u_k satisfies the heat equation and the boundary value in (3.3.8). In order to get a solution satisfying the equation, the boundary value and the initial value in (3.3.8), we consider an infinite linear combination of u_k and choose coefficients appropriately.

We emphasize that we identified an eigenvalue problem (3.3.10) from the initial/boundary-value problem (3.3.8). We note that $-\frac{d^2}{dx^2}$ in (3.3.10) originates from the term evolving spatial derivative in the equation in (3.3.8) and that the boundary condition in (3.3.10) is the same as that in (3.3.8).

Now, let us suppose that

$$u(x, t) = \sqrt{\frac{2}{\pi}} \sum_{k=1}^{\infty} a_k e^{-k^2 t} \sin kx \tag{3.3.11}$$

solves (3.3.8). In order to identify the coefficients a_k, $k = 1, 2, \cdots$, we calculate formally:

$$u(x,0) = \sqrt{\frac{2}{\pi}} \sum_{k=1}^{\infty} a_k \sin kx,$$

but we are given the initial condition $u(x,0) = u_0(x)$ for $x \in (0,\pi)$. Thus we take the constants a_k, $k = 1, 2, \cdots$, to be the Fourier coefficients of u_0 with respect to the basis $\{\sqrt{\frac{2}{\pi}} \sin kx\}$ of $L^2(0,\pi)$, i.e.,

$$a_k = \sqrt{\frac{2}{\pi}} \int_0^{\pi} u_0(x) \sin kx \, dx \quad \text{for } k = 1, 2, \cdots . \tag{3.3.12}$$

Next we prove that u in (3.3.11) indeed solves (3.3.8). To do this, we need to prove that u is at least C^2 in x and C^1 in t and satisfies (3.3.8) under appropriate conditions on u_0. We first have the following result.

Theorem 3.3.5. *Suppose $u_0 \in L^2(0,\pi)$ and u is given by (3.3.11) and (3.3.12). Then u is smooth in $[0,\pi] \times (0,\infty)$ and*

$$\begin{aligned} u_t - u_{xx} &= 0 \quad \text{in } (0,\pi) \times (0,\infty), \\ u(0,t) = u(\pi,t) &= 0 \quad \text{for } t \in (0,\infty). \end{aligned}$$

Moreover,

$$\lim_{t \to 0} \|u(\cdot,t) - u_0\|_{L^2(0,\pi)} = 0.$$

Proof. Let i and j be nonnegative integers. For any $x \in [0,\pi]$ and $t \in (0,\infty)$, we have formally

$$\partial_x^i \partial_t^j u(x,t) = \sqrt{\frac{2}{\pi}} \sum_{k=1}^{\infty} a_k \frac{d^j}{dt^j}(e^{-k^2 t}) \frac{d^i}{dx^i}(\sin kx).$$

In order to justify the interchange of the order of differentiation and summation, we need to prove that the series in the right-hand side is convergent absolutely and uniformly for any $(x,t) \in [0,\pi] \times [t_0,\infty)$, for an arbitrarily fixed $t_0 > 0$. Set

$$S_{ij}(x,t) = \sum_{k=1}^{\infty} \left| a_k \frac{d^j}{dt^j}(e^{-k^2 t}) \frac{d^i}{dx^i}(\sin kx) \right| . \tag{3.3.13}$$

Fix $t_0 > 0$. Then for any $(x,t) \in [0,\pi] \times [t_0,\infty)$,

$$S_{ij}(x,t) \le \sum_{k=1}^{\infty} |a_k| \frac{k^{i+2j}}{e^{k^2 t}} \le \sum_{k=1}^{\infty} |a_k| \frac{k^{i+2j}}{e^{k^2 t_0}}.$$

Since $u_0 \in L^2(0,\pi)$, we have

$$\|u_0\|_{L^2(0,\pi)}^2 = \sum_{k=1}^{\infty} a_k^2 < \infty.$$

Then the Cauchy inequality implies, for any $(x,t) \in [0,\pi] \times [t_0,\infty)$,

$$S_{ij}(x,t) \leq \left(\sum_{k=1}^{\infty} a_k^2\right)^{\frac{1}{2}} \left(\sum_{k=1}^{\infty} \frac{k^{2i+4j}}{e^{2k^2 t_0}}\right)^{\frac{1}{2}} \leq C_{i,j,t_0} \|u_0\|_{L^2(0,\pi)}, \tag{3.3.14}$$

where C_{i,j,t_0} is a positive constant depending only on i, j and t_0. This verifies that the series defining $\partial_x^i \partial_t^j u(x,t)$ is convergent absolutely and uniformly for $(x,t) \in [0,\pi] \times [t_0,\infty)$, for any nonnegative integers i and j. Hence u is smooth in $[0,\pi] \times [t_0,\infty)$ for any $t_0 > 0$. Therefore, all derivatives of u can be obtained from term-by-term differentiation in (3.3.11). It is then easy to conclude that u satisfies the heat equation and the boundary condition in (3.3.8).

We now prove the L^2-convergence. First, from the series expansions of u and u_0, we see that

$$u(x,t) - u_0(x) = \sqrt{\frac{2}{\pi}} \sum_{k=1}^{\infty} a_k \big(e^{-k^2 t} - 1\big) \sin kx,$$

and then

$$\int_0^{\pi} |u(x,t) - u_0(x)|^2 \, dx = \sum_{k=1}^{\infty} a_k^2 \big(e^{-k^2 t} - 1\big)^2.$$

We note that $e^{-k^2 t} \to 1$ as $t \to 0$ for each fixed $k \geq 1$. For a positive integer K to be determined, we write

$$\int_0^{\pi} |u(x,t) - u_0(x)|^2 \, dx = \sum_{k=1}^{K} a_k^2 \big(e^{-k^2 t} - 1\big)^2 + \sum_{k=K+1}^{\infty} a_k^2 \big(e^{-k^2 t} - 1\big)^2.$$

For any $\varepsilon > 0$, there exists a positive integer $K = K(\varepsilon)$ such that

$$\sum_{k=K+1}^{\infty} a_k^2 < \varepsilon.$$

Then there exists a $\delta > 0$, depending on ε and K, such that

$$\sum_{k=1}^{K} a_k^2 \big(e^{-k^2 t} - 1\big)^2 < \varepsilon \quad \text{for any } t \in (0,\delta),$$

since the series in the left-hand side consists of finitely many terms. Therefore, we obtain

$$\int_0^{\pi} |u(x,t) - u_0(x)|^2 \, dx < 2\varepsilon \quad \text{for any } t \in (0,\delta).$$

This implies the desired L^2-convergence as $t \to 0$. □

In fact, (3.3.14) implies the following estimate. For any integer $m \geq 0$ and any $t_0 > 0$,

$$\|u\|_{C^m([0,\pi]\times[t_0,\infty))} \leq C_{m,t_0}\|u_0\|_{L^2(0,\pi)},$$

where C_{m,t_0} is a positive constant depending only on m and t_0. This estimate controls the C^m-norm of u in $[0,\pi] \times [t_0,\infty)$ in terms of the L^2-norm of u_0 on $(0,\pi)$. It is referred to as an *interior estimate* (with respect to t). We note that u becomes smooth instantly after $t = 0$ even if the initial value u_0 is only L^2.

Naturally, we ask whether u in Theorem 3.3.5 is continuous up to $\{t = 0\}$, or, more generally, whether u is smooth up to $\{t = 0\}$. First, we assume that u is continuous up to $\{t = 0\}$. Then $u_0 \in C[0,\pi]$. By comparing the initial value with the homogeneous boundary value at corners, we have

$$u_0(0) = 0, \quad u_0(\pi) = 0.$$

Next, we assume that u is C^2 in x and C^1 in t up to $\{t = 0\}$. Then $u_0 \in C^2[0,\pi]$. By the homogeneous boundary condition and differentiation with respect to t, we have

$$u_t(0,t) = 0, \quad u_t(\pi,t) = 0 \quad \text{for } t \geq 0.$$

Evaluating at $t = 0$ yields

$$u_t(0,0) = 0, \quad u_t(\pi,0) = 0.$$

Then by the heat equation, we get

$$u_{xx}(0,0) = 0, \quad u_{xx}(\pi,0) = 0,$$

and hence

$$u_0''(0) = 0, \quad u_0''(\pi) = 0.$$

If u is smooth up to $\{t = 0\}$, we can continue this process. Then we have a necessary condition

$$u_0^{(2\ell)}(0) = 0, \quad u_0^{(2\ell)}(\pi) = 0 \quad \text{for any } \ell = 0, 1, \cdots. \tag{3.3.15}$$

Now, we prove that this is also a sufficient condition.

Theorem 3.3.6. *Suppose $u_0 \in C^\infty[0,\pi]$ and u is given by* (3.3.11) *and* (3.3.12). *If* (3.3.15) *holds, then u is smooth in $[0,\pi] \times [0,\infty)$, and $u(\cdot, 0) = u_0$.*

Proof. Let i and j be nonnegative integers. We need to prove that the series defining $\partial_x^i \partial_t^j u(x,t)$ converges absolutely and uniformly for $(x,t) \in$

$[0, \pi] \times [0, \infty)$. Let S_{ij} be the series defined in (3.3.13). Then for any $x \in [0, \pi]$ and $t \geq 0$,

$$S_{ij}(x,t) \leq \sum_{k=1}^{\infty} k^{i+2j}|a_k|.$$

To prove that the series in the right-hand side is convergent, we need to improve estimates of a_k, the coefficients of Fourier series of u_0. With (3.3.15) for $\ell = 0$, we have, upon simple integrations by parts,

$$\begin{aligned} a_k &= \sqrt{\frac{2}{\pi}} \int_0^{\pi} u_0(x) \sin kx \, dx = \sqrt{\frac{2}{\pi}} \int_0^{\pi} u_0'(x) \frac{\cos kx}{k} \, dx \\ &= -\sqrt{\frac{2}{\pi}} \int_0^{\pi} u_0''(x) \frac{\sin kx}{k^2} \, dx. \end{aligned}$$

We note that values at the endpoints are not present since $u_0(0) = u_0(\pi) = 0$ in the first integration by parts and $\sin kx = 0$ at $x = 0$ and $x = \pi$ in the second integration by parts. Hence for any $m \geq 1$, we continue this process with the help of (3.3.15) for $\ell = 0, \cdots, [(m-1)/2]$ and obtain

$$a_k = (-1)^{\frac{m-1}{2}} \sqrt{\frac{2}{\pi}} \int_0^{\pi} u_0^{(m)}(x) \frac{\cos kx}{k^m} \, dx \quad \text{if } m \text{ is odd},$$

and

$$a_k = (-1)^{\frac{m}{2}} \sqrt{\frac{2}{\pi}} \int_0^{\pi} u_0^{(m)}(x) \frac{\sin kx}{k^m} \, dx \quad \text{if } m \text{ is even}.$$

In other words, $\{k^m a_k\}$ is the sequence of coefficients of the Fourier series of $\pm u_0^{(m)}$ with respect to $\{\sqrt{\frac{2}{\pi}} \sin kx\}$ or $\{\sqrt{\frac{2}{\pi}} \cos kx\}$, where m determines uniquely the choice of positive or negative sign and the choice of the sine or the cosine function. Then, we have

$$\sum_{k=1}^{\infty} k^{2m} a_k^2 \leq \|u_0^{(m)}\|_{L^2(0,\pi)}^2.$$

Hence, by the Cauchy inequality, we obtain that, for any $(x,t) \in [0, \pi] \times [0, \infty)$ and any m,

$$S_{ij}(x,t) \leq \sum_{k=1}^{\infty} k^{i+2j}|a_k| \leq \left(\sum_{k=1}^{\infty} k^{2m} a_k^2 \right)^{\frac{1}{2}} \left(\sum_{k=1}^{\infty} k^{2(i+2j-m)} \right)^{\frac{1}{2}}.$$

By taking $m = i + 2j + 1$, we get

$$S_{ij}(x,t) \leq C_{ij} \|u_0^{(m)}\|_{L^2(0,\pi)},$$

where C_{ij} is a positive constant depending only on i and j. This implies that the series defining $\partial_x^i \partial_t^j u(x,t)$ converges absolutely and uniformly for $(x,t) \in [0, \pi] \times [0, \infty)$. Therefore, $\partial_x^i \partial_t^j u$ is continuous in $[0, \pi] \times [0, \infty)$. □

If we are interested only in the continuity of u up to $t = 0$, we have the following result.

Corollary 3.3.7. *Suppose $u_0 \in C^1[0,\pi]$ and u is given by (3.3.11) and (3.3.12). If $u_0(0) = u_0(\pi) = 0$, then u is smooth in $[0,\pi]\times(0,\infty)$, continuous in $[0,\pi]\times[0,\infty)$ and satisfies (3.3.8).*

Proof. It follows from Theorem 3.3.5 that u is smooth in $[0,\pi]\times(0,\infty)$ and satisfies the heat equation and the homogeneous boundary condition in (3.3.8). The continuity of u up to $t = 0$ follows from the proof of Theorem 3.3.6 with $i = j = 0$ and $m = 1$. □

The regularity assumption on u_0 in Corollary 3.3.7 does not seem to be optimal. It is natural to ask whether it suffices to assume that u_0 is in $C[0,\pi]$ instead of in $C^1[0,\pi]$. To answer this question, we need to analyze pointwise convergence of Fourier series. We will not pursue this issue in this book.

Now we provide another expression of u in (3.3.11). With explicit expressions of a_k in terms of u_0 in (3.3.12), we can write

$$u(x,t) = \int_0^\pi G(x,y;t)u_0(y)\,dy, \tag{3.3.16}$$

where

$$G(x,y;t) = \frac{2}{\pi}\sum_{k=1}^\infty e^{-k^2t}\sin kx\sin ky,$$

for any $x, y \in [0,\pi]$ and $t > 0$. The function G is called the *Green's function* of the initial/boundary-value problem (3.3.8). For each fixed $t > 0$, the series for G is convergent absolutely and uniformly for any $x, y \in [0,\pi]$. In fact, this uniform convergence justifies the interchange of the order of summation and integration in obtaining (3.3.16). The Green's function G satisfies the following properties:

(1) Symmetry: $G(x,y;t) = G(y,x;t)$.

(2) Smoothness: $G(x,y;t)$ is smooth in $x, y \in [0,\pi]$ and $t > 0$.

(3) Solution of the heat equation: $G_t - G_{xx} = 0$.

(4) Homogeneous boundary values: $G(0,y;t) = G(\pi,y;t) = 0$.

These properties follow easily from the explicit expression for G. They imply that u in (3.3.16) is a smooth function in $[0,\pi]\times(0,\infty)$ and satisfies the heat equation with homogeneous boundary values. We can prove directly with the help of the explicit expression of G that u in (3.3.16) is continuous up to $t = 0$ and satisfies $u(\cdot,0) = u_0$ under appropriate assumptions on u_0. We point out that G can also be expressed in terms of the fundamental solution of the heat equation. See Chapter 5 for discussions of the fundamental solution.

Next we discuss initial/boundary-value problems for the 1-dimensional wave equation. Let u_0 and u_1 be continuous functions on $[0, \pi]$. Consider

$$
\begin{aligned}
&u_{tt} - u_{xx} = 0 \quad \text{in } (0, \pi) \times (0, \infty), \\
&u(x, 0) = u_0(x), \ u_t(x, 0) = u_1(x) \quad \text{for } x \in (0, \pi), \\
&u(0, t) = u(\pi, t) = 0 \quad \text{for } t \in (0, \infty).
\end{aligned} \tag{3.3.17}
$$

We proceed as for the heat equation, first considering the problem

$$
\begin{aligned}
&u_{tt} - u_{xx} = 0 \quad \text{in } (0, \pi) \times (0, \infty), \\
&u(0, t) = u(\pi, t) = 0 \quad \text{for } t \in (0, \infty),
\end{aligned} \tag{3.3.18}
$$

and asking for solutions of the form

$$u(x, t) = c(t)w(x).$$

An argument similar to that given for the heat equation shows that w must be a solution of the homogeneous eigenvalue problem for $-\frac{d^2}{dx^2}$ on $(0, \pi)$. The eigenvalues of this problem are $\lambda_k = k^2$, $k = 1, 2, \cdots$, and the corresponding normalized eigenfunctions

$$w_k(x) = \sqrt{\frac{2}{\pi}} \sin kx$$

form a complete orthonormal set in $L^2(0, \pi)$. For $k = 1, 2, \cdots$, let

$$u_k(x, t) = c_k(t)w_k(x)$$

be a solution of (3.3.18). Then $c_k(t)$ satisfies the ordinary differential equation

$$c_k''(t) + k^2 c_k(t) = 0.$$

Thus, $c_k(t)$ has the form

$$c_k(t) = a_k \cos kt + b_k \sin kt,$$

where a_k and b_k are constants. Therefore, for $k = 1, 2, \cdots$, we have

$$u_k(x, t) = \sqrt{\frac{2}{\pi}} (a_k \cos kt + b_k \sin kt) \sin kx.$$

Now, let us suppose that

$$u(x, t) = \sqrt{\frac{2}{\pi}} \sum_{k=1}^{\infty} (a_k \cos kt + b_k \sin kt) \sin kx \tag{3.3.19}$$

solves (3.3.17). In order to identify the coefficients a_k and b_k, $k = 1, 2, \cdots$, we calculate formally:

$$u(x,0) = \sqrt{\frac{2}{\pi}} \sum_{k=1}^{\infty} a_k \sin kx,$$

but we are given the initial condition $u(x,0) = u_0(x)$ for $x \in (0,\pi)$. Thus we take the constants a_k, $k = 1, 2, \cdots$, to be the Fourier coefficients of u_0 with respect to the basis $\left\{\sqrt{\frac{2}{\pi}} \sin kx\right\}$ of $L^2(0,\pi)$, i.e.,

$$a_k = \sqrt{\frac{2}{\pi}} \int_0^{\pi} u_0(x) \sin kx \, dx \quad \text{for } k = 1, 2, \cdots. \tag{3.3.20}$$

Differentiating (3.3.19) term by term, we find

$$u_t(x,t) = \sqrt{\frac{2}{\pi}} \sum_{k=1}^{\infty} (-ka_k \sin kt + kb_k \cos kt) \sin kx,$$

and evaluating at $t = 0$ gives

$$u_t(x,0) = \sqrt{\frac{2}{\pi}} \sum_{k=1}^{\infty} kb_k \sin kx.$$

From the initial condition $u_t(x,0) = u_1(x)$, we see that kb_k, for $k = 1, 2, \cdots$, are the Fourier coefficients of u_1 with respect to the basis $\left\{\sqrt{\frac{2}{\pi}} \sin kx\right\}$ of $L^2(0,\pi)$, i.e.,

$$b_k = \sqrt{\frac{2}{\pi}} \int_0^{\pi} u_1(x) \frac{\sin kx}{k} \, dx \quad \text{for } k = 1, 2, \cdots. \tag{3.3.21}$$

We now discuss the regularity of u in (3.3.19). Unlike the case of the heat equation, in order to get differentiability of u now, we need to impose similar differentiability assumptions on initial values. Proceding as for the heat equation, we note that if u is a C^2-solution, then

$$\begin{aligned} u_0(0) &= 0, \quad u_1(0) = 0, \quad u_0''(0) = 0, \\ u_0(\pi) &= 0, \quad u_1(\pi) = 0, \quad u_0''(\pi) = 0. \end{aligned} \tag{3.3.22}$$

Theorem 3.3.8. *Suppose $u_0 \in C^3[0,\pi]$, $u_1 \in C^2[0,\pi]$ and u is defined by (3.3.19), (3.3.20) and (3.3.21). If u_0, u_1 satisfy (3.3.22), then u is C^2 in $[0,\pi] \times [0,\infty)$ and is a solution of (3.3.17).*

Proof. Let i and j be two nonnegative integers with $0 \le i + j \le 2$. For any $x \in [0,\pi]$ and $t \in (0,\infty)$, we have formally

$$\partial_x^i \partial_t^j u(x,t) = \sqrt{\frac{2}{\pi}} \sum_{k=1}^{\infty} \frac{d^j}{dt^j}(a_k \cos kt + b_k \sin kt) \frac{d^i}{dx^i}(\sin kx).$$

In order to justify the interchange of the order of differentiation and summation, we need to prove that the series in the right-hand side is convergent absolutely and uniformly for any $(x,t) \in [0,\pi]\times[0,\infty)$. Set

$$T_{ij}(x,t) = \sum_{k=1}^{\infty} \left| \frac{d^j}{dt^j}(a_k \cos kt + b_k \sin kt)\frac{d^i}{dx^i}(\sin kx) \right|.$$

Hence, for any $(x,t) \in [0,\pi]\times[0,\infty)$,

$$T_{ij}(x,t) \le \sum_{k=1}^{\infty} k^{i+j}\left(a_k^2 + b_k^2\right)^{\frac{1}{2}}.$$

To prove the convergence of the series in the right-hand side, we need to improve estimates for a_k and b_k. By (3.3.22) and integration by parts, we have

$$a_k = \sqrt{\frac{2}{\pi}}\int_0^{\pi} u_0(x)\sin kx\,dx = -\sqrt{\frac{2}{\pi}}\int_0^{\pi} u_0'''(x)\frac{\cos kx}{k^3}\,dx,$$

$$b_k = \sqrt{\frac{2}{\pi}}\int_0^{\pi} u_1(x)\frac{\sin kx}{k}\,dx = -\sqrt{\frac{2}{\pi}}\int_0^{\pi} u_1''(x)\frac{\sin kx}{k^3}\,dx.$$

In other words, $\{k^3 a_k\}$ is the sequence of Fourier coefficients of $-u_0'''(x)$ with respect to $\{\sqrt{\frac{2}{\pi}}\cos kx\}$, and $\{k^3 b_k\}$ is the sequence of Fourier coefficients of $-u_1''(x)$ with respect to $\{\sqrt{\frac{2}{\pi}}\sin kx\}$. Hence

$$\sum_{k=1}^{\infty}(k^6 a_k^2 + k^6 b_k^2) \le \|u_0'''\|^2_{L^2(0,\pi)} + \|u_1''\|^2_{L^2(0,\pi)}.$$

By the Cauchy inequality, we obtain that, for any $(x,t) \in [0,\pi]\times[0,\infty)$,

$$T_{ij}(x,t) \le \left(\sum_{k=1}^{\infty} k^{2(i+j+1)}\left(a_k^2 + b_k^2\right)\right)^{\frac{1}{2}} \cdot \left(\sum_{k=1}^{\infty}\frac{1}{k^2}\right)^{\frac{1}{2}} < \infty.$$

Therefore, u is C^2 in $[0,\pi]\times[0,\infty)$ and any derivative of u up to order two may be calculated by a simple term-by-term differentiation. Thus u satisfies (3.3.17). □

By examining the proof, we have

$$\|u\|_{C^2([0,\pi]\times[0,\infty))} \le C\left(\sum_{i=0}^{3}\|u_0^{(i)}\|_{L^2(0,\pi)} + \sum_{i=0}^{2}\|u_1^{(i)}\|_{L^2(0,\pi)}\right),$$

where C is a positive constant independent of u.

In fact, in order to get a C^2-solution of (3.3.17), it suffices to assume $u_0 \in C^2[0,\pi]$, $u_1 \in C^1[0,\pi]$ and the compatibility condition (3.3.22). We will prove this for a more general initial/boundary-value problem for the wave equation in Section 6.1. See Theorem 6.1.3.

Now, we compare the regularity results for solutions of initial/boundary-value problems in Theorems 3.3.5, 3.3.6 and 3.3.8. For the heat equation in Theorem 3.3.5, solutions become smooth immediately after $t = 0$, even for L^2-initial values. This is the *interior smoothness* (with respect to time). We also proved in Theorem 3.3.6 that solutions are smooth up to $\{t = 0\}$ if initial values are smooth with a compatibility condition. This property is called the *global smoothness*. However, solutions of the wave equation exhibit a different property. We proved in Theorem 3.3.8 that solutions have a similar degree of regularity as initial values. In general, solutions of the wave equation do not have better regularity than initial values, and in higher dimensions they are less regular than initial values. We will discuss in Chapter 6 how solutions of the wave equation depend on initial values.

To conclude, we point out that the methods employed in this section to solve initial/boundary-value problems for the 1-dimensional heat equation and wave equation can actually be generalized to higher dimensions. We illustrate this by the heat equation. Let Ω be a bounded smooth domain in $\mathbb{R}^n$ and u_0 be an L^2-function in Ω. We consider

$$\begin{aligned} u_t - \Delta u &= 0 \quad \text{in } \Omega \times (0,\infty), \\ u(\cdot,0) &= u_0 \quad \text{in } \Omega, \\ u &= 0 \quad \text{on } \partial\Omega \times (0,\infty). \end{aligned} \tag{3.3.23}$$

To solve (3.3.23) by separation of variables, we need to solve the eigenvalue problem of $-\Delta$ in Ω with homogeneous boundary values, i.e.,

$$\begin{aligned} \Delta\varphi + \lambda\varphi &= 0 \quad \text{in } \Omega, \\ \varphi &= 0 \quad \text{on } \partial\Omega. \end{aligned} \tag{3.3.24}$$

This is much harder to solve than its 1-dimensional counterpart (3.3.10). Nevertheless, a similar result still holds. In fact, solutions of (3.3.24) are given by a sequence (λ_k, φ_k), where λ_k is a nondecreasing sequence of positive numbers such that $\lambda_k \to \infty$ as $k \to \infty$ and φ_k is a sequence of smooth functions in $\bar{\Omega}$ which forms a basis in $L^2(\Omega)$. Then we can use a similar method to find a solution of (3.3.23) of the form

$$u(x,t) = \sum_{k=1}^{\infty} a_k e^{-\lambda_k t}\varphi_k(x) \quad \text{for any } (x,t) \in \Omega \times (0,\infty).$$

We should remark that solving (3.3.24) is a complicated process. We need to work in Sobolev spaces, spaces of functions with L^2-integrable derivatives. A brief discussion of Sobolev spaces can be found in Subsection 4.4.2.

3.4. Exercises

Exercise 3.1. Classify the following second-order PDEs:

(1) $\displaystyle\sum_{i=1}^{n} u_{x_i x_i} + \sum_{1\le i<j\le n} u_{x_i x_j} = 0.$

(2) $\displaystyle\sum_{1\le i<j\le n} u_{x_i x_j} = 0.$

Exercise 3.2. (1) Let (r,θ) be polar coordinates in $\mathbb{R}^2$, i.e.,

$$x = r\cos\theta, \quad y = r\sin\theta.$$

Prove that the Laplace operator Δ can be expressed by

$$\Delta u = u_{rr} + \frac{1}{r}u_r + \frac{1}{r^2}u_{\theta\theta}.$$

(2) Let (r,θ,ϕ) be spherical coordinates in $\mathbb{R}^3$, i.e.,

$$x = r\sin\theta\cos\varphi, \quad y = r\sin\theta\sin\varphi, \quad z = r\cos\theta.$$

Prove that the Laplace operator Δ can be expressed by

$$\Delta u = \frac{1}{r^2}\frac{\partial}{\partial r}\left(r^2\frac{\partial u}{\partial r}\right) + \frac{1}{r^2\sin\theta}\frac{\partial}{\partial\theta}\left(\sin\theta\frac{\partial u}{\partial\theta}\right) + \frac{1}{r^2\sin^2\theta}\frac{\partial^2 u}{\partial\varphi^2}.$$

Exercise 3.3. Discuss the uniqueness of the following problems using energy methods:

(1) $\begin{cases} \Delta u - u^3 = f & \text{in } \Omega, \\ u = \varphi & \text{on } \partial\Omega; \end{cases}$

(2) $\begin{cases} \Delta u - u\int_\Omega u^2(y)\,dy = f & \text{in } \Omega, \\ u = \varphi & \text{on } \partial\Omega. \end{cases}$

Exercise 3.4. Let Ω be a bounded C^1-domain in $\mathbb{R}^n$ and u be a C^2-function in $\bar\Omega\times[0,T]$ satisfying

$$\begin{aligned} u_t - \Delta u &= f \quad \text{in } \Omega\times(0,\infty), \\ u(\cdot,0) &= u_0 \quad \text{in } \Omega, \\ u &= 0 \quad \text{on } \partial\Omega\times(0,\infty). \end{aligned}$$

Prove

$$\begin{aligned} &\sup_{0\le t\le T}\int_\Omega |\nabla u(\cdot,t)|^2\,dx + \int_0^T\int_\Omega u_t^2\,dxdt \\ &\qquad \le C\Big(\int_\Omega |\nabla u_0|^2\,dx + \int_0^T\int_\Omega f^2\,dxdt\Big), \end{aligned}$$

where C is a positive constant depending only on Ω.

Exercise 3.5. Prove that the Poisson kernel in (3.3.6) is given by (3.3.7).

Exercise 3.6. For any $u_0 \in L^2(0,\pi)$, let u be given by (3.3.11). For any nonnegative integers i and j, prove

$$\sup_{[0,\pi]} |\partial_x^i \partial_t^j u(\cdot,t)| \to 0 \quad \text{as } t \to \infty.$$

Exercise 3.7. Let G be defined as in (3.3.16). Prove

$$|G(x,y;t)| \le \frac{1}{\sqrt{\pi t}} \quad \text{for any } x, y \in [0,\pi] \text{ and } t > 0.$$

Exercise 3.8. For any $u_0 \in L^2(0,\pi)$, solve the following problem by separation of variables:

$$\begin{aligned} &u_t - u_{xx} = 0 \quad \text{in } (0,\pi)\times(0,\infty),\\ &u(x,0) = u_0(x) \quad \text{for any } x \in (0,\pi),\\ &u_x(0,t) = u_x(\pi,t) = 0 \quad \text{for any } t \in (0,\infty). \end{aligned}$$

Exercise 3.9. For any $u_0 \in L^2(0,\pi)$ and $f \in L^2((0,\pi)\times(0,\infty))$, find a formal explicit expression of a solution of the problem

$$\begin{aligned} &u_t - u_{xx} = f \quad \text{in } (0,\pi)\times(0,\infty),\\ &u(x,0) = u_0(x) \quad \text{for any } x \in (0,\pi),\\ &u(0,t) = u(\pi,t) = 0 \quad \text{for any } t \in (0,\infty). \end{aligned}$$

Exercise 3.10. For any $u_0, u_1 \in L^2(0,\pi)$ and $f \in L^2((0,\pi)\times(0,\infty))$, find a formal explicit expression of a solution of the problem

$$\begin{aligned} &u_{tt} - u_{xx} = f \quad \text{in } (0,\pi)\times(0,\infty),\\ &u(x,0) = u_0(x),\ u_t(x,0) = u_1(x) \quad \text{for any } x \in (0,\pi),\\ &u(0,t) = u(\pi,t) = 0 \quad \text{for any } t \in (0,\infty). \end{aligned}$$

Exercise 3.11. Let T be a positive constant, Ω be a bounded C^1-domain in $\mathbb{R}^n$ and u be C^2 in x and C^1 in t in $\bar\Omega\times[0,T]$. Suppose u satisfies

$$\begin{aligned} u_t - \Delta u &= 0 \quad \text{in } \Omega\times(0,T),\\ u(\cdot,T) &= 0 \quad \text{in } \Omega,\\ u &= 0 \quad \text{on } \partial\Omega\times(0,T). \end{aligned}$$

Prove that $u = 0$ in $\Omega\times(0,T)$.

Hint: The function $J(t) = \log\int_\Omega u^2(x,t)\,dx$ is a decreasing convex function.

Exercise 3.12. Classify homogeneous harmonic polynomials in $\mathbb{R}^3$ by following the steps outlined below. Let (r,θ,ϕ) be spherical coordinates in $\mathbb{R}^3$. (Refer to Exercise 3.2.) Suppose u is a homogeneous harmonic polynomial of degree m in $\mathbb{R}^3$ and set $u = r^m Q_m(\theta,\varphi)$ for some function Q_m defined in $\mathbb{S}^2$.

(1) Prove that Q_m satisfies

$$m(m+1)Q_m + \frac{1}{\sin\theta}\frac{\partial}{\partial\theta}\left(\sin\theta\frac{\partial Q_m}{\partial\theta}\right) + \frac{1}{\sin^2\theta}\frac{\partial^2 Q_m}{\partial\varphi^2} = 0.$$

(2) Prove that, if Q_m is of the form $f(\theta)g(\varphi)$, then

$$Q_m(\theta,\varphi) = \big(A\cos k\varphi + B\sin k\varphi\big)f_{m,k}(\cos\theta),$$

where

$$f_{m,k}(\mu) = (1-\mu^2)^{\frac{k}{2}}\frac{d^{m+k}}{d\mu^{m+k}}(1-\mu^2)^m \quad \text{for } \mu\in[-1,1],$$

for $k = 0, 1, \cdots, m$.

(3) Sketch the zero set of Q_m on $\mathbb{S}^2$ according to $k=0$, $1\le k\le m-1$ and $k=m$.

Chapter 4

Laplace Equations

The Laplace operator is probably the most important differential operator and has a wide range of important applications.

In Section 4.1, we discuss the fundamental solution of the Laplace equation and its applications. First, we introduce the important notion of Green's functions, which are designed to solve Dirichlet boundary-value problems. Due to the simple geometry of balls, we are able to find Green's functions in balls and derive an explicit expression of solutions of the Dirichlet problem in balls, the so-called Poisson integral formula. Second, we discuss regularity of harmonic functions using the fundamental solution. We derive interior gradient estimates and prove that harmonic functions are analytic.

In Section 4.2, we study the mean-value property of harmonic functions. First, we demonstrate that the mean-value property presents an equivalent description of harmonic functions. Due to this equivalence, the mean-value property provides another tool to study harmonic functions. To illustrate this, we derive the maximum principle for harmonic functions from the mean-value property.

In Section 4.3, we discuss harmonic functions using the maximum principle. This section is independent of Section 4.1 and Section 4.2. The maximum principle is an important tool in studying harmonic functions, or in general, solutions of second-order elliptic differential equations. In this section, the maximum principle is proved based on the algebraic structure of the Laplace equation. As an application, we derive a priori estimates for solutions of the Dirichlet boundary-value problem. We also derive interior gradient estimates and the differential Harnack inequality. As a final application, we solve the Dirichlet problem for the Laplace equation in a large class of bounded domains by Perron's method.

We point out that several results in this chapter are proved by multiple methods. For example, interior gradient estimates are proved by three methods: the fundamental solution, the mean-value property and the maximum principle.

In Section 4.4, we discuss the Poisson equation. We first discuss regularity of classical solutions using the fundamental solution. Then we discuss weak solutions and solve the Dirichlet problem in the weak sense. The method is from functional analysis, and the Riesz representation theorem plays an essential role. The presentation in this part is brief. The main purpose is to introduce notions of weak solutions and Sobolev spaces.

4.1. Fundamental Solutions

The *Laplace operator* Δ is defined on C^2-functions u in a domain in $\mathbb{R}^n$ by

$$\Delta u = \sum_{i=1}^{n} u_{x_i x_i}.$$

The equation $\Delta u = 0$ is called the *Laplace equation* and its C^2-solutions are called *harmonic functions*.

4.1.1. Green's Identities. One of the important properties of the Laplace equation is its spherical symmetry. As discussed in Example 3.1.7, the Laplace equation is preserved by rotations about some point in $\mathbb{R}^n$, say the origin. Hence, it is plausible that there exist special solutions that are invariant under rotations. We now seek harmonic functions u in $\mathbb{R}^n$ which are *radial*, i.e., functions depending only on $r = |x|$. Set

$$v(r) = u(x).$$

For any $i = 1, \cdots, n$ and $x \neq 0$, we get

$$u_{x_i} = v'(r)\frac{x_i}{r},$$

and

$$u_{x_i x_i} = v''(r)\frac{x_i^2}{r^2} + v'(r)\left(\frac{1}{r} - \frac{x_i^2}{r^3}\right).$$

Hence

$$\Delta u = v'' + \frac{n-1}{r}v' = 0.$$

If $v' \neq 0$, then

$$(\log v')' + \frac{n-1}{r} = 0,$$

or

$$\big(\log(r^{n-1}v')\big)' = 0.$$

A simple integration then yields, for $n = 2$,

$$v(r) = c_1 + c_2 \log r \quad \text{for any } r > 0,$$

and for $n \geq 3$,

$$v(r) = c_3 + c_4 r^{2-n} \quad \text{for any } r > 0,$$

where c_i are constants for $i = 1, 2, 3, 4$. We note that $v(r)$ has a singularity at $r = 0$ as long as it is not constant. For reasons to be apparent soon, we are interested in solutions with a singularity such that

$$\int_{\partial B_r} \frac{\partial u}{\partial \nu}\, dS = 1 \quad \text{for any } r > 0.$$

In the following, we set $c_1 = c_3 = 0$ and choose c_2 and c_4 accordingly. In fact, we have

$$c_2 = \frac{1}{2\pi},$$

and

$$c_4 = \frac{1}{(2-n)\omega_n},$$

where ω_n is the surface area of the unit sphere in $\mathbb{R}^n$.

Definition 4.1.1. Let Γ be defined for $x \in \mathbb{R}^n \setminus \{0\}$ by

$$\Gamma(x) = \frac{1}{2\pi} \log |x| \quad \text{for } n = 2,$$

and

$$\Gamma(x) = \frac{1}{(2-n)\omega_n} |x|^{2-n} \quad \text{for } n \geq 3.$$

The function Γ is called the *fundamental solution* of the Laplace operator.

We note that Γ is harmonic in $\mathbb{R}^n \setminus \{0\}$, i.e.,

$$\Delta \Gamma = 0 \quad \text{in } \mathbb{R}^n \setminus \{0\},$$

and

$$\int_{\partial B_r} \frac{\partial \Gamma}{\partial \nu}\, dS = 1 \quad \text{for any } r > 0.$$

Moreover, Γ has a singularity at the origin. By a simple calculation, we have, for any $i, j = 1, \cdots, n$ and any $x \neq 0$,

$$\Gamma_{x_i}(x) = \frac{1}{\omega_n} \cdot \frac{x_i}{|x|^n},$$

and

$$\Gamma_{x_i x_j} = \frac{1}{\omega_n} \left(\frac{\delta_{ij}}{|x|^n} - \frac{n x_i x_j}{|x|^{n+2}} \right).$$

We note that Γ and its first derivatives are integrable in any neighborhood of the origin, even though Γ has a singularity there. However, the second derivatives of Γ are not integrable near the origin.

To proceed, we review several integral formulas. Let Ω be a C^1-domain in $\mathbb{R}^n$ and $\nu = (\nu_1, \cdots, \nu_n)$ be the unit exterior normal to $\partial\Omega$. Then for any $u, v \in C^1(\Omega) \cap C(\bar{\Omega})$ and $i = 1, \cdots, n$,

$$\int_\Omega u_{x_i} v \, dx = \int_{\partial\Omega} uv\nu_i \, dS - \int_\Omega uv_{x_i} \, dx.$$

This is the *integration by parts* in higher-dimensional Euclidean space. Now for any $w \in C^2(\Omega) \cap C^1(\bar{\Omega})$ and $v \in C^1(\Omega) \cap C(\bar{\Omega})$, substitute w_{x_i} for u to get

$$\int_\Omega (vw_{x_i x_i} + v_{x_i} w_{x_i}) \, dx = \int_{\partial\Omega} vw_{x_i}\nu_i \, dS.$$

By summing up for $i = 1, \cdots, n$, we get *Green's formula*,

$$\int_\Omega \big(v\Delta w + \nabla v \cdot \nabla w\big) \, dx = \int_{\partial\Omega} v\frac{\partial w}{\partial \nu} \, dS.$$

For any $v, w \in C^2(\Omega) \cap C^1(\bar{\Omega})$, we interchange v and w and subtract to get a second version of Green's formula,

$$\int_\Omega (v\Delta w - w\Delta v) \, dx = \int_{\partial\Omega} \left(v\frac{\partial w}{\partial \nu} - w\frac{\partial v}{\partial \nu}\right) dS.$$

Taking $v \equiv 1$ in either version of Green's formula, we get

$$\int_\Omega \Delta w \, dx = \int_{\partial\Omega} \frac{\partial w}{\partial \nu} \, dS.$$

We note that all these integral formulas hold if Ω is only a piecewise C^1-domain.

Now we prove *Green's identity*, which plays an important role in discussions of harmonic functions.

Theorem 4.1.2. *Suppose Ω is a bounded C^1-domain in $\mathbb{R}^n$ and that $u \in C^1(\bar{\Omega}) \cap C^2(\Omega)$. Then for any $x \in \Omega$,*

$$\begin{aligned} u(x) = &\int_\Omega \Gamma(x-y)\Delta_y u(y) \, dy \\ &- \int_{\partial\Omega} \left(\Gamma(x-y)\frac{\partial u}{\partial \nu_y}(y) - u(y)\frac{\partial \Gamma}{\partial \nu_y}(x-y)\right) dS_y. \end{aligned}$$

Proof. We fix an $x \in \Omega$ and write $\Gamma = \Gamma(x - \cdot)$ for brevity. For any $r > 0$ such that $B_r(x) \subset \Omega$, the function Γ is smooth in $\Omega \setminus B_r(x)$. By applying Green's formula to u and Γ in $\Omega \setminus B_r(x)$, we get

$$\begin{aligned} \int_{\Omega\setminus B_r(x)} (\Gamma\Delta u - u\Delta\Gamma) \, dy = &\int_{\partial\Omega} \left(\Gamma\frac{\partial u}{\partial \nu} - u\frac{\partial \Gamma}{\partial \nu_y}\right) dS_y \\ &+ \int_{\partial B_r(x)} \left(\Gamma\frac{\partial u}{\partial \nu} - u\frac{\partial \Gamma}{\partial \nu_y}\right) dS_y, \end{aligned}$$

where ν is the unit exterior normal to $\partial(\Omega\backslash B_r(x))$. Now $\Delta\Gamma = 0$ in $\Omega\backslash B_r(x)$, so letting $r \to 0$, we have

$$\int_\Omega \Gamma\Delta u\, dy = \int_{\partial\Omega}\left(\Gamma\frac{\partial u}{\partial\nu} - u\frac{\partial\Gamma}{\partial\nu_y}\right) dS_y + \lim_{r\to 0}\int_{\partial B_r(x)}\left(\Gamma\frac{\partial u}{\partial\nu} - u\frac{\partial\Gamma}{\partial\nu_y}\right) dS_y.$$

For $n \geq 3$, by the definition of Γ, we get

$$\begin{aligned}\left|\int_{\partial B_r(x)} \Gamma\frac{\partial u}{\partial\nu}\, dS_y\right| &= \left|\frac{r^{2-n}}{(2-n)\omega_n}\int_{\partial B_r(x)}\frac{\partial u}{\partial\nu}\, dS_y\right| \\ &\leq \frac{r}{n-2}\max_{\partial B_r(x)}|\nabla u| \to 0 \quad \text{as} \quad r \to 0,\end{aligned}$$

and

$$-\int_{\partial B_r(x)} u\frac{\partial\Gamma}{\partial\nu_y}\, dS_y = \frac{1}{\omega_n r^{n-1}}\int_{\partial B_r(x)} u\, dS_y \to u(x) \quad \text{as} \quad r \to 0,$$

where ν is normal to $\partial B_r(x)$ and points to x. This implies the desired result for $n \geq 3$. We proceed similarly for $n = 2$. □

Remark 4.1.3. We note that

$$\int_{\partial\Omega}\frac{\partial\Gamma}{\partial\nu_y}(x-y)\, dS_y = 1,$$

for any $x \in \Omega$. This can be obtained by taking $u \equiv 1$ in Theorem 4.1.2.

If u has a compact support in Ω, then Theorem 4.1.2 implies

$$u(x) = \int_\Omega \Gamma(x-y)\Delta u(y)\, dy.$$

By computing formally, we have

$$u(x) = \int_\Omega \Delta_y\Gamma(x-y)u(y)\, dy.$$

In the sense of distributions, we write

$$\Delta_y\Gamma(x-y) = \delta_x.$$

Here δ_x is the Dirac measure at x, which assigns unit mass to x. The term "fundamental solution" is reflected in this identity. We will not give a formal definition of distribution in this book.

4.1.2. Green's Functions. Now we discuss the Dirichlet boundary-value problem using Theorem 4.1.2. Let f be a continuous function in Ω and φ a continuous function on $\partial\Omega$. Consider

$$\begin{aligned}\Delta u &= f \quad \text{in } \Omega,\\ u &= \varphi \quad \text{on } \partial\Omega.\end{aligned} \tag{4.1.1}$$

Lemma 3.2.1 asserts the uniqueness of a solution in $C^2(\Omega)\cap C^1(\bar\Omega)$. An alternative method to obtain the uniqueness is by the maximum principle, which will be discussed later in this chapter. Let $u \in C^2(\Omega) \cap C^1(\bar{\Omega})$ be a solution of (4.1.1). By Theorem 4.1.2, u can be expressed in terms of f and φ, with one *unknown term* $\frac{\partial u}{\partial \nu}$ on $\partial\Omega$. We intend to eliminate this term by adjusting Γ. We emphasize that we cannot prescribe $\frac{\partial u}{\partial \nu}$ on $\partial\Omega$ together with u on $\partial\Omega$.

For each fixed $x \in \Omega$, we consider a function $\Phi(x, \cdot) \in C^2(\Omega) \cap C^1(\bar{\Omega})$ with $\Delta_y \Phi(x, y) = 0$ in Ω. Green's formula implies

$$0 = \int_\Omega \Phi(x,y)\Delta u(y)\,dy - \int_{\partial\Omega} \left(\Phi(x,y)\frac{\partial u}{\partial \nu_y}(y) - u(y)\frac{\partial \Phi}{\partial \nu_y}(x,y) \right) dS_y.$$

Set

$$\gamma(x,y) = \Gamma(x-y) - \Phi(x,y).$$

By a substraction from Green's identity in Theorem 4.1.2, we obtain, for any $x \in \Omega$,

$$u(x) = \int_\Omega \gamma(x,y)\Delta u(y)\,dy - \int_{\partial\Omega} \left(\gamma(x,y)\frac{\partial u}{\partial \nu_y}(y) - u(y)\frac{\partial \gamma}{\partial \nu_y}(x,y) \right) dS_y.$$

We will choose Φ appropriately so that $\gamma(x, \cdot) = 0$ on $\partial\Omega$. Then, $\frac{\partial u}{\partial \nu}$ on $\partial\Omega$ is eliminated from the boundary integral. The process described above leads to the important concept of Green's functions.

To summarize, for each fixed $x \in \Omega$, we consider $\Phi(x, \cdot) \in C^1(\bar{\Omega}) \cap C^2(\Omega)$ such that

$$\begin{aligned} \Delta_y \Phi(x,y) &= 0 \quad \text{for any } y \in \Omega, \\ \Phi(x,y) &= \Gamma(x-y) \quad \text{for any } y \in \partial\Omega. \end{aligned} \tag{4.1.2}$$

The existence of Φ in general domains is not the main issue in our discussion here. We will prove later that $\Phi(x, \cdot)$ is smooth in Ω for each fixed x if it exists. (See Theorem 4.1.10.)

Definition 4.1.4. The *Green's function* G for the domain Ω is defined by

$$G(x,y) = \Gamma(x-y) - \Phi(x,y),$$

for any $x, y \in \Omega$ with $x \neq y$.

In other words, for each fixed $x \in \Omega$, $G(x, \cdot)$ differs from $\Gamma(x - \cdot)$ by a harmonic function in Ω and vanishes on $\partial\Omega$. If such a G exists, then the solution u of the Dirichlet problem (4.1.1) can be expressed by

$$u(x) = \int_\Omega G(x,y)f(y)\,dy + \int_{\partial\Omega} \varphi(y)\frac{\partial G}{\partial \nu_y}(x,y)\,dS_y. \tag{4.1.3}$$

We note that the Green's function $G(x, y)$ is defined as a function of $y \in \bar{\Omega} \setminus \{x\}$ for each fixed $x \in \Omega$. Now we discuss properties of G as a function of x and y. As was mentioned, we will not discuss the existence of the Green's function in general domains. However, we should point out that the Green's function is unique if it exists. This follows from Lemma 3.2.1 or Corollary 4.2.9, since the difference of any two Green's functions is harmonic, with vanishing boundary values.

Lemma 4.1.5. *Let G be the Green's function in Ω. Then $G(x,y)=G(y,x)$ for any $x,y\in\Omega$ with $x\neq y$.*

Proof. For any $x_1,x_2\in\Omega$ with $x_1\neq x_2$, take $r>0$ small enough that $B_r(x_1)\subset\Omega$, $B_r(x_2)\subset\Omega$ and $B_r(x_1)\cap B_r(x_2)=\emptyset$. Set $G_i(y)=G(x_i,y)$ and $\Gamma_i(y)=\Gamma(x_i-y)$ for $i=1,2$. By Green's formula in $\Omega\setminus(B_r(x_1)\cup B_r(x_2))$, we get

$$\int_{\Omega\setminus(B_r(x_1)\cup B_r(x_2))}(G_1\Delta G_2-G_2\Delta G_1)\,dy=\int_{\partial\Omega}\left(G_1\frac{\partial G_2}{\partial\nu}-G_2\frac{\partial G_1}{\partial\nu}\right)dS_y$$
$$+\int_{\partial B_r(x_1)}\left(G_1\frac{\partial G_2}{\partial\nu}-G_2\frac{\partial G_1}{\partial\nu}\right)dS_y+\int_{\partial B_r(x_2)}\left(G_1\frac{\partial G_2}{\partial\nu}-G_2\frac{\partial G_1}{\partial\nu}\right)dS_y,$$

where ν is the unit exterior normal to $\partial\big(\Omega\setminus(B_r(x_1)\cup B_r(x_2))\big)$. Since $G_i(y)$ is harmonic for $y\neq x_i$, $i=1,2$, and vanishes on $\partial\Omega$, we have

$$\int_{\partial B_r(x_1)}\left(G_1\frac{\partial G_2}{\partial\nu}-G_2\frac{\partial G_1}{\partial\nu}\right)dS_y+\int_{\partial B_r(x_2)}\left(G_1\frac{\partial G_2}{\partial\nu}-G_2\frac{\partial G_1}{\partial\nu}\right)dS_y=0.$$

Now we replace G_1 in the first integral by Γ_1 and replace G_2 in the second integral by Γ_2. Since $G_1-\Gamma_1$ is C^2 in Ω and G_2 is C^2 in $\Omega\setminus B_r(x_2)$, we have

$$\int_{\partial B_r(x_1)}\left((G_1-\Gamma_1)\frac{\partial G_2}{\partial\nu}-G_2\frac{\partial(G_1-\Gamma_1)}{\partial\nu}\right)dS_y\to 0\quad\text{as } r\to 0.$$

Similarly,

$$\int_{\partial B_r(x_2)}\left(G_1\frac{\partial(G_2-\Gamma_2)}{\partial\nu}-(G_2-\Gamma_2)\frac{\partial G_1}{\partial\nu}\right)dS_y\to 0\quad\text{as } r\to 0.$$

Therefore, we obtain

$$\int_{\partial B_r(x_1)}\left(\Gamma_1\frac{\partial G_2}{\partial\nu}-G_2\frac{\partial\Gamma_1}{\partial\nu}\right)dS_y+\int_{\partial B_r(x_2)}\left(G_1\frac{\partial\Gamma_2}{\partial\nu}-\Gamma_2\frac{\partial G_1}{\partial\nu}\right)dS_y\to 0,$$

as $r\to 0$. On the other hand, by explicit expressions of Γ_1 and Γ_2, we have

$$\int_{\partial B_r(x_1)}\Gamma_1\frac{\partial G_2}{\partial\nu}\,dS_y\to 0,\quad \int_{\partial B_r(x_2)}\Gamma_2\frac{\partial G_1}{\partial\nu}\,dS_y\to 0,$$

and

$$-\int_{\partial B_r(x_1)}G_2\frac{\partial\Gamma_1}{\partial\nu}\,dS_y\to G_2(x_1),\quad -\int_{\partial B_r(x_2)}G_1\frac{\partial\Gamma_2}{\partial\nu}\,dS_y\to G_1(x_2),$$

as $r\to 0$. These limits can be proved similarly as in the proof of Theorem 4.1.2. We point out that ν points to x_i on $\partial B_r(x_i)$, for $i=1,2$. We then obtain $G_2(x_1)-G_1(x_2)=0$ and hence $G(x_2,x_1)=G(x_1,x_2)$. □

Finding a Green's function involves solving a Dirichlet problem for the Laplace equation. Meanwhile, Green's functions are introduced to yield an explicit expression of solutions of the Dirichlet problem. It turns out that we can construct Green's functions for some special domains.

4.1.3. Poisson Integral Formula. In the next result, we give an explicit expression of Green's functions in balls. We exploit the geometry of balls in an essential way.

Theorem 4.1.6. *Let G be the Green's function in the ball $B_R \subset \mathbb{R}^n$.*

(1) *In case $n \geq 3$,*

$$G(0,y) = \frac{1}{(2-n)\omega_n}\left(|y|^{2-n} - R^{2-n}\right),$$

for any $y \in B_R \setminus \{0\}$, and

$$G(x,y) = \frac{1}{(2-n)\omega_n}\left(|y-x|^{2-n} - \left(\frac{R}{|x|}\right)^{n-2}\left|y - \frac{R^2}{|x|^2}x\right|^{2-n}\right),$$

for any $x \in B_R \setminus \{0\}$ and $y \in B_R \setminus \{x\}$.

(2) *In case $n = 2$,*

$$G(0,y) = \frac{1}{2\pi}\left(\log|y| - \log R\right),$$

for any $y \in B_R \setminus \{0\}$, and

$$G(x,y) = \frac{1}{2\pi}\left(\log|y-x| - \log\left(\frac{|x|}{R}\left|y - \frac{R^2}{|x|^2}x\right|\right)\right),$$

for any $x \in B_R \setminus \{0\}$ and $y \in B_R \setminus \{x\}$.

Proof. By Definition 4.1.4, we need to find Φ in (4.1.2). We consider $n \geq 3$ first. For $x = 0$,

$$\Gamma(0-y) = \frac{1}{(2-n)\omega_n}|y|^{2-n}.$$

Hence we take

$$\Phi(0,y) = \frac{1}{(2-n)\omega_n}R^{2-n},$$

for any $y \in \bar{B}_R$. Next, we fix an $x \in B_R \setminus \{0\}$ and let $X = R^2x/|x|^2$. Obviously, we have $X \notin \bar{B}_R$ and hence $\Gamma(y-X)$ is harmonic for $y \in B_R$. For any $y \in \partial B_R$, by

$$\frac{|x|}{R} = \frac{R}{|X|},$$

we have $\Delta Oxy \sim \Delta OyX$. Then for any $y \in \partial B_R$,

$$\frac{|x|}{R} = \frac{|y-x|}{|y-X|},$$

and hence,

$$|y - x| = \frac{|x|}{R}|y - X|. \tag{4.1.4}$$

This implies

$$\Gamma(y - x) = \left(\frac{R}{|x|}\right)^{n-2} \Gamma(y - X),$$

for any $x \in B_R \setminus \{0\}$ and $y \in \partial B_R$. Then we take

$$\Phi(x, y) = \left(\frac{R}{|x|}\right)^{n-2} \Gamma(y - X),$$

for any $x \in B_R \setminus \{0\}$ and $y \in B_R \setminus \{x\}$. The proof for $n = 2$ is similar and is omitted. □

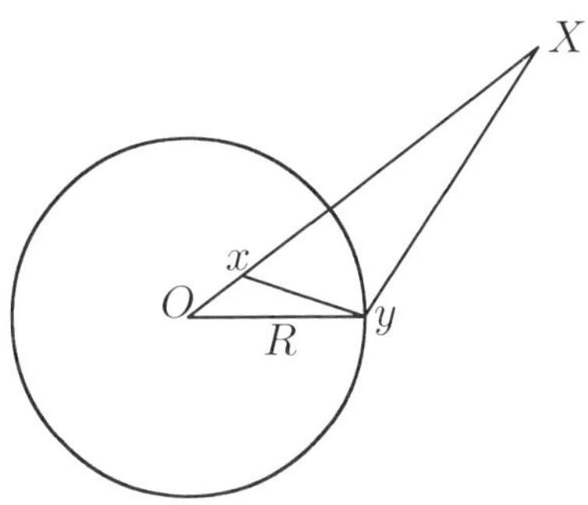

Figure 4.1.1. The reflection about the sphere.

Next, we calculate normal derivatives of the Green's function on spheres.

Corollary 4.1.7. *Let G be the Green's function in B_R. Then*

$$\frac{\partial G}{\partial \nu_y}(x, y) = \frac{R^2 - |x|^2}{\omega_n R|x - y|^n},$$

for any $x \in B_R$ and $y \in \partial B_R$.

Proof. We first consider $n \geq 3$. With $X = R^2 x/|x|^2$ as in the proof of Theorem 4.1.6, we have

$$G(x, y) = \frac{1}{(2 - n)\omega_n}\left(|y - x|^{2-n} - \left(\frac{R}{|x|}\right)^{n-2} |y - X|^{2-n}\right),$$

for any $x \in B_R \setminus \{0\}$ and $y \in B_R \setminus \{x\}$. Hence we get, for such x and y,

$$G_{y_i}(x, y) = \frac{1}{\omega_n}\left(\frac{y_i - x_i}{|y - x|^n} - \left(\frac{R}{|x|}\right)^{n-2} \cdot \frac{y_i - X_i}{|y - X|^n}\right).$$

By (4.1.4) in the proof of Theorem 4.1.6, we have, for any $x \in B_R \setminus \{0\}$ and $y \in \partial B_R$,

$$G_{y_i}(x,y) = \frac{y_i}{\omega_n R^2} \frac{R^2 - |x|^2}{|x-y|^n}.$$

This formula also holds when $x = 0$. With $\nu_i = y_i/R$ for any $y \in \partial B_R$, we obtain

$$\frac{\partial G}{\partial \nu_y}(x,y) = \sum_{i=1}^n \nu_i G_{y_i}(x,y) = \frac{1}{\omega_n R} \cdot \frac{R^2 - |x|^2}{|x-y|^n}.$$

This yields the desired result for $n \geq 3$. The proof for $n = 2$ is similar and is omitted. □

Denote by $K(x,y)$ the function in Corollary 4.1.7, i.e.,

$$K(x,y) = \frac{R^2 - |x|^2}{\omega_n R |x-y|^n}, \tag{4.1.5}$$

for any $x \in B_R$ and $y \in \partial B_R$. It is called the *Poisson kernel.*

Lemma 4.1.8. *Let K be the Poisson kernel defined by (4.1.5). Then*

(1) *$K(x,y)$ is smooth for any $x \in B_R$ and $y \in \partial B_R$;*

(2) *$K(x,y) > 0$ for any $x \in B_R$ and $y \in \partial B_R$;*

(3) *for any fixed $x_0 \in \partial B_R$ and $\delta > 0$,*

$$\lim_{x \to x_0, |x|<R} K(x,y) = 0 \quad \textit{uniformly in } y \in \partial B_R \setminus B_\delta(x_0);$$

(4) *$\Delta_x K(x,y) = 0$ for any $x \in B_R$ and $y \in \partial B_R$;*

(5) *$\int_{\partial B_R} K(x,y)\, dS_y = 1$ for any $x \in B_R$.*

Proof. First, (1), (2) and (3) follow easily from the explicit expression for K as in (4.1.5), and (4) follows easily from the definition $K(x,y) = \frac{\partial}{\partial \nu_y} G(x,y)$ and the fact that $G(x,y)$ is harmonic in x. Of course, we can also verify (4) by a straightforward calculation. An easy derivation of (5) is based on (4.1.3). By taking a $C^2(\bar{B}_R)$ harmonic function u in (4.1.3), we conclude that

$$u(x) = \int_{\partial B_R} K(x,y) u(y)\, dS_y \quad \text{for any } x \in B_R.$$

Then we have (5) by taking $u \equiv 1$. □

Now we are ready to solve the Laplace equation in balls, with prescribed Dirichlet boundary values.

Theorem 4.1.9. *Let φ be a continuous function on ∂B_R and u be defined by*

$$u(x) = \int_{\partial B_R} K(x,y) \varphi(y)\, dS_y \quad \textit{for any } x \in B_R, \tag{4.1.6}$$

where K is the Poisson kernel given by (4.1.5). *Then u is smooth in B_R and $\Delta u = 0$ in B_R. Moreover, for any $x_0 \in \partial B_R$,*

$$\lim_{x \to x_0} u(x) = \varphi(x_0).$$

Proof. By Lemma 4.1.8(1) and (4), we conclude easily that u defined by (4.1.6) is smooth and harmonic in B_R. We need only prove the convergence of u up to the boundary ∂B_R. We fix $x_0 \in \partial B_R$ and take an $x \in B_R$. By Lemma 4.1.8(5), we have

$$\varphi(x_0) = \int_{\partial B_R} K(x,y)\varphi(x_0)\, dS_y.$$

Then

$$u(x) - \varphi(x_0) = \int_{\partial B_R} K(x,y)\big(\varphi(y) - \varphi(x_0)\big)\, dS_y = I_1 + I_2,$$

where

$$I_1 = \int_{\partial B_R \cap B_\delta(x_0)} \cdots, \quad I_2 = \int_{\partial B_R \setminus B_\delta(x_0)} \cdots,$$

for a positive constant δ to be determined. For any $\varepsilon > 0$, we can choose $\delta = \delta(\varepsilon) > 0$ small so that

$$|\varphi(y) - \varphi(x_0)| < \varepsilon$$

for any $y \in \partial B_R \cap B_\delta(x_0)$, because φ is continuous at x_0. Then by Lemma 4.1.8(2) and (5),

$$|I_1| \le \int_{\partial B_R \cap B_\delta(x_0)} K(x,y)\big|\varphi(y) - \varphi(x_0)\big|\, dS_y < \varepsilon.$$

Next, set $M = \max_{\partial B_R} |\varphi|$. By Lemma 4.1.8(3), we can find a $\delta' > 0$ such that

$$K(x,y) \le \frac{\varepsilon}{2M\omega_n R^{n-1}},$$

for any $x \in B_R \cap B_{\delta'}(x_0)$ and any $y \in \partial B_R \setminus B_\delta(x_0)$. We note that δ' depends on ε and $\delta = \delta(\varepsilon)$, and hence only on ε. Then

$$|I_2| \le \int_{\partial B_R \setminus B_\delta(x_0)} K(x,y)\big(|\varphi(y)| + |\varphi(x_0)|\big)\, dS_y \le \varepsilon.$$

Hence

$$|u(x) - \varphi(x_0)| < 2\varepsilon,$$

for any $x \in B_R \cap B_{\delta'}(x_0)$. This implies the convergence of u at $x_0 \in \partial B_R$. □

We note that the function u in (4.1.6) is defined only in B_R. We can extend u to ∂B_R by defining $u = \varphi$ on ∂B_R. Then $u \in C^\infty(B_R) \cap C(\bar{B}_R)$. Therefore, u is a solution of

$$\begin{aligned} \Delta u &= 0 \quad \text{in } B_R, \\ u &= \varphi \quad \text{on } \partial B_R. \end{aligned}$$

The formula (4.1.6) is called the *Poisson integral formula*, or simply the *Poisson formula*.

For $n = 2$, with $x = (r\cos\theta, r\sin\theta)$ and $y = (R\cos\eta, R\sin\eta)$ in (4.1.6), we have

$$u(r\cos\theta, r\sin\theta) = \frac{1}{2\pi}\int_0^{2\pi} K(r,\theta,\eta)\varphi(R\cos\eta, R\sin\eta)\,d\eta,$$

where

$$K(r,\theta,\eta) = \frac{R^2 - r^2}{R^2 - 2Rr\cos(\theta-\eta) + r^2}.$$

Compare with (3.3.6) and (3.3.7) in Section 3.3.

Now we discuss properties of the function defined in (4.1.6). First, $u(x)$ in (4.1.6) is smooth for $|x| < R$, even if the boundary value φ is simply continuous on ∂B_R. In fact, any harmonic function is smooth. We will prove this result later in this section.

Next, by letting $x = 0$ in (4.1.6), we have

$$u(0) = \frac{1}{\omega_n R^{n-1}} \int_{\partial B_R} u(y)\,dS_y.$$

We note that $\omega_n R^{n-1}$ is the surface area of the sphere ∂B_R. Hence, values of harmonic functions at the center of spheres are equal to their average over spheres. This is the *mean-value property*.

Moreover, by Lemma 4.1.8(2) and (5), u in (4.1.6) satisfies

$$\min_{\partial B_R} \varphi \le u \le \max_{\partial B_R} \varphi \quad \text{in } B_R.$$

In other words, harmonic functions in balls are bounded from above by their maximum on the boundary and bounded from below by their minimum on the boundary. Such a result is referred to as the *maximum principle*. Again, this is a general fact, and we will prove it for any harmonic function in any bounded domain.

The mean-value property and the maximum principle are the main topics in Section 4.2 and Section 4.3, respectively.

4.1.4. Regularity of Harmonic Functions. In the following, we discuss regularity of harmonic functions using the fundamental solution of the Laplace equation. First, as an application of Theorem 4.1.2, we prove that harmonic functions are smooth.

Theorem 4.1.10. *Let Ω be a domain in $\mathbb{R}^n$ and $u \in C^2(\Omega)$ be a harmonic function in Ω. Then u is smooth in Ω.*

Proof. We take an arbitrary bounded C^1-domain Ω' in Ω such that $\bar{\Omega}' \subset \Omega$. Obviously, u is C^1 in $\bar{\Omega}'$ and $\Delta u = 0$ in Ω'. By Theorem 4.1.2, we have

$$u(x) = -\int_{\partial\Omega'} \left(\Gamma(x-y)\frac{\partial u}{\partial \nu_y}(y) - u(y)\frac{\partial \Gamma}{\partial \nu_y}(x-y) \right) dS_y,$$

for any $x \in \Omega'$. There is no singularity in the integrand, since $x \in \Omega'$ and $y \in \partial\Omega'$. This implies easily that u is smooth in Ω'. □

We note that, in its definition, a harmonic function is required only to be C^2. Theorem 4.1.10 asserts that the simple algebraic relation $\Delta u = 0$ among *some* of second derivatives of u implies that all partial derivatives of u exist. We will prove a more general result later in Theorem 4.4.2 that u is smooth if Δu is smooth.

Harmonic functions are not only smooth but also analytic. We will prove the analyticity by estimating the radius of convergence for Taylor series of harmonic functions. As the first step, we estimate derivatives of harmonic functions. For convenience, we consider harmonic functions in balls. The following result is referred to as an *interior gradient estimate.* It asserts that first derivatives of a harmonic function at any point are controlled by its maximum absolute value in a ball centered at this point.

Theorem 4.1.11. *Suppose $u \in C\big(\bar{B}_R(x_0)\big)$ is harmonic in $B_R(x_0) \subset \mathbb{R}^n$. Then*

$$|\nabla u(x_0)| \le \frac{C}{R} \max_{\bar{B}_R(x_0)} |u|,$$

where C is a positive constant depending only on n.

Proof. Without loss of generality, we may assume $x_0 = 0$.

We first consider $R = 1$ and employ a local version of Green's identity. Take a cutoff function $\varphi \in C_0^\infty(B_{3/4})$ such that $\varphi = 1$ in $B_{1/2}$ and $0 \le \varphi \le 1$. For any $x \in B_{1/4}$, we write $\Gamma = \Gamma(x - \cdot)$ temporarily. For any $r > 0$ small enough, applying Green's formula to u and $\varphi\Gamma$ in $B_1 \setminus B_r(x)$, we get

$$\int_{B_1\setminus B_r(x)} \big(\varphi\Gamma\Delta u - u\Delta(\varphi\Gamma)\big)\,dy = \int_{\partial B_1} \left(\varphi\Gamma\frac{\partial u}{\partial\nu} - u\frac{\partial(\varphi\Gamma)}{\partial\nu}\right) dS_y + \int_{\partial B_r(x)} \left(\varphi\Gamma\frac{\partial u}{\partial\nu} - u\frac{\partial(\varphi\Gamma)}{\partial\nu}\right) dS_y,$$

where ν is the unit exterior normal to $\partial(B_1 \setminus B_r(x))$. The boundary integral over ∂B_1 is zero since $\varphi = \frac{\partial\varphi}{\partial\nu} = 0$ on ∂B_1. In the boundary integral over $\partial B_r(x)$, we may replace φ by 1 since $B_r(x) \subset B_{1/2}$ if $r < 1/4$. As shown in the proof of Theorem 4.1.2, we have

$$u(x) = \lim_{r\to 0}\int_{\partial B_r(x)} \left(\Gamma\frac{\partial u}{\partial\nu} - u\frac{\partial\Gamma}{\partial\nu}\right) dS_y,$$

where ν is normal to $\partial B_r(x)$ and points toward x. For the domain integral, the first term is zero since $\Delta u = 0$ in B_1. For the second term, we have

$$\Delta(\varphi\Gamma) = \Delta\varphi\Gamma + 2\nabla\varphi\cdot\nabla\Gamma + \varphi\Delta\Gamma.$$

We note that $\Delta\Gamma = 0$ in $B_1 \setminus B_r(x)$ and that the derivatives of φ are zero for $|y| < 1/2$ and $3/4 < |y| < 1$ since φ is constant there. Then we obtain

$$u(x) = -\int_{\frac{1}{2}<|y|<\frac{3}{4}} u(y)\big(\Delta_y\varphi(y)\Gamma(x-y) + 2\nabla_y\varphi(y)\cdot\nabla_y\Gamma(x-y)\big)\,dy,$$

for any $x \in B_{1/4}$. We note that there is no singularity in the integrand for $|x| < 1/4$ and $1/2 < |y| < 3/4$. (This also gives an alternative proof of the smoothness of u in $B_{1/4}$.) Therefore,

$$\nabla u(x) = -\int_{\frac{1}{2}<|y|<\frac{3}{4}} u(y)\big(\Delta_y\varphi(y)\nabla_x\Gamma(x-y) + 2\nabla_y\varphi(y)\cdot\nabla_x\nabla_y\Gamma(x-y)\big)\,dy,$$

for any $x \in B_{1/4}$. Hence, we obtain

$$|\nabla u(x)| \le C\sup_{B_1}|u| \quad \text{for any } x \in B_{\frac{1}{4}},$$

where C is a positive constant depending only on n. We obtain the desired result by taking $x = 0$.

The general case follows from a simple dilation. Define

$$\tilde{u}(x) = u(Rx) \quad \text{for any } x \in B_1.$$

Then $\tilde{u}$ is a harmonic function in B_1. By applying the result we just proved to $\tilde{u}$, we obtain

$$|\nabla\tilde{u}(0)| \le C\sup_{B_1}|\tilde{u}|.$$

Since $\nabla\tilde{u}(0) = R\nabla u(0)$, we have the desired result. □

We note that the proof above consists of two steps. We first prove the desired estimate for $R = 1$ and then extend such an estimate to arbitrary R by a simple scaling. Such a scaling argument is based on the following fact: If u is a harmonic function in B_R, then $\tilde{u}(x) = u(Rx)$ is a harmonic function in B_1. We point out that this scaling argument is commonly used in studying elliptic and parabolic differential equations.

Next, we estimate derivatives of harmonic functions of arbitrary order.

Theorem 4.1.12. *Suppose $u \in C(\bar{B}_R(x_0))$ is harmonic in $B_R(x_0) \subset \mathbb{R}^n$. Then for any multi-index α with $|\alpha| = m$,*

$$|\partial^\alpha u(x_0)| \le \frac{C^m e^{m-1} m!}{R^m} \max_{\bar{B}_R(x_0)} |u|,$$

where C is a positive constant depending only on n.

Proof. The proof is by induction on $m \ge 1$. The case of $m = 1$ holds by Theorem 4.1.11. We assume it holds for m and consider $m + 1$. Let v be an arbitrary derivative of u of order m. Obviously, it is harmonic in $B_R(x_0)$. For any $\theta \in (0, 1)$, by applying Theorem 4.1.11 to v in $B_{(1-\theta)R}(x_0)$, we get

$$|\nabla v(x_0)| \le \frac{C}{(1-\theta)R} \max_{\bar{B}_{(1-\theta)R}(x_0)} |v|.$$

For any $x \in B_{(1-\theta)R}(x_0)$, we have $B_{\theta R}(x) \subset B_R(x_0)$. By the induction assumption, we obtain

$$|v(x)| \le \frac{C^m e^{m-1} m!}{(\theta R)^m} \max_{\bar{B}_{\theta R}(x)} |u|,$$

for any $x \in B_{(1-\theta)R}(x_0)$, and hence

$$\max_{\bar{B}_{(1-\theta)R}(x_0)} |v| \le \frac{C^m e^{m-1} m!}{(\theta R)^m} \max_{\bar{B}_R(x_0)} |u|.$$

Therefore,

$$|\nabla v(x_0)| \le \frac{C^{m+1} e^{m-1} m!}{(1-\theta)\theta^m R^{m+1}} \max_{\bar{B}_R(x_0)} |u|.$$

By taking $\theta = \frac{m}{m+1}$, we have

$$\frac{1}{(1-\theta)\theta^m} = \left(1 + \frac{1}{m}\right)^m (m+1) < e(m+1).$$

This implies

$$|\nabla v(x_0)| \le \frac{C^{m+1} e^m (m+1)!}{R^{m+1}} \max_{\bar{B}_R(x_0)} |u|.$$

Hence the desired result is established for any derivatives of u of order $m + 1$. $\square$

As a consequence of the interior estimate on derivatives, we prove the following compactness result.

Corollary 4.1.13. *Let Ω be a bounded domain in $\mathbb{R}^n$, M be a positive constant and $\{u_k\}$ be a sequence of harmonic functions in Ω such that*

$$\sup_{\Omega} |u_k| \le M \quad \text{for any } k.$$

Then there exist a harmonic function u in Ω and a subsequence $\{u_{k'}\}$ such that

$$u_{k'} \to u \quad \text{uniformly in } \Omega' \text{ as } k' \to \infty,$$

for any Ω' with $\bar{\Omega}' \subset \Omega$.

Proof. Take any Ω' with $\bar{\Omega}' \subset \Omega$ and set $d = \text{dist}(\Omega', \partial\Omega)$. For any $x \in \Omega'$, we have $B_d(x) \subset \Omega$. By applying Theorem 4.1.12 to u_k in $B_d(x)$, we get, for any integer $m \ge 1$,

$$|\nabla^m u_k(x)| \le \frac{C_m}{d^m} \sup_{B_d(x)} |u_k| \le C_m d^{-m} M,$$

where C_m is a positive constant depending only on n and m. Hence

$$\max_{\Omega'} |\nabla^m u_k| \le C_m d^{-m} M.$$

For any $\ell = 0, 1, \cdots$, the mean-value theorem implies

$$|\nabla^\ell u_k(x) - \nabla^\ell u_k(y)| \le C_{\ell+1} d^{-\ell-1} M |x - y|,$$

for any $k = 1, 2, \cdots$, and any $x, y \in \Omega'$.

Next, we take a sequence of domains $\{\Omega_j\}$ with $\bar{\Omega}_j \subset \Omega_{j+1} \subset \cdots \subset \Omega$ and $d_j = \text{dist}(\Omega_j, \partial\Omega) \le 1/j$. Then

$$|\nabla^\ell u_k(x) - \nabla^\ell u_k(y)| \le C_{\ell+1} d_j^{-\ell-1} M |x - y|,$$

for any $\ell = 0, 1, \cdots$, any $k = 1, 2, \cdots$, and any $x, y \in \Omega_j$. By Arzelà's theorem and diagonalization, we can find a function u in Ω and a subsequence $\{u_{k'}\}$ such that

$$u_{k'} \to u \quad \text{in the } C^\ell\text{-norm in } \Omega_j \text{ as } k' \to \infty,$$

for any $j = 1, 2, \cdots$ and any $\ell = 0, 1, \cdots$. By taking $\ell = 2$, we then get $\Delta u = 0$ in each Ω_j from $\Delta u_{k'} = 0$. □

As shown in the proof, $u_{k'}$ converges to u in $C^\ell(\Omega')$ for any Ω' with $\bar{\Omega}' \subset \Omega$ and any $\ell = 0, 1, \cdots$.

Now we are ready to prove that harmonic functions are analytic. Real analytic functions will be studied in Section 7.2. Now we simply introduce the notion. Let u be a (real-valued) function defined in a neighborhood of

$x_0 \in \mathbb{R}^n$. Then u is *analytic* near x_0 if its Taylor series about x_0 is convergent to u in a neighborhood of x_0, i.e., for some $r > 0$,

$$u(x) = \sum_{\alpha} \frac{1}{\alpha!} \partial^{\alpha} u(x_0)(x - x_0)^{\alpha},$$

for any $x \in B_r(x_0)$.

Theorem 4.1.14. *Harmonic functions are analytic.*

Proof. Let Ω be a domain in $\mathbb{R}^n$ and u be a harmonic function in Ω. For any fixed $x_0 \in \Omega$, we prove that u is equal to its Taylor series about x_0 in a neighborhood of x_0. To do this, we take $B_{2R}(x_0) \subset \Omega$ and $h \in \mathbb{R}^n$ with $|h| \leq R$. For any integer $m \geq 1$, we have, by the Taylor expansion,

$$u(x_0 + h) = u(x_0) + \sum_{i=1}^{m-1} \frac{1}{i!} \left[(h_1 \partial_{x_1} + \cdots + h_n \partial_{x_n})^i \, u \right] (x_0) + R_m(h),$$

where

$$R_m(h) = \frac{1}{m!} \left[(h_1 \partial_{x_1} + \cdots + h_n \partial_{x_n})^m \, u \right] (x_0 + \theta h),$$

for some $\theta \in (0, 1)$. Note that $x_0 + h \in B_R(x_0)$ for $|h| < R$ and that $R_m(h)$ consists of n^m terms of derivatives of u of order m. By applying Theorem 4.1.12 to u in $B_R(x_0 + \theta h)$, we obtain

$$\begin{aligned} |R_m(h)| &\leq \frac{1}{m!} |h|^m \cdot n^m \cdot \frac{C^m e^{m-1} m!}{R^m} \max_{\bar{B}_{2R}(x_0)} |u| \\ &\leq \left(\frac{Cne|h|}{R} \right)^m \max_{\bar{B}_{2R}(x_0)} |u|. \end{aligned}$$

Then for any h with $Cne|h| < R/2$, $R_m(h) \to 0$ as $m \to \infty$. Hence,

$$u(x_0 + h) = \sum_{i=0}^{\infty} \frac{1}{i!} \left[(h_1 \partial_{x_1} + \cdots + h_n \partial_{x_n})^i \, u \right] (x_0),$$

for any h with $|h| < (2Cne)^{-1} R$. □

4.2. Mean-Value Properties

It is a simple consequence of the Poisson integral formula that the mean value of a harmonic function over a sphere is equal to its value at the center. Indeed, this *mean-value property* is equivalent to harmonicity. In this section, we briefly discuss harmonic functions using the mean-value property. The fundamental solution and the Poisson integral formula are not used to prove the equivalence of harmonicity and the mean-value property. We point out that the mean-value property is special and cannot be generalized to solutions of general elliptic differential equations. Many results in this section were either proved in the previous section or will be proved in the next section.

We first define the mean-value property. There are two versions of the mean-value property, mean values over spheres and mean values over balls.

Definition 4.2.1. Let Ω be a domain in $\mathbb{R}^n$ and u be a continuous function in Ω. Then

(i) u satisfies the mean-value property over spheres if for any $B_r(x) \subset \Omega$,

$$u(x) = \frac{1}{\omega_n r^{n-1}} \int_{\partial B_r(x)} u(y)\, dS_y;$$

(ii) u satisfies the mean-value property over balls if for any $B_r(x) \subset \Omega$,

$$u(x) = \frac{n}{\omega_n r^n} \int_{B_r(x)} u(y)\, dy,$$

where ω_n is the surface area of the unit sphere in $\mathbb{R}^n$.

We note that $\omega_n r^{r-1}$ is the surface area of the sphere $\partial B_r(x)$ and that $\omega_n r^n/n$ is the volume of the ball $B_r(x)$.

These two versions of the mean-value property are equivalent. In fact, if we write (i) as

$$u(x) r^{n-1} = \frac{1}{\omega_n} \int_{\partial B_r(x)} u(y)\, dS_y,$$

we can integrate with respect to r to get (ii). If we write (ii) as

$$u(x) r^n = \frac{n}{\omega_n} \int_{B_r(x)} u(y)\, dy,$$

we can differentiate with respect to r to get (i).

By a change of variables, we also write mean-value properties in the following equivalent forms: for any $B_r(x) \subset \Omega$,

$$u(x) = \frac{1}{\omega_n} \int_{\partial B_1} u(x + ry)\, dS_y$$

or

$$u(x) = \frac{n}{\omega_n} \int_{B_1} u(x + ry)\, dy.$$

A function satisfying mean-value properties is required only to be continuous to start with. However, a harmonic function is required to be C^2. We now prove that these two requirements are actually equivalent.

Theorem 4.2.2. *Let Ω be a domain in $\mathbb{R}^n$ and u be a function in Ω.*

(i) *If $u \in C^2(\Omega)$ is harmonic in Ω, then u satisfies the mean-value property in Ω.*

(ii) *If* $u \in C(\Omega)$ *satisfies the mean-value property in* Ω, *then* u *is smooth and harmonic in* Ω.

Proof. Take any ball $B_r(x) \subset \Omega$. Then for any $u \in C^2(\Omega)$ and any $\rho \in (0, r)$, we have

$$\begin{aligned}\int_{B_\rho(x)} \Delta u \, dy &= \int_{\partial B_\rho(x)} \frac{\partial u}{\partial \nu} \, dS = \rho^{n-1} \int_{\partial B_1} \frac{\partial u}{\partial \rho}(x + \rho w) \, dS_w \\ &= \rho^{n-1} \frac{\partial}{\partial \rho} \int_{\partial B_1} u(x + \rho w) \, dS_w.\end{aligned} \tag{4.2.1}$$

(i) Let $u \in C^2(\Omega)$ be harmonic in Ω. Then for any $\rho \in (0, r)$,

$$\frac{\partial}{\partial \rho} \int_{\partial B_1} u(x + \rho w) \, dS_w = 0.$$

Integrating from 0 to r, we obtain

$$\int_{\partial B_1} u(x + rw) \, dS_w = \int_{\partial B_1} u(x) \, dS_w = u(x)\omega_n,$$

and hence

$$u(x) = \frac{1}{\omega_n} \int_{\partial B_1} u(x + rw) \, dS_w.$$

This yields the desired mean-value property.

(ii) Let $u \in C(\Omega)$ satisfy the mean-value property. For the smoothness, we prove that u is equal to the convolution of itself with some smooth function. To this end, we choose a smooth function ψ in $[0, 1]$ such that ψ is constant in $[0, \varepsilon]$ and $\psi = 0$ in $[1 - \varepsilon, 1]$ for some $\varepsilon \in (0, 1/2)$, and

$$\omega_n \int_0^1 r^{n-1} \psi(r) \, dr = 1.$$

The existence of such a function can be verified easily. Define $\varphi(x) = \psi(|x|)$. Then $\varphi \in C_0^\infty(B_1)$ and

$$\int_{B_1} \varphi \, dx = 1.$$

Next, we define $\varphi_\varepsilon(x) = \frac{1}{\varepsilon^n} \varphi\left(\frac{x}{\varepsilon}\right)$ for any $\varepsilon > 0$. Then $\operatorname{supp} \varphi_\varepsilon \subset B_\varepsilon$. We claim that

$$u(x) = \int_\Omega u(y) \varphi_\varepsilon(y - x) \, dy,$$

for any $x \in \Omega$ with $\operatorname{dist}(x, \partial\Omega) > \varepsilon$. Then it follows easily that u is smooth. Moreover, by (4.2.1) and the mean-value property, we have, for any $B_r(x) \subset \Omega$,

$$\int_{B_r(x)} \Delta u \, dy = r^{n-1} \frac{\partial}{\partial r} \int_{\partial B_1} u(x + rw) \, dS_w = r^{n-1} \frac{\partial}{\partial r} \big(\omega_n u(x)\big) = 0.$$

This implies $\Delta u = 0$ in Ω.

Now we prove the claim. For any $x \in \Omega$ and $\varepsilon < \text{dist}(x, \partial\Omega)$, we have, by a change of variables and the mean-value property,

$$\begin{aligned}\int_\Omega u(y)\varphi_\varepsilon(y-x)\,dy &= \int_{B_\varepsilon} u(x+y)\varphi_\varepsilon(y)\,dy = \frac{1}{\varepsilon^n}\int_{B_\varepsilon} u(x+y)\varphi\left(\frac{y}{\varepsilon}\right)dy\\ &= \int_{B_1} u(x+\varepsilon z)\varphi(z)\,dz\\ &= \int_0^1 \int_{\partial B_1} u(x+\varepsilon rw)\varphi(rw)r^{n-1}\,dS_w dr\\ &= \int_0^1 \psi(r)r^{n-1}\int_{\partial B_1} u(x+\varepsilon rw)\,dS_w\,dr\\ &= u(x)\omega_n\int_0^1 \psi(r)r^{n-1}\,dr = u(x).\end{aligned}$$

This proves the claim. □

By combining both parts of Theorem 4.2.2, we have the following result.

Corollary 4.2.3. *Harmonic functions are smooth and satisfy the mean-value property.*

Next, we prove an *interior gradient estimate* using the mean-value property.

Theorem 4.2.4. *Suppose $u \in C\big(\bar{B}_R(x_0)\big)$ is harmonic in $B_R(x_0) \subset \mathbb{R}^n$. Then*

$$|\nabla u(x_0)| \le \frac{n}{R}\max_{\bar{B}_R(x_0)}|u|.$$

We note that Theorem 4.2.4 gives an explicit expression of the constant C in Theorem 4.1.11.

Proof. Without loss of generality, we assume $u \in C^1(\bar{B}_R(x_0))$. Otherwise, we consider u in $B_r(x_0)$ for any $r < R$, derive the desired estimate in $B_r(x_0)$ and then let $r \to R$. Since u is smooth, $\Delta(u_{x_i}) = 0$. In other words, u_{x_i} is also harmonic in $B_R(x_0)$. Hence u_{x_i} satisfies the mean-value property. Upon a simple integration by parts, we obtain

$$u_{x_i}(x_0) = \frac{n}{\omega_n R^n}\int_{B_R(x_0)} u_{x_i}(y)\,dy = \frac{n}{\omega_n R^n}\int_{\partial B_R(x_0)} u(y)\nu_i\,dS_y,$$

and hence

$$\begin{aligned}|u_{x_i}(x_0)| &\le \frac{n}{\omega_n R^n}\int_{\partial B_R(x_0)} |u(y)|\,dS_y\\ &\le \frac{n}{\omega_n R^n}\max_{\partial B_R(x_0)}|u|\cdot\omega_n R^{n-1} \le \frac{n}{R}\max_{\bar{B}_R(x_0)}|u|.\end{aligned}$$

This yields the desired result. □

When harmonic functions are nonnegative, we can improve Theorem 4.2.4.

Theorem 4.2.5. *Suppose $u \in C(\bar{B}_R(x_0))$ is a nonnegative harmonic function in $B_R(x_0) \subset \mathbb{R}^n$. Then*

$$|\nabla u(x_0)| \leq \frac{n}{R} u(x_0).$$

This result is referred to as the *differential Harnack inequality*. It has many important consequences.

Proof. As in the proof of Theorem 4.2.4, from integration by parts and the nonnegativeness of u, we have

$$|u_{x_i}(x_0)| \leq \frac{n}{\omega_n R^n} \int_{\partial B_R(x_0)} u(y)\, dS_y = \frac{n}{R} u(x_0),$$

where in the last equality we used the mean-value property. □

As an application, we prove the *Liouville theorem*.

Corollary 4.2.6. *Any harmonic function in $\mathbb{R}^n$ bounded from above or below is constant.*

Proof. Suppose u is a harmonic function in $\mathbb{R}^n$ with $u \geq c$ for some constant c. Then $v = u - c$ is a nonnegative harmonic function in $\mathbb{R}^n$. Let $x \in \mathbb{R}^n$ be an arbitrary point. By applying Theorem 4.2.5 to v in $B_R(x)$ for any $R > 0$, we have

$$|\nabla v(x)| \leq \frac{n}{R} v(x).$$

By letting $R \to \infty$, we conclude that $\nabla v(x) = 0$. This holds for any $x \in \mathbb{R}^n$. Hence v is constant and so is u. □

As another application, we prove the *Harnack inequality*, which asserts that nonnegative harmonic functions have comparable values in compact subsets.

Corollary 4.2.7. *Let u be a nonnegative harmonic function in $B_R(x_0) \subset \mathbb{R}^n$. Then*

$$u(x) \leq C u(y) \quad \text{for any } x, y \in B_{\frac{R}{2}}(x_0),$$

where C is a positive constant depending only on n.

Proof. Without loss of generality, we assume that u is positive in $B_R(x_0)$. Otherwise, we consider $u + \varepsilon$ for any constant $\varepsilon > 0$, derive the desired estimate for $u+\varepsilon$ and then let $\varepsilon \to 0$. For any $x \in B_{R/2}(x_0)$, we have $B_{R/2}(x) \subset B_R(x_0)$. By applying Theorem 4.2.5 to u in $B_{R/2}(x)$, we get

$$|\nabla u(x)| \le \frac{2n}{R} u(x),$$

or

$$|\nabla \log u(x)| \le \frac{2n}{R}.$$

For any $x, y \in B_{R/2}(x_0)$, a simple integration yields

$$\begin{aligned} \log \frac{u(x)}{u(y)} &= \int_0^1 \frac{d}{dt} \log u(tx + (1-t)y)\, dt \\ &= (x-y) \cdot \int_0^1 \nabla \log u(tx + (1-t)y)\, dt. \end{aligned}$$

Since $tx + (1-t)y \in B_{R/2}(x_0)$ for any $t \in [0,1]$ and $|x-y| \le R$, we obtain

$$\log \frac{u(x)}{u(y)} \le |x-y| \int_0^1 |\nabla \log u(tx + (1-t)y)|\, dt \le \frac{2n}{R}|x-y| \le 2n.$$

Therefore

$$u(x) \le e^{2n} u(y).$$

This is the desired result. □

In fact, Corollary 4.2.7 can be proved directly by the mean-value property.

Another proof of Corollary 4.2.7. First, we take any $B_{4r}(\bar{x}) \subset B_R(x_0)$ and claim that

$$u(x) \le 3^n u(\tilde{x}),$$

for any $x, \tilde{x} \in B_r(\bar{x})$. To see this, we note that $B_r(x) \subset B_{3r}(\tilde{x}) \subset B_{4r}(\bar{x})$ for any $x, \tilde{x} \in B_r(\bar{x})$. Then the mean-value property implies

$$u(x) = \frac{n}{\omega_n r^n} \int_{B_r(x)} u\, dy \le \frac{n}{\omega_n r^n} \int_{B_{3r}(\tilde{x})} u\, dy = 3^n u(\tilde{x}).$$

Next we take $r = R/8$ and choose finitely many $\bar{x}_1, \cdots, \bar{x}_N \in B_{R/2}(x_0)$ such that $\{B_r(\bar{x}_i)\}_{i=1}^N$ covers $B_{R/2}(x_0)$. We note that $B_{4r}(\bar{x}_i) \subset B_R(x_0)$, for any $i = 1, \cdots, N$, and that N is a constant depending only on n.

For any $x, y \in B_{R/2}(x_0)$, we can find $\tilde{x}_1, \cdots, \tilde{x}_k \in B_{R/2}$, for some $k \le N$, such that any two consecutive points in the ordered collection of $x, \tilde{x}_1, \cdots, \tilde{x}_k, y$ belong to a ball in $\{B_r(\bar{x}_i)\}_{i=1}^N$. Then we obtain

$$u(x) \le 3^n u(\tilde{x}_1) \le 3^{2n} u(\tilde{x}_2) \le \cdots \le 3^{nk} u(\tilde{x}_k) \le 3^{n(k+1)} u(y).$$

Then we have the desired result by taking $C = 3^{n(N+1)}$. □

As the final application of the mean-value property, we prove the *strong maximum principle* for harmonic functions.

Theorem 4.2.8. *Let Ω be a bounded domain in $\mathbb{R}^n$ and $u \in C(\bar{\Omega})$ be harmonic in Ω. Then u attains its maximum and minimum only on $\partial\Omega$ unless u is constant. In particular,*

$$\inf_{\partial\Omega} u \leq u \leq \sup_{\partial\Omega} u \quad \text{in } \Omega.$$

Proof. We only discuss the maximum of u. Set $M = \max_{\bar{\Omega}} u$ and

$$D = \{x \in \Omega : \ u(x) = M\}.$$

It is obvious that D is relatively closed; namely, for any sequence $\{x_m\} \subset D$, if $x_m \to x \in \Omega$, then $x \in D$. This follows easily from the continuity of u. Next we show that D is open. For any $x_0 \in D$, we take $r > 0$ such that $B_r(x_0) \subset \Omega$. By the mean-value property, we have

$$M = u(x_0) = \frac{n}{\omega_n r^n} \int_{B_r(x_0)} u \, dy \leq \frac{n}{\omega_n r^n} \int_{B_r(x_0)} M \, dy = M.$$

This implies $u = M$ in $B_r(x_0)$ and hence $B_r(x_0) \subset D$. In conclusion, D is both relatively closed and open in Ω. Therefore either $D = \emptyset$ or $D = \Omega$. In other words, u either attains its maximum only on $\partial\Omega$ or u is constant. □

A consequence of the maximum principle is the uniqueness of solutions of the Dirichlet problem in a bounded domain.

Corollary 4.2.9. *Let Ω be a bounded domain in $\mathbb{R}^n$. Then for any $f \in C(\Omega)$ and $\varphi \in C(\partial\Omega)$, there exists at most one solution $u \in C^2(\Omega) \cap C(\bar{\Omega})$ of the problem*

$$\begin{aligned} \Delta u &= f \quad \text{in } \Omega, \\ u &= \varphi \quad \text{on } \partial\Omega. \end{aligned}$$

Proof. Let w be the difference of any two solutions. Then $\Delta w = 0$ in Ω and $w = 0$ on $\partial\Omega$. Theorem 4.2.8 implies $w = 0$ in Ω. □

Compare Corollary 4.2.9 with Lemma 3.2.1, where the uniqueness was proved by energy estimates for solutions $u \in C^2(\Omega) \cap C^1(\bar{\Omega})$.

The maximum principle is an important tool in studying harmonic functions. We will study it in detail in Section 4.3, where we will prove the maximum principle using the algebraic structure of the Laplace equation and discuss its applications.

4.3. The Maximum Principle

One of the important methods in studying harmonic functions is the maximum principle. In this section, we discuss the maximum principle for a class of elliptic differential equations slightly more general than the Laplace equation. As applications of the maximum principle, we derive a priori estimates for solutions of the Dirichlet problem, and interior gradient estimates and the Harnack inequality for harmonic functions.

4.3.1. The Weak Maximum Principle. In the following, we assume Ω is a bounded domain in $\mathbb{R}^n$. We first prove the maximum principle for subharmonic functions without using the mean-value property.

Definition 4.3.1. Let u be a C^2-function in Ω. Then u is a *subharmonic* (or *superharmonic*) function in Ω if $\Delta u \geq$ (or $\leq$) 0 in Ω.

Theorem 4.3.2. *Let Ω be a bounded domain in $\mathbb{R}^n$ and $u \in C^2(\Omega) \cap C(\bar{\Omega})$ be subharmonic in Ω. Then u attains on $\partial\Omega$ its maximum in $\bar{\Omega}$, i.e.,*

$$\max_{\bar{\Omega}} u = \max_{\partial\Omega} u.$$

Proof. If u has a local maximum at a point x_0 in Ω, then the Hessian matrix $\big(\nabla^2 u(x_0)\big)$ is negative semi-definite. Thus,

$$\Delta u(x_0) = \operatorname{tr}(\nabla^2 u(x_0)) \leq 0.$$

Hence, in the special case that $\Delta u > 0$ in Ω, the maximum value of u in $\bar{\Omega}$ is attained only on $\partial\Omega$.

We now consider the general case and assume that Ω is contained in the ball B_R for some $R > 0$. For any $\varepsilon > 0$, consider

$$u_\varepsilon(x) = u(x) - \varepsilon(R^2 - |x|^2).$$

Then

$$\Delta u_\varepsilon = \Delta u + 2n\varepsilon \geq 2n\varepsilon > 0 \quad \text{in } \Omega.$$

By the special case we just discussed, u_ε attains its maximum only on $\partial\Omega$ and hence

$$\max_{\bar{\Omega}} u_\varepsilon = \max_{\partial\Omega} u_\varepsilon.$$

Then

$$\max_{\bar{\Omega}} u \leq \max_{\bar{\Omega}} u_\varepsilon + \varepsilon R^2 = \max_{\partial\Omega} u_\varepsilon + \varepsilon R^2 \leq \max_{\partial\Omega} u + \varepsilon R^2.$$

We have the desired result by letting $\varepsilon \to 0$ and using the fact that $\partial\Omega \subset \bar{\Omega}$. □

A continuous function in $\bar{\Omega}$ always attains its maximum in $\bar{\Omega}$. Theorem 4.3.2 asserts that any subharmonic function continuous up to the boundary attains its maximum on the boundary $\partial\Omega$, but possibly also in Ω. Theorem 4.3.2 is referred to as the *weak maximum principle*. A stronger version asserts that subharmonic functions attain their maximum only on the boundary. We will prove the strong maximum principle later.

Next, we discuss a class of elliptic equations slightly more general than the Laplace equation. Let c and f be continuous functions in Ω. We consider

$$\Delta u + cu = f \quad \text{in } \Omega.$$

Here, we require $u \in C^2(\Omega)$. The function c is referred to as the coefficient of the zeroth-order term. It is obvious that u is harmonic if $c = f = 0$.

A C^2-function u is called a *subsolution* (or *supersolution*) if $\Delta u + cu \geq f$ (or $\Delta u + cu \leq f$). If $c = 0$ and $f = 0$, subsolutions (or supersolutions) are subharmonic (or superharmonic).

Now we prove the *weak maximum principle* for subsolutions.

Recall that u^+ is the nonnegative part of u defined by $u^+ = \max\{0, u\}$.

Theorem 4.3.3. *Let Ω be a bounded domain in $\mathbb{R}^n$ and c be a continuous function in Ω with $c \leq 0$. Suppose $u \in C^2(\Omega) \cap C(\bar{\Omega})$ satisfies*

$$\Delta u + cu \geq 0 \quad \text{in } \Omega.$$

Then u attains on $\partial\Omega$ its nonnegative maximum in $\bar{\Omega}$, i.e.,

$$\max_{\bar{\Omega}} u \leq \max_{\partial\Omega} u^+.$$

Proof. We can proceed as in the proof of Theorem 4.3.2 with simple modifications. In the following, we provide an alternative proof based on Theorem 4.3.2. Set $\Omega_+ = \{x \in \Omega; u(x) > 0\}$. If $\Omega_+ = \emptyset$, then $u \leq 0$ in Ω, so $u^+ \equiv 0$. If $\Omega_+ \neq \emptyset$, then

$$\Delta u = \Delta u + cu - cu \geq -cu \geq 0 \quad \text{in } \Omega_+.$$

Theorem 4.3.2 implies

$$\max_{\bar{\Omega}_+} u = \max_{\partial\Omega_+} u = \max_{\partial\Omega} u^+.$$

This yields the desired result. □

If $c \equiv 0$ in Ω, Theorem 4.3.3 reduces to Theorem 4.3.2 and we can draw conclusions about the maximum of u rather than its nonnegative maximum. A similar remark holds for the strong maximum principle to be proved later.

We point out that Theorem 4.3.3 holds for general elliptic differential equations. Let a_{ij}, b_i and c be continuous functions in Ω with $c \leq 0$. We assume

$$\sum_{i,j=1}^{n} a_{ij}(x)\xi_i\xi_j \geq \lambda|\xi|^2 \quad \text{for any } x \in \Omega \text{ and any } \xi \in \mathbb{R}^n,$$

for some positive constant λ. In other words, we have a uniform positive lower bound for the eigenvalues of (a_{ij}) in Ω. For $u \in C^2(\Omega) \cap C(\bar{\Omega})$ and $f \in C(\Omega)$, consider the uniformly elliptic equation

$$Lu \equiv \sum_{i,j=1}^{n} a_{ij}u_{x_ix_j} + \sum_{i=1}^{n} b_iu_{x_i} + cu = f \quad \text{in } \Omega.$$

Many results in this section hold for uniformly elliptic equations.

As a simple consequence of Theorem 4.3.3, we have the following result.

Corollary 4.3.4. *Let Ω be a bounded domain in $\mathbb{R}^n$ and c be a continuous function in Ω with $c \leq 0$. Suppose $u \in C^2(\Omega) \cap C(\bar{\Omega})$ satisfies*

$$\begin{aligned} \Delta u + cu &\geq 0 \quad \text{in } \Omega, \\ u &\leq 0 \quad \text{on } \partial\Omega. \end{aligned}$$

Then $u \leq 0$ in Ω.

More generally, we have the following *comparison principle*.

Corollary 4.3.5. *Let Ω be a bounded domain in $\mathbb{R}^n$ and c be a continuous function in Ω with $c \leq 0$. Suppose $u, v \in C^2(\Omega) \cap C(\bar{\Omega})$ satisfy*

$$\begin{aligned} \Delta u + cu &\geq \Delta v + cv \quad \text{in } \Omega, \\ u &\leq v \quad \text{on } \partial\Omega. \end{aligned}$$

Then $u \leq v$ in Ω.

Proof. The difference $w = u - v$ satisfies $\Delta w + cw \geq 0$ in Ω and $w \leq 0$ on $\partial\Omega$. Then Corollary 4.3.4 implies $w \leq 0$ in Ω. □

The comparison principle provides a reason that functions u satisfying $\Delta u + cu \geq f$ are called subsolutions. They are less than a solution v of $\Delta v + cv = f$ with the same boundary values.

In the following, we simply say *by the maximum principle* when we apply Theorem 4.3.3, Corollary 4.3.4 or Corollary 4.3.5.

A consequence of the maximum principle is the uniqueness of solutions of Dirichlet problems.

Corollary 4.3.6. *Let Ω be a bounded domain in $\mathbb{R}^n$ and c be a continuous function in Ω with $c \leq 0$. For any $f \in C(\Omega)$ and $\varphi \in C(\partial\Omega)$, there exists at* *most one solution* $u \in C^2(\Omega) \cap C(\bar{\Omega})$ *of*

$$\begin{aligned} \Delta u + cu &= f \quad \text{in } \Omega, \\ u &= \varphi \quad \text{on } \partial\Omega. \end{aligned}$$

Proof. Let $u_1, u_2 \in C^2(\Omega) \cap C(\bar{\Omega})$ be two solutions. Then $w = u_1 - u_2$ satisfies

$$\begin{aligned} \Delta w + cw &= 0 \quad \text{in } \Omega, \\ w &= 0 \quad \text{on } \partial\Omega. \end{aligned}$$

By the maximum principle (applied to w and $-w$), we obtain $w = 0$ and hence $u_1 = u_2$ in Ω. □

The boundedness of the domain Ω is essential, since it guarantees the existence of the maximum and minimum of u in $\bar{\Omega}$. The uniqueness may not hold if the domain is unbounded. Consider the Dirichlet problem

$$\begin{aligned} \Delta u &= 0 \quad \text{in } \Omega \\ u &= 0 \quad \text{on } \partial\Omega, \end{aligned}$$

where $\Omega = \mathbb{R}^n \setminus B_1$. Then a nontrivial solution u is given by

$$u(x) = \begin{cases} \log|x| & \text{for } n = 2; \\ |x|^{2-n} - 1 & \text{for } n \geq 3. \end{cases}$$

Note that $u(x) \to \infty$ as $|x| \to \infty$ for $n = 2$ and u is bounded for $n \geq 3$. Next, we consider the same problem in the upper half-space $\Omega = \{x \in \mathbb{R}^n : x_n > 0\}$. Then $u(x) = x_n$ is a nontrivial solution, which is unbounded. These examples demonstrate that uniqueness may not hold for the Dirichlet problem in unbounded domains. Equally important for uniqueness is the condition $c \leq 0$. For example, we consider $\Omega = (0, \pi) \times \cdots \times (0, \pi) \subset \mathbb{R}^n$, and

$$u(x) = \prod_{i=1}^{n} \sin x_i.$$

Then u is a nontrivial solution of the problem

$$\begin{aligned} \Delta u + nu &= 0 \quad \text{in } \Omega, \\ u &= 0 \quad \text{on } \partial\Omega. \end{aligned}$$

In fact, such a u is an eigenfunction of Δ in Ω with zero boundary values.

4.3.2. The Strong Maximum Principle. The weak maximum principle asserts that subsolutions of elliptic differential equations attain their nonnegative maximum on the boundary if the coefficients of the zeroth-order term is nonpositive. In fact, these subsolutions can attain their nonnegative maximum only on the boundary, unless they are constant. This is the strong maximum principle. To prove this, we need the following Hopf lemma.

For any C^1-function u in $\bar{\Omega}$ that attains its maximum on $\partial\Omega$, say at $x_0 \in \partial\Omega$, we have $\frac{\partial u}{\partial \nu}(x_0) \geq 0$. The *Hopf lemma* asserts that the normal derivative is in fact positive if u is a subsolution in Ω.

Lemma 4.3.7. *Let B be an open ball in $\mathbb{R}^n$ with $x_0 \in \partial B$ and c be a continuous function in $\bar{B}$ with $c \leq 0$. Suppose $u \in C^2(B) \cap C^1(\bar{B})$ satisfies*

$$\Delta u + cu \geq 0 \quad \text{in } B.$$

Assume $u(x) < u(x_0)$ for any $x \in B$ and $u(x_0) \geq 0$. Then

$$\frac{\partial u}{\partial \nu}(x_0) > 0,$$

where ν is the exterior unit normal to B at x_0.

Proof. Without loss of generality, we assume $B = B_R$ for some $R > 0$. By the continuity of u up to ∂B_R, we have

$$u(x) \leq u(x_0) \quad \text{for any } x \in \bar{B}_R.$$

For positive constants α and ε to be determined, we set

$$w(x) = e^{-\alpha|x|^2} - e^{-\alpha R^2},$$

and

$$v(x) = u(x) - u(x_0) + \varepsilon w(x).$$

We consider w and v in $D = B_R \setminus \bar{B}_{R/2}$.

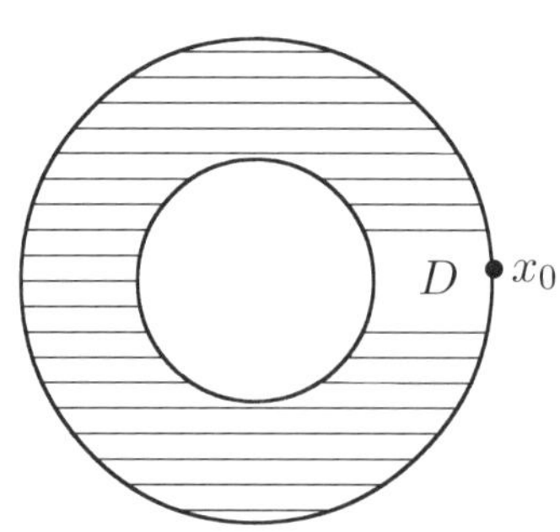

Figure 4.3.1. The domain D.

A direct calculation yields

$$\begin{aligned}\Delta w + cw &= e^{-\alpha|x|^2}\big(4\alpha^2|x|^2 - 2n\alpha + c\big) - ce^{-\alpha R^2}\\ &\geq e^{-\alpha|x|^2}\big(4\alpha^2|x|^2 - 2n\alpha + c\big),\end{aligned}$$

where we used $c \leq 0$ in B_R. Since $R/2 \leq |x| \leq R$ in D, we have

$$\Delta w + cw \geq e^{-\alpha|x|^2}(\alpha^2 R^2 - 2n\alpha + c) > 0 \quad \text{in } D,$$

if we choose α sufficiently large. By $c \leq 0$ and $u(x_0) \geq 0$, we obtain, for any $\varepsilon > 0$,

$$\Delta v + cv = \Delta u + cu + \varepsilon(\Delta w + cw) - cu(x_0) \geq 0 \quad \text{in } D.$$

We discuss v on ∂D in two cases. First, on $\partial B_{R/2}$, we have $u - u(x_0) < 0$, and hence $u - u(x_0) < -\varepsilon$ for some $\varepsilon > 0$. Note that $w < 1$ on $\partial B_{R/2}$. Then for such an ε, we obtain $v < 0$ on $\partial B_{R/2}$. Second, for $x \in \partial B_R$, we have $w(x) = 0$ and $u(x) \leq u(x_0)$. Hence $v(x) \leq 0$ for any $x \in \partial B_R$ and $v(x_0) = 0$. Therefore, $v \leq 0$ on ∂D.

In conclusion,

$$\begin{aligned}\Delta v + cv \geq 0 \quad &\text{in } D,\\ v \leq 0 \quad &\text{on } \partial D.\end{aligned}$$

By the maximum principle, we have

$$v \leq 0 \quad \text{in } D.$$

In view of $v(x_0) = 0$, then v attains at x_0 its maximum in $\bar{D}$. Hence, we obtain

$$\frac{\partial v}{\partial \nu}(x_0) \geq 0,$$

and then

$$\frac{\partial u}{\partial \nu}(x_0) \geq -\varepsilon \frac{\partial w}{\partial \nu}(x_0) = 2\varepsilon\alpha R e^{-\alpha R^2} > 0.$$

This is the desired result. □

Remark 4.3.8. Lemma 4.3.7 still holds if we substitute for B any bounded C^1-domain which satisfies an *interior sphere condition* at $x_0 \in \partial\Omega$, namely, if there exists a ball $B \subset \Omega$ with $x_0 \in \partial B$. This is because such a ball B is tangent to $\partial\Omega$ at x_0. We note that the interior sphere condition always holds for C^2-domains.

Now, we are ready to prove the *strong maximum principle.*

Theorem 4.3.9. *Let Ω be a bounded domain in $\mathbb{R}^n$ and c be a continuous function in Ω with $c \leq 0$. Suppose $u \in C^2(\Omega) \cap C(\bar{\Omega})$ satisfies*

$$\Delta u + cu \geq 0 \quad \text{in } \Omega.$$

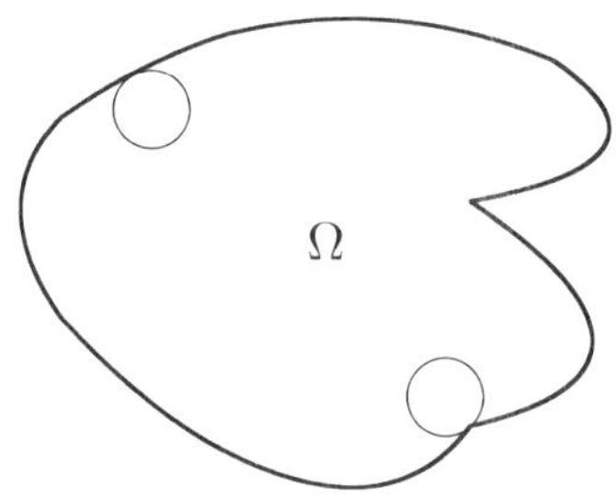

Figure 4.3.2. Interior sphere conditions.

Then u attains only on $\partial\Omega$ its nonnegative maximum in $\bar{\Omega}$ unless u is a constant.

Proof. Let M be the nonnegative maximum of u in $\bar{\Omega}$ and set

$$D = \{x \in \Omega : \ u(x) = M\}.$$

We prove either $D = \emptyset$ or $D = \Omega$ by a contradiction argument. Suppose D is a nonempty proper subset of Ω. It follows from the continuity of u that D is relatively closed in Ω. Then $\Omega \setminus D$ is open and we can find an open ball $B \subset \Omega \setminus D$ such that $\partial B \cap D \neq \emptyset$. In fact, we may choose a point $x_* \in \Omega \setminus D$ with $\operatorname{dist}(x_*, D) < \operatorname{dist}(x_*, \partial\Omega)$ and then take the ball centered at x_* with radius $\operatorname{dist}(x_*, D)$. Suppose $x_0 \in \partial B \cap D$.

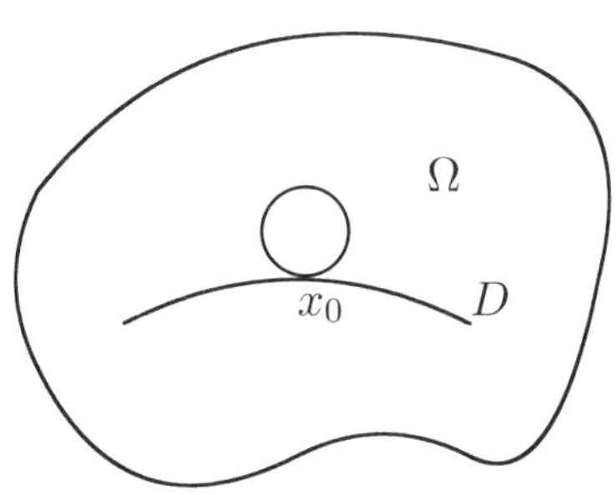

Figure 4.3.3. The domain Ω and its subset D.

Obviously, we have

$$\Delta u + cu \geq 0 \quad \text{in } B,$$

and

$$u(x) < u(x_0) \quad \text{for any } x \in B \text{ and } u(x_0) = M \geq 0.$$

By Lemma 4.3.7, we have

$$\frac{\partial u}{\partial \nu}(x_0) > 0,$$

where ν is the exterior unit normal to B at x_0. On the other hand, x_0 is an interior maximum point of Ω. This implies $\nabla u(x_0) = 0$, which leads to a contradiction. Therefore, either $D = \emptyset$ or $D = \Omega$. In the first case, u attains only on $\partial\Omega$ its nonnegative maximum in $\bar{\Omega}$; while in the second case, u is constant in Ω. □

The following result improves Corollary 4.3.5.

Corollary 4.3.10. *Let Ω be a bounded domain in $\mathbb{R}^n$ and c be a continuous function in Ω with $c \le 0$. Suppose $u \in C^2(\Omega) \cap C(\bar{\Omega})$ satisfies*

$$\begin{aligned} \Delta u + cu &\ge 0 \quad \text{in } \Omega, \\ u &\le 0 \quad \text{on } \partial\Omega. \end{aligned}$$

Then either $u < 0$ in Ω or u is a nonpositive constant in Ω.

We now consider the Neumann problem.

Corollary 4.3.11. *Let Ω be a bounded C^1-domain in $\mathbb{R}^n$ satisfying the interior sphere condition at every point of $\partial\Omega$ and c be a continuous function in Ω with $c \le 0$. Suppose $u \in C^2(\Omega) \cap C^1(\bar{\Omega})$ is a solution of the boundary-value problem*

$$\begin{aligned} \Delta u + cu &= f \quad \text{in } \Omega, \\ \frac{\partial u}{\partial \nu} &= \varphi \quad \text{on } \partial\Omega, \end{aligned}$$

for some $f \in C(\bar{\Omega})$ and $\varphi \in C(\partial\Omega)$. Then u is unique if $c \not\equiv 0$ and is unique up to additive constants if $c \equiv 0$.

Proof. We assume $f = 0$ in Ω and $\varphi = 0$ on $\partial\Omega$ and consider

$$\begin{aligned} \Delta u + cu &= 0 \quad \text{in } \Omega, \\ \frac{\partial u}{\partial \nu} &= 0 \quad \text{on } \partial\Omega. \end{aligned}$$

We will prove that $u = 0$ if $c \not\equiv 0$ and that u is constant if $c \equiv 0$.

We first consider the case $c \not\equiv 0$ and prove $u = 0$ by contradiction. Suppose u has a positive maximum at $x_0 \in \bar{\Omega}$. If u is a positive constant, then $c \equiv 0$ in Ω, which leads to a contradiction. If u is not a constant, then $x_0 \in \partial\Omega$ and $u(x) < u(x_0)$ for any $x \in \Omega$ by Theorem 4.3.9. Then Lemma 4.3.7 implies $\frac{\partial u}{\partial \nu}(x_0) > 0$, which contradicts the homogeneous boundary condition. Therefore, u has no positive maximum and hence $u \le 0$ in Ω. Similarly, $-u$ has no positive maximum and then $u \ge 0$ in Ω. In conclusion, $u = 0$ in Ω.

We now consider the case $c \equiv 0$. Suppose u is a nonconstant solution. Then its maximum in $\bar{\Omega}$ is attained only on $\partial\Omega$ by Theorem 4.3.9, say at $x_0 \in$

$\partial\Omega$. Lemma 4.3.7 implies $\frac{\partial u}{\partial \nu}(x_0) > 0$, which contradicts the homogeneous boundary value. This contradiction shows that u is constant. □

4.3.3. A Priori Estimates. As we have seen, an important application of the maximum principle is to prove the uniqueness of solutions of boundary-value problems. Equally or more important is to derive a priori estimates. In derivations of a priori estimates, it is essential to construct *auxiliary functions*. We will explain in the proof of the next result what auxiliary functions are and how they are used to yield necessary estimates by the maximum principle. We point out that we need only the weak maximum principle in the following discussion.

We now derive an a priori estimate for solutions of the Dirichlet problem.

Theorem 4.3.12. *Let Ω be a bounded domain in $\mathbb{R}^n$, c and f be continuous functions in $\bar{\Omega}$ with $c \leq 0$ and φ be a continuous function on $\partial\Omega$. Suppose $u \in C^2(\Omega) \cap C(\bar{\Omega})$ satisfies*

$$\begin{aligned} \Delta u + cu &= f \quad \text{in } \Omega, \\ u &= \varphi \quad \text{on } \partial\Omega. \end{aligned}$$

Then

$$\sup_{\Omega} |u| \leq \sup_{\partial\Omega} |\varphi| + C \sup_{\Omega} |f|,$$

where C is a positive constant depending only on n and $\operatorname{diam}(\Omega)$.

Proof. Set

$$F = \sup_{\Omega} |f|, \quad \Phi = \sup_{\partial\Omega} |\varphi|.$$

Then

$$\begin{aligned} (\Delta + c)(\pm u) &= \pm f \geq -F \quad \text{in } \Omega, \\ \pm u &= \pm\varphi \leq \Phi \quad \text{on } \partial\Omega. \end{aligned}$$

Without loss of generality, we assume $\Omega \subset B_R$, for some $R > 0$. Set

$$v(x) = \Phi + \frac{F}{2n}(R^2 - |x|^2) \quad \text{for any } x \in \Omega.$$

We note that $v \geq 0$ in Ω since $B_R \subset \Omega$. Then, by the property $c \leq 0$ in Ω, we have

$$\Delta v + cv = -F + cv \leq -F.$$

We also have $v \geq \Phi$ on $\partial\Omega$. Hence v satisfies

$$\begin{aligned} \Delta v + cv &\leq -F \quad \text{in } \Omega, \\ v &\geq \Phi \quad \text{on } \partial\Omega. \end{aligned}$$

Therefore,

$$(\Delta + c)(\pm u) \geq (\Delta + c)v \quad \text{in } \Omega,$$
$$\pm u \leq v \quad \text{on } \partial\Omega.$$

By the maximum principle, we obtain

$$\pm u \leq v \quad \text{in } \Omega,$$

and hence $|u| \leq v$ in Ω. Therefore,

$$|u(x)| \leq \Phi + \frac{1}{2n}(R^2 - |x|^2)F \quad \text{for any } x \in \Omega.$$

This yields the desired result. □

If $\Omega = B_R(x_0)$, then we have

$$\sup_{B_R(x_0)} |u| \leq \sup_{\partial B_R(x_0)} |\varphi| + \frac{R^2}{2n} \sup_{B_R(x_0)} |f|.$$

This follows easily from the proof.

The function v in the proof above is what we called an *auxiliary function.* In fact, auxiliary functions were already used in the proof of Lemma 4.3.7.

4.3.4. Gradient Estimates. In the following, we derive gradient estimates, estimates of first derivatives. The basic method is to derive a differential equation for $|\nabla u|^2$ and then apply the maximum principle. This is the *Bernstein method.*

There are two classes of gradient estimates, global gradient estimates and interior gradient estimates. Global gradient estimates yield estimates of gradients ∇u in Ω in terms of ∇u on $\partial\Omega$, as well as u in Ω, while interior gradient estimates yield estimates of ∇u in compact subsets of Ω in terms of u in Ω. In the next result, we will prove the *interior gradient estimate* for harmonic functions. Compare with Theorem 4.1.11 and Theorem 4.2.4.

Theorem 4.3.13. *Suppose $u \in C(\bar{B}_1)$ is harmonic in B_1. Then*

$$\sup_{B_{\frac{1}{2}}} |\nabla u| \leq C \sup_{\partial B_1} |u|,$$

where C is a positive constant depending only on n.

Proof. Recall that u is smooth in B_1 by Theorem 4.1.10. A direct calculation yields

$$\Delta(|\nabla u|^2) = 2\sum_{i,j=1}^{n} u_{x_i x_j}^2 + 2\sum_{i=1}^{n} u_{x_i}(\Delta u)_{x_i} = 2\sum_{i,j=1}^{n} u_{x_i x_j}^2,$$

where we used $\Delta u = 0$ in B_1. We note that $|\nabla u|^2$ is a subharmonic function. Hence we can easily obtain an estimate of ∇u in B_1 in terms of ∇u on ∂B_1.

This is the global gradient estimate. To get interior estimates, we need to introduce a cutoff function. For any nonnegative function $\varphi \in C_0^2(B_1)$, we have

$$\Delta(\varphi|\nabla u|^2) = (\Delta\varphi)|\nabla u|^2 + 4\sum_{i,j=1}^n \varphi_{x_i} u_{x_j} u_{x_i x_j} + 2\varphi \sum_{i,j=1}^n u_{x_i x_j}^2.$$

By the Cauchy inequality, we get

$$4|\varphi_{x_i} u_{x_j} u_{x_i x_j}| \le 2\varphi u_{x_i x_j}^2 + \frac{2}{\varphi}\varphi_{x_i}^2 u_{x_j}^2.$$

Then

$$\Delta(\varphi|\nabla u|^2) \ge \left(\Delta\varphi - \frac{2|\nabla\varphi|^2}{\varphi}\right)|\nabla u|^2.$$

We note that the ratio $|\nabla\varphi|^2/\varphi$ makes sense only when $\varphi \neq 0$. To interpret this ratio in B_1, we take $\varphi = \eta^2$ for some $\eta \in C_0^2(B_1)$. Then

$$\Delta(\eta^2|\nabla u|^2) \ge \left(2\eta\Delta\eta - 6|\nabla\eta|^2\right)|\nabla u|^2 \ge -C|\nabla u|^2,$$

where C is a positive constant depending only on η and n. Note that

$$\Delta(u^2) = 2|\nabla u|^2 + 2u\Delta u = 2|\nabla u|^2,$$

since u is harmonic. By taking a constant α large enough, we obtain

$$\Delta(\eta^2|\nabla u|^2 + \alpha u^2) \ge (2\alpha - C)|\nabla u|^2 \ge 0.$$

By the maximum principle, we obtain

$$\sup_{B_1}(\eta^2|\nabla u|^2 + \alpha u^2) \le \sup_{\partial B_1}(\eta^2|\nabla u|^2 + \alpha u^2).$$

In choosing $\eta \in C_0^2(B_1)$, we require in addition that $\eta \equiv 1$ in $B_{1/2}$. With $\eta = 0$ on ∂B_1, we get

$$\sup_{B_{1/2}} |\nabla u|^2 \le \alpha \sup_{\partial B_1} u^2.$$

This is the desired estimate. □

As consequences of interior gradient estimates, we have interior estimates on derivatives of arbitrary order as in Theorem 4.1.12 and the compactness as in Corollary 4.1.13. The compactness result will be used later in Perron's method.

Next we derive the *differential Harnack inequality* for positive harmonic functions using the maximum principle. Compare this with Theorem 4.2.5.

Theorem 4.3.14. *Suppose u is a positive harmonic function in B_1. Then*

$$\sup_{B_{1/2}} |\nabla \log u| \le C,$$

where C is a positive constant depending only on n.

Proof. Set $v = \log u$. A direct calculation yields

$$\Delta v = -|\nabla v|^2.$$

Next, we prove an interior gradient estimate for v. By setting $w = |\nabla v|^2$, we get

$$\Delta w + 2\sum_{i=1}^n v_{x_i} w_{x_i} = 2\sum_{i,j=1}^n v_{x_ix_j}^2.$$

As in Theorem 4.3.13, we need to introduce a cutoff function. First, by

$$\left(\sum_{i=1}^n v_{x_ix_i}\right)^2 \leq n\sum_{i=1}^n v_{x_ix_i}^2,$$

we have

$$\sum_{i,j=1}^n v_{x_ix_j}^2 \geq \sum_{i=1}^n v_{x_ix_i}^2 \geq \frac{1}{n}(\Delta v)^2 = \frac{|\nabla v|^4}{n} = \frac{w^2}{n}. \tag{4.3.1}$$

Take a nonnegative function $\varphi \in C_0^2(B_1)$. A straightforward calculation yields

$$\begin{aligned}\Delta(\varphi w) + 2\sum_{i=1}^n v_{x_i}(\varphi w)_{x_i} = 2\varphi \sum_{i,j=1}^n v_{x_ix_j}^2 + 4\sum_{i,j=1}^n \varphi_{x_i} v_{x_j} v_{x_ix_j} \\ + 2w\sum_{i=1}^n \varphi_{x_i} v_{x_i} + (\Delta\varphi)w.\end{aligned}$$

The Cauchy inequality implies

$$4|\varphi_{x_i} v_{x_j} v_{x_ix_j}| \leq \varphi v_{x_ix_j}^2 + \frac{4\varphi_{x_i}^2}{\varphi} v_{x_j}^2.$$

Then

$$\begin{aligned}\Delta(\varphi w) + 2\sum_{i=1}^n v_{x_i}(\varphi w)_{x_i} \geq \varphi \sum_{i,j=1}^n v_{x_ix_j}^2 - 2|\nabla\varphi||\nabla v|^3 \\ + \left(\Delta\varphi - \frac{4|\nabla\varphi|^2}{\varphi}\right)|\nabla v|^2.\end{aligned}$$

Here we keep one term of $\varphi v_{x_ix_j}^2$ in the right-hand side instead of dropping it entirely as in the proof of Theorem 4.3.13. To make sense of $|\nabla\varphi|^2/\varphi$ in B_1, we take $\varphi = \eta^4$ for some $\eta \in C_0^2(B_1)$. In addition, we require that $\eta = 1$ in $B_{1/2}$. We obtain, by (4.3.1),

$$\begin{aligned}&\Delta(\eta^4 w) + 2\sum_{i=1}^n v_{x_i}(\eta^4 w)_{x_i} \\ &\quad \geq \frac{1}{n}\eta^4|\nabla v|^4 - 8\eta^3|\nabla\eta||\nabla v|^3 + 4\eta^2(\eta\Delta\eta - 13|\nabla\eta|^2)|\nabla v|^2.\end{aligned}$$

We note that the right-hand side can be regarded as a polynomial of degree 4 in $\eta|\nabla v|$ with a positive leading coefficient. Other coefficients depend on η and hence are bounded functions of x. For the leading term, we save half of it for a later purpose. Now,

$$\frac{1}{2n}t^4 - 8|\nabla\eta|t^3 + 4(\eta\Delta\eta - 13|\nabla\eta|^2)t^2 \geq -C \quad \text{for any } t \in \mathbb{R},$$

where C is a positive constant depending only on n and η. Hence with $t = \eta|\nabla v|$, we get

$$\Delta(\eta^4 w) + 2\sum_{i=1}^{n} v_{x_i}(\eta^4 w)_{x_i} \geq \frac{1}{2n}\eta^4 w^2 - C.$$

We note that $\eta^4 w$ is nonnegative in B_1 and zero near ∂B_1. Next, we assume that $\eta^4 w$ attains its maximum at $x_0 \in B_1$. Then $\nabla(\eta^4 w) = 0$ and $\Delta(\eta^4 w) \leq 0$ at x_0. Hence

$$\eta^4 w^2(x_0) \leq C.$$

If $w(x_0) \geq 1$, then $\eta^4 w(x_0) \leq C$. Otherwise $\eta^4 w(x_0) \leq \eta^4(x_0)$. By combining these two cases, we obtain

$$\eta^4 w \leq C_* \quad \text{in } B_1,$$

where C_* is a positive constant depending only on n and η. With the definition of w and $\eta = 1$ in $B_{1/2}$, we obtain the desired result. □

The following result is referred to as the *Harnack inequality*. Compare it with Corollary 4.2.7.

Corollary 4.3.15. *Suppose u is a nonnegative harmonic function in B_1. Then*

$$u(x_1) \leq Cu(x_2) \quad \textit{for any } x_1, x_2 \in B_{1/2},$$

where C is a positive constant depending only on n.

The proof is identical to the first proof of Corollary 4.2.7 and is omitted. We note that u is required to be positive in Theorem 4.3.14 since $\log u$ is involved, while u is only nonnegative in Corollary 4.3.15.

The Harnack inequality describes an important property of harmonic functions. Any nonnegative harmonic functions have comparable values in a proper subdomain. We point out that the Harnack inequality in fact implies the strong maximum principle: *Any nonnegative harmonic function in a domain is identically zero if it is zero somewhere in the domain.*

4.3.5. Removable Singularity. Next, we discuss isolated singularity of harmonic functions. We note that the fundamental solution of the Laplace operator has an isolated singularity and is harmonic elsewhere. The next result asserts that an isolated singularity of harmonic functions can be removed, if it is "better" than that of the fundamental solution.

Theorem 4.3.16. *Suppose u is harmonic in $B_R \setminus \{0\} \subset \mathbb{R}^n$ and satisfies*

$$u(x) = \begin{cases} o(\log|x|), & n = 2, \\ o(|x|^{2-n}), & n \geq 3 \end{cases} \quad \text{as } |x| \to 0.$$

Then u can be defined at 0 so that it is harmonic in B_R.

Proof. Without loss of generality, we assume that u is continuous in $0 < |x| \leq R$. Let v solve

$$\begin{aligned} \Delta v &= 0 \quad \text{in } B_R, \\ v &= u \quad \text{on } \partial B_R. \end{aligned}$$

The existence of v is guaranteed by the Poisson integral formula in Theorem 4.1.9. Set $M = \max_{\partial B_R} |u|$. We note that the constant functions $\pm M$ are obviously harmonic and $-M \leq v \leq M$ on ∂B_R. By the maximum principle, we have $-M \leq v \leq M$ in B_R and hence,

$$|v| \leq M \quad \text{in } B_R.$$

Next, we prove $u = v$ in $B_R \setminus \{0\}$. Set $w = v - u$ in $B_R \setminus \{0\}$ and $M_r = \max_{\partial B_r} |w|$ for any $r < R$. We only consider the case $n \geq 3$. First, we have

$$-M_r \cdot \frac{r^{n-2}}{|x|^{n-2}} \leq w(x) \leq M_r \cdot \frac{r^{n-2}}{|x|^{n-2}},$$

for any $x \subset \partial B_r \cup \partial B_R$. It holds on ∂B_r by the definition of M_r and on ∂B_R since $w = 0$ on ∂B_R. Note that w and $|x|^{2-n}$ are harmonic in $B_R \setminus B_r$. Then the maximum principle implies

$$-M_r \cdot \frac{r^{n-2}}{|x|^{n-2}} \leq w(x) \leq M_r \cdot \frac{r^{n-2}}{|x|^{n-2}},$$

and hence

$$|w(x)| \leq M_r \cdot \frac{r^{n-2}}{|x|^{n-2}},$$

for any $x \in B_R \setminus B_r$. With

$$M_r = \max_{\partial B_r} |v - u| \leq \max_{\partial B_r} |v| + \max_{\partial B_r} |u| \leq M + \max_{\partial B_r} |u|,$$

we then have

$$|w(x)| \leq \frac{r^{n-2}}{|x|^{n-2}} M + \frac{1}{|x|^{n-2}} \cdot \left(r^{n-2} \max_{\partial B_r} |u|\right),$$

for any $x \in B_R \setminus B_r$. Now for each fixed $x \neq 0$, we take $r < |x|$ and then let $r \to 0$. By the assumption on u, we obtain $w(x) = 0$. This implies $w = 0$ and hence $u = v$ in $B_R \setminus \{0\}$. □

4.3.6. Perron's Method. In this subsection, we solve the Dirichlet problem for the Laplace equation in bounded domains by *Perron's method.* Essentially used are the maximum principle and the Poisson integral formula. The latter provides the solvability of the Dirichlet problem for the Laplace equation in balls.

We first discuss subharmonic functions. By Definition 4.3.1, a C^2-function v is subharmonic if $\Delta v \geq 0$.

Lemma 4.3.17. *Let Ω be a domain in $\mathbb{R}^n$ and v be a C^2-function in Ω. Then $\Delta v \geq 0$ in Ω if and only if for any ball $B \subset \Omega$ and any harmonic function $w \in C(\bar{B})$,*

$$v \leq w \text{ on } \partial B \text{ implies } v \leq w \text{ in } B.$$

Proof. We first prove the *only if* part. For any ball $B \subset \Omega$ and any harmonic function $w \in C(\bar{B})$ with $v \leq w$ on ∂B, we have

$$\begin{aligned} \Delta v &\geq \Delta w \quad \text{in } B, \\ v &\leq w \quad \text{on } \partial B. \end{aligned}$$

By the maximum principle, we have $v \leq w$ in B.

Now we prove the *if* part by a contradiction argument. If $\Delta v < 0$ somewhere in Ω, then $\Delta v < 0$ in B for some ball B with $\bar{B} \subset \Omega$. Let w solve

$$\begin{aligned} \Delta w &= 0 \quad \text{in } B, \\ w &= v \quad \text{on } \partial B. \end{aligned}$$

The existence of w in B is implied by the Poisson integral formula in Theorem 4.1.9. We have $v \leq w$ in B by the assumption. Next, we note that

$$\begin{aligned} \Delta w = 0 &> \Delta v \quad \text{in } B, \\ w &= v \quad \text{on } \partial B. \end{aligned}$$

We have $w \leq v$ in B by the maximum principle. Hence $v = w$ in B, which contradicts $\Delta w > \Delta v$ in B. Therefore, $\Delta v \geq 0$ in Ω. □

Lemma 4.3.17 leads to the following definition.

Definition 4.3.18. Let Ω be a domain in $\mathbb{R}^n$ and v be a continuous function in Ω. Then v is *subharmonic (superharmonic)* in Ω if for any ball $B \subset \Omega$ and any harmonic function $w \in C(\bar{B})$,

$$v \leq (\geq)\ w \text{ on } \partial B \text{ implies } v \leq (\geq)\ w \text{ in } B.$$

We point out that in Definition 4.3.18 subharmonic (superharmonic) functions are defined for continuous functions. We now prove a maximum principle for such subharmonic and superharmonic functions.

Lemma 4.3.19. *Let Ω be a bounded domain in $\mathbb{R}^n$ and $u, v \in C(\bar{\Omega})$. Suppose u is subharmonic in Ω and v is superharmonic in Ω with $u \le v$ on $\partial\Omega$. Then $u \le v$ in Ω.*

Proof. We first note that $u - v \le 0$ on $\partial\Omega$. Set $M = \max_{\bar{\Omega}}(u - v)$ and

$$D = \big\{x \in \Omega : \; u(x) - v(x) = M\big\} \subset \Omega.$$

Then D is relatively closed by the continuity of u and v. Next we prove that D is open. For any $x_0 \in D$, we take $r < \operatorname{dist}(x_0, \partial\Omega)$. Let $\bar{u}$ and $\bar{v}$ solve, respectively,

$$\begin{aligned} \Delta\bar{u} &= 0 \quad \text{in } B_r(x_0),\\ \bar{u} &= u \quad \text{on } \partial B_r(x_0), \end{aligned}$$

and

$$\begin{aligned} \Delta\bar{v} &= 0 \quad \text{in } B_r(x_0),\\ \bar{v} &= v \quad \text{on } \partial B_r(x_0). \end{aligned}$$

The existence of $\bar{u}$ and $\bar{v}$ in $B_r(x_0)$ is implied by the Poisson integral formula in Theorem 4.1.9. Definition 4.3.18 implies

$$u \le \bar{u}, \quad \bar{v} \le v \quad \text{in } B_r(x_0).$$

Hence,

$$\bar{u} - \bar{v} \ge u - v \quad \text{in } B_r(x_0).$$

Next,

$$\begin{aligned} \Delta(\bar{u} - \bar{v}) &= 0 \quad \text{in } B_r(x_0),\\ \bar{u} - \bar{v} &= u - v \quad \text{on } \partial B_r(x_0). \end{aligned}$$

With $u - v \le M$ on $\partial B_r(x_0)$, the maximum principle implies $\bar{u} - \bar{v} \le M$ in $B_r(x_0)$. In particular,

$$M \ge (\bar{u} - \bar{v})(x_0) \ge (u - v)(x_0) = M.$$

Hence, $(\bar{u} - \bar{v})(x_0) = M$ and then $\bar{u} - \bar{v}$ has an interior maximum at x_0. By the strong maximum principle, $\bar{u} - \bar{v} \equiv M$ in $\bar{B}_r(x_0)$. Therefore, $u - v = M$ on $\partial B_r(x_0)$. This holds for any $r < \operatorname{dist}(x_0, \partial\Omega)$. Then, $u - v = M$ in $B_r(x_0)$ and hence $B_r(x_0) \subset D$. In conclusion, D is both relatively closed and open in Ω. Therefore either $D = \emptyset$ or $D = \Omega$. In other words, $u - v$ either attains its maximum only on $\partial\Omega$ or $u - v$ is constant. Since $u \le v$ in $\partial\Omega$, we have $u \le v$ in Ω in both cases. $\square$

The proof in fact yields the strong maximum principle: Either $u < v$ in Ω or $u - v$ is constant in Ω.

Next, we describe Perron's method. Let Ω be a bounded domain in $\mathbb{R}^n$ and φ be a continuous function on $\partial\Omega$. We will find a function $u \in C^\infty(\Omega) \cap C(\bar{\Omega})$ such that

$$\begin{aligned} \Delta u &= 0 \quad \text{in } \Omega, \\ u &= \varphi \quad \text{on } \partial\Omega. \end{aligned} \tag{4.3.2}$$

Suppose there exists a solution $u = u_\varphi$. Then for any $v \in C(\bar{\Omega})$ which is subharmonic in Ω with $v \leq \varphi$ on $\partial\Omega$, we obtain, by Lemma 4.3.19,

$$v \leq u_\varphi \quad \text{in } \Omega.$$

Hence for any $x \in \Omega$

$$u_\varphi(x) = \sup\{v(x) : \ v \in C(\bar{\Omega}) \text{ is subharmonic in } \Omega \text{ and } v \leq \varphi \text{ on } \partial\Omega\}. \tag{4.3.3}$$

We note that the equality holds since u_φ is obviously an element of the set in the right-hand side. Here, we assumed the existence of the solution u_φ. In Perron's method, we will prove that the function u_φ defined in (4.3.3) is indeed a solution of (4.3.2) under appropriate assumptions on the domain. The proof consists of two steps. In the first step, we prove that u_φ is harmonic in Ω. This holds for arbitrary bounded domains. We note that u_φ in (4.3.3) is defined only in Ω. So in the second step, we prove that u_φ has a limit on $\partial\Omega$ and this limit is precisely φ. For this, we need appropriate assumptions on $\partial\Omega$.

Before we start our discussion of Perron's method, we demonstrate how to generate greater subharmonic functions from given subharmonic functions.

Lemma 4.3.20. *Let $v \in C(\bar{\Omega})$ be a subharmonic function in Ω and B be a ball in Ω with $\bar{B} \subset \Omega$. Let w be defined by*

$$w = v \quad \text{in } \bar{\Omega} \setminus B,$$

and

$$\Delta w = 0 \quad \text{in } B.$$

Then w is a subharmonic function in Ω and $v \leq w$ in $\bar{\Omega}$.

The function w is called the *harmonic lifting* of v (in B).

Proof. The existence of w in B is implied by the Poisson integral formula in Theorem 4.1.9. Then w is smooth in B and is continuous in $\bar{\Omega}$. We also have $v \leq w$ in B by Definition 4.3.18.

Next, we take any B' with $\bar{B}' \subset \Omega$ and consider a harmonic function $u \in C(\bar{B}')$ with $w \leq u$ on $\partial B'$. By $v \leq w$ on $\partial B'$, we have $v \leq u$ on $\partial B'$. Then, v is subharmonic and u is harmonic in B' with $v \leq u$ on $\partial B'$. By Lemma 4.3.19, we have $v \leq u$ in B'. Hence $w \leq u$ in $B' \setminus B$. Next, both w and u are harmonic in $B \cap B'$ and $w \leq u$ on $\partial(B \cap B')$. By the maximum principle, we have $w \leq u$ in $B \cap B'$. Hence $w \leq u$ in B'. Therefore, w is subharmonic in Ω by Definition 4.3.18. □

Lemma 4.3.20 asserts that we obtain greater subharmonic functions if we preserve the values of subharmonic functions outside the balls and extend them inside the balls by the Poisson integral formula.

Now we are ready to prove that u_φ in (4.3.3) is a harmonic function in Ω.

Lemma 4.3.21. *Let Ω be a bounded domain in $\mathbb{R}^n$ and φ be a continuous function on $\partial\Omega$. Then u_φ defined in* (4.3.3) *is harmonic in Ω.*

Proof. Set

$$\mathcal{S}_\varphi = \{v : \ v \in C(\bar{\Omega}) \text{ is subharmonic in } \Omega \text{ and } v \leq \varphi \text{ on } \partial\Omega\}.$$

Then for any $x \in \Omega$,

$$u_\varphi(x) = \sup\{v(x) : v \in \mathcal{S}_\varphi\}.$$

In the following, we simply write $\mathcal{S} = \mathcal{S}_\varphi$.

Step 1. We prove that u_φ is well defined. To do this, we set

$$m = \min_{\partial\Omega} \varphi, \quad M = \max_{\partial\Omega} \varphi.$$

We note that the constant function m is in $\mathcal{S}$ and hence the set $\mathcal{S}$ is not empty. Next, the constant function M is obviously harmonic in Ω with $\varphi \leq M$ on $\partial\Omega$. By Lemma 4.3.19, for any $v \in \mathcal{S}$,

$$v \leq M \quad \text{in } \bar{\Omega}.$$

Thus u_φ is well defined and $u_\varphi \leq M$ in Ω.

Step 2. We prove that $\mathcal{S}$ is closed by taking maximum among finitely many functions in $\mathcal{S}$. We take arbitrary $v_1, \cdots, v_k \in \mathcal{S}$ and set

$$v = \max\{v_1, \cdots, v_k\}.$$

It follows easily from Definition 4.3.18 that v is subharmonic in Ω. In fact, we take any ball $B \subset \Omega$ and any harmonic function $w \in C(\bar{B})$ with $v \leq w$ on ∂B. Then $v_i \leq w$ on ∂B, for $i = 1, \cdots, k$. Since v_i is subharmonic, we get $v_i \leq w$ in B, so $v \leq w$ in B. We conclude that v is subharmonic in Ω. Hence $v \in \mathcal{S}$.

Step 3. For any $B_r(x_0) \subset \Omega$, we prove that u_φ is harmonic in $B_r(x_0)$. First, by the definition of u_φ, there exists a sequence of functions $v_i \in \mathcal{S}$ such that

$$\lim_{i\to\infty} v_i(x_0) = u_\varphi(x_0).$$

We point out that the sequence $\{v_i\}$ depends on x_0. We may replace v_i above by any $\tilde{v}_i \in \mathcal{S}$ with $\tilde{v}_i \geq v_i$ since

$$v_i(x_0) \leq \tilde{v}_i(x_0) \leq u_\varphi(x_0).$$

Replacing, if necessary, v_i by $\max\{m, v_i\} \in \mathcal{S}$, we may also assume

$$m \leq v_i \leq u_\varphi \quad \text{in } \Omega.$$

For the fixed $B_r(x_0)$ and each v_i, we let w_i be the harmonic lifting in Lemma 4.3.20. In other words, $w_i = v_i$ in $\Omega \setminus B_r(x_0)$ and $\Delta w_i = 0$ in $B_r(x_0)$. By Lemma 4.3.20, $w_i \in \mathcal{S}$ and $v_i \leq w_i$ in Ω. Hence,

$$\lim_{i\to\infty} w_i(x_0) = u_\varphi(x_0),$$

and

$$m \leq w_i \leq u_\varphi \quad \text{in } \Omega,$$

for any $i = 1, 2, \cdots$. In particular, $\{w_i\}$ is a bounded sequence of harmonic functions in $B_r(x_0)$. By Corollary 4.1.13, there exists a harmonic function w in $B_r(x_0)$ such that a subsequence of $\{w_i\}$, still denoted by $\{w_i\}$, converges to w in any compact subset of $B_r(x_0)$. We then conclude easily that

$$w \leq u_\varphi \text{ in } B_r(x_0) \text{ and } w(x_0) = u_\varphi(x_0).$$

We now claim $u_\varphi = w$ in $B_r(x_0)$. To see this, we take any $\bar{x} \in B_r(x_0)$ and proceed similarly as before, with $\bar{x}$ replacing x_0. By the definition of u_φ, there exists a sequence of functions $\bar{v}_i \in \mathcal{S}$ such that

$$\lim_{i\to\infty} \bar{v}_i(\bar{x}) = u_\varphi(\bar{x}).$$

Replacing, if necessary, $\bar{v}_i$ by $\max\{\bar{v}_i, w_i\} \in \mathcal{S}$, we may also assume

$$w_i \leq \bar{v}_i \leq u_\varphi \quad \text{in } \Omega.$$

For the fixed $B_r(x_0)$ and each $\bar{v}_i$, we let $\bar{w}_i$ be the harmonic lifting in Lemma 4.3.20. Then, $\bar{w}_i \in \mathcal{S}$ and $\bar{v}_i \leq \bar{w}_i$ in Ω. Moreover, $\bar{w}_i$ is harmonic in $B_r(x_0)$ and satisfies

$$\lim_{i\to\infty} \bar{w}_i(\bar{x}) = u_\varphi(\bar{x}),$$

and

$$m \leq w_i \leq \bar{v}_i \leq \bar{w}_i \leq u_\varphi \quad \text{in } \Omega,$$

for any $i = 1, 2, \cdots$. By Corollary 4.1.13, there exists a harmonic function $\bar{w}$ in $B_r(x_0)$ such that a subsequence of $\bar{w}_i$ converges to $\bar{w}$ in any compact subset of $B_r(x_0)$. We then conclude easily that

$$w \leq \bar{w} \leq u_\varphi \text{ in } B_r(x_0) \text{ and } w(x_0) = \bar{w}(x_0) = u_\varphi(x_0),$$

and

$$\bar{w}(\bar{x}) = u_\varphi(\bar{x}).$$

Next, we note that $w-\bar{w}$ is a harmonic function in $B_r(x_0)$ with a maximum attained at x_0. By applying the strong maximum principle to $w-\bar{w}$ in $B_r(x_0)$, we conclude that $w-\bar{w}$ is constant, which is obviously zero. This implies $w=\bar{w}$ in $B_r(x_0)$, and in particular, $w(\bar{x})=\bar{w}(\bar{x})=u_\varphi(\bar{x})$. We then have $w=u_\varphi$ in $B_r(x_0)$ since $\bar{x}$ is arbitrary in $B_r(x_0)$. Therefore, u_φ is harmonic in $B_r(x_0)$. □

We note that u_φ in Lemma 4.3.21 is defined only in Ω. We have to discuss limits of $u_\varphi(x)$ as x approaches the boundary. For this, we need to impose additional assumptions on the boundary $\partial\Omega$.

Lemma 4.3.22. *Let φ be a continuous function on $\partial\Omega$ and u_φ be the function defined in* (4.3.3)*. For some $x_0 \in \partial\Omega$, suppose $w_{x_0} \in C(\bar{\Omega})$ is a subharmonic function in Ω such that*

$$w_{x_0}(x_0)=0, \quad w_{x_0}(x)<0 \quad \text{for any } x\in\partial\Omega\setminus\{x_0\}. \tag{4.3.4}$$

Then

$$\lim_{x\to x_0} u_\varphi(x)=\varphi(x_0).$$

Proof. As in the proof of Lemma 4.3.21, we set

$$\mathcal{S}_\varphi=\{v:\ v\in C(\bar{\Omega}) \text{ is subharmonic in } \Omega \text{ and } v\le\varphi \text{ on } \partial\Omega\}.$$

We simply write $w=w_{x_0}$ and set $M=\max_{\partial\Omega}|\varphi|$.

Let ε be an arbitrary positive constant. By the continuity of φ at x_0, there exists a positive constant δ such that

$$|\varphi(x)-\varphi(x_0)|\le\varepsilon,$$

for any $x\in\partial\Omega\cap B_\delta(x_0)$. We then choose K sufficiently large so that

$$-Kw(x)\ge 2M,$$

for any $x\in\partial\Omega\setminus B_\delta(x_0)$. Hence,

$$|\varphi-\varphi(x_0)|\le\varepsilon-Kw \quad \text{on } \partial\Omega.$$

Since $\varphi(x_0)-\varepsilon+Kw(x)$ is a subharmonic function in Ω with $\varphi(x_0)-\varepsilon+Kw\le\varphi$ on $\partial\Omega$, we have $\varphi(x_0)-\varepsilon+Kw\in\mathcal{S}_\varphi$. The definition of u_φ implies

$$\varphi(x_0)-\varepsilon+Kw\le u_\varphi \quad \text{in } \Omega.$$

On the other hand, $\varphi(x_0)+\varepsilon-Kw$ is a superharmonic in Ω with $\varphi(x_0)+\varepsilon-Kw\ge\varphi$ on Ω. Hence for any $v\in\mathcal{S}_\varphi$, we obtain, by Lemma 4.3.19,

$$v\le\varphi(x_0)+\varepsilon-Kw \quad \text{in } \Omega.$$

Again by the definition of u_φ, we have

$$u_\varphi \le \varphi(x_0) + \varepsilon - Kw \quad \text{in } \Omega.$$

Therefore,

$$|u_\varphi - \varphi(x_0)| \le \varepsilon - Kw \quad \text{in } \Omega.$$

This implies

$$\limsup_{x \to x_0} |u_\varphi(x) - \varphi(x_0)| \le \varepsilon.$$

We obtain the desired result by letting $\varepsilon \to 0$. □

The function w_{x_0} satisfying (4.3.4) is called a *barrier function*. As shown in the proof, w_{x_0} provides a barrier for the function u_φ near x_0.

We note that u_φ in Lemma 4.3.21 is defined only in Ω. It is natural to extend u_φ to $\partial\Omega$ by defining $u_\varphi(x_0) = \varphi(x_0)$ for $x_0 \in \partial\Omega$. If (4.3.4) is satisfied for x_0, Lemma 4.3.22 asserts that u_φ is continuous at x_0. If (4.3.4) is satisfied for any $x_0 \in \partial\Omega$, we then obtain a continuous function u_φ in $\bar{\Omega}$.

Barrier functions can be constructed for a large class of domains Ω. Take, for example, the case where Ω satisfies the *exterior sphere condition* at $x_0 \in \partial\Omega$ in the sense that there exists a ball $B_{r_0}(y_0)$ such that

$$\Omega \cap B_{r_0}(y_0) = \emptyset, \quad \bar{\Omega} \cap \bar{B}_{r_0}(y_0) = \{x_0\}.$$

To construct a barrier function at x_0, we set

$$w_{x_0}(x) = \Gamma(x - y_0) - \Gamma(x_0 - y_0) \qquad \text{for any } x \in \bar{\Omega},$$

where Γ is the fundamental solution of the Laplace operator. It is easy to see that w_{x_0} is a harmonic function in Ω and satisfies (4.3.4). We note that the exterior sphere condition always holds for C^2-domains.

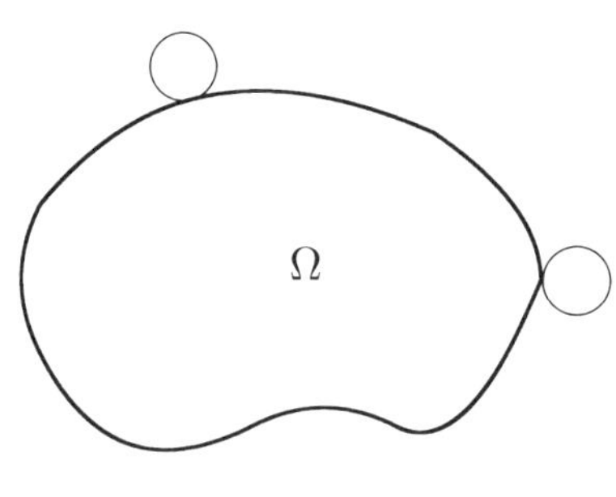

Figure 4.3.4. Exterior sphere conditions.

Theorem 4.3.23. *Let Ω be a bounded domain in $\mathbb{R}^n$ satisfying the exterior sphere condition at every boundary point. Then for any $\varphi \in C(\partial\Omega)$, (4.3.2) admits a solution $u \in C^\infty(\Omega) \cap C(\bar{\Omega})$.*

In summary, Perron's method yields a solution of the Dirichlet problem for the Laplace equation. This method depends essentially on the maximum principle and the solvability of the Dirichlet problem in balls. An important feature here is that the interior existence problem is separated from the boundary behavior of solutions, which is determined by the local geometry of domains.

4.4. Poisson Equations

In this section, we discuss briefly the Poisson equation. We first discuss regularity of classical solutions using the fundamental solution. Then we discuss weak solutions and introduce Sobolev spaces.

4.4.1. Classical Solutions. Let Ω be a domain in $\mathbb{R}^n$ and f be a continuous function in Ω. The Poisson equation has the form

$$\Delta u = f \quad \text{in } \Omega. \tag{4.4.1}$$

If u is a smooth solution of (4.4.1) in Ω, then obviously f is smooth. Conversely, we ask whether u is smooth if f is smooth. At first glance, this does not seem to be a reasonable question. We note that Δu is just a linear combination of second derivatives of u. Essentially, we are asking whether all partial derivatives exist and are continuous if a special combination of second derivatives is smooth. This question turns out to have an affirmative answer.

To proceed, we define

$$w_f(x) = \int_\Omega \Gamma(x-y)f(y)\,dy, \tag{4.4.2}$$

where Γ is the fundamental solution of the Laplace operator as in Definition 4.1.1. The function w_f is called the *Newtonian potential* of f in Ω. We will write $w_{f,\Omega}$ to emphasize the dependence on the domain Ω. It is easy to see that w_f is well defined in $\mathbb{R}^n$ if Ω is a bounded domain and f is a bounded function, although Γ has a singularity. We recall that the derivatives of Γ have asymptotic behavior of the form

$$\nabla\Gamma(x-y) \sim \frac{1}{|x-y|^{n-1}}, \quad \nabla^2\Gamma(x-y) \sim \frac{1}{|x-y|^n},$$

as $y \to x$. By differentiating under the integral sign formally, we have

$$\partial_{x_i} w_f(x) = \int_\Omega \partial_{x_i}\Gamma(x-y)f(y)\,dy,$$

for any $x \in \mathbb{R}^n$ and $i = 1, \cdots, n$. We note that the right-hand side is a well-defined integral and defines a continuous function of x. We will not use this identity directly in the following and leave its proof as an exercise. Assuming its validity, we cannot simply differentiate the expression of $\partial_{x_i} w_f$

to get second derivatives of w_f due to the singularity of $\partial_{x_ix_j}\Gamma$. In fact, extra conditions are needed in order to infer that w_f is C^2. If w_f is indeed C^2 and $\Delta w_f = f$ in Ω, then any solution of (4.4.1) differs from w_f by an addition of a harmonic function. Since harmonic functions are smooth, regularity of arbitrary solutions of (4.4.1) is determined by that of w_f.

Lemma 4.4.1. *Let Ω be a bounded domain in Ω, f be a bounded function in Ω and w_f be defined in (4.4.2). Assume that $f \in C^{k-1}(\Omega)$ for some integer $k \geq 2$. Then $w_f \in C^k(\Omega)$ and $\Delta w_f = f$ in Ω. Moreover, if f is smooth in Ω, then w_f is smooth in Ω.*

Proof. For brevity, we write $w = w_f$.

We first consider a special case where f has a compact support in Ω. For any $x \in \Omega$, we write

$$w(x) = \int_{\mathbb{R}^n} \Gamma(x-y)f(y)\,dy.$$

We point out that the integration is in fact over a bounded region. Note that Γ is evaluated as a function of $|x-y|$. By the change of variables $z = y - x$, we have

$$w(x) = \int_{\mathbb{R}^n} \Gamma(z)f(z+x)\,dz.$$

By the assumption, f is at least C^1. By a simple differentiation under the integral sign and an integration by parts, we obtain

$$\begin{aligned} w_{x_i}(x) &= \int_{\mathbb{R}^n} \Gamma(z)f_{x_i}(z+x)\,dz = \int_{\mathbb{R}^n} \Gamma(z)f_{z_i}(z+x)\,dz \\ &= -\int_{\mathbb{R}^n} \Gamma_{z_i}(z)f(z+x)\,dz. \end{aligned}$$

For $f \in C^{k-1}(\Omega)$ for some $k \geq 2$, we can differentiate under the integral sign to get

$$\partial^\alpha \partial_{x_i} w(x) = -\int_{\mathbb{R}^n} \Gamma_{z_i}(z)\partial_z^\alpha f(z+x)\,dz,$$

for any $\alpha \in \mathbb{Z}_+^n$ with $|\alpha| \leq k-1$. Hence, w is C^k in Ω. Moreover, if f is smooth in Ω, then w is smooth in Ω. Next, we calculate Δw if f is at least C^1. For any $x \in \Omega$, we have

$$\begin{aligned} \Delta w(x) &= \sum_{i=1}^n w_{x_ix_i}(x) = -\int_{\mathbb{R}^n} \sum_{i=1}^n \Gamma_{z_i}(z)f_{z_i}(z+x)\,dz \\ &= -\lim_{\varepsilon\to 0} \int_{\mathbb{R}^n\setminus B_\varepsilon} \sum_{i=1}^n \Gamma_{z_i}(z)f_{z_i}(z+x)\,dz. \end{aligned}$$

We note that $f(\cdot + x)$ has a compact support in $\mathbb{R}^n$. An integration by parts implies

$$\Delta w(x) = -\lim_{\varepsilon\to 0}\int_{\partial B_\varepsilon} \frac{\partial \Gamma}{\partial \nu}(z) f(z+x)\, dS_z,$$

where ν is the unit exterior normal to the boundary ∂B_ε of the domain $\mathbb{R}^n \setminus B_\varepsilon$, which points toward the origin. With $r = |z|$, we obtain

$$\begin{aligned}\Delta w(x) &= \lim_{\varepsilon\to 0}\int_{\partial B_\varepsilon} \frac{\partial \Gamma}{\partial r}(z) f(z+x)\, dS_z\\ &= \lim_{\varepsilon\to 0}\frac{1}{\omega_n \varepsilon^{n-1}}\int_{\partial B_\varepsilon} f(z+x)\, dS_z = f(x),\end{aligned}$$

by the explicit expression of Γ.

Next, we consider the general case. For any $x_0 \in \Omega$, we prove that w is C^k and $\Delta w = f$ in a neighborhood of x_0. To this end, we take $r < \mathrm{dist}(x_0, \partial\Omega)$ and a function $\varphi \in C_0^\infty(B_r(x_0))$ with $\varphi \equiv 1$ in $B_{r/2}(x_0)$. Then we write

$$\begin{aligned}w(x) &= \int_\Omega \Gamma(x-y)\big(1-\varphi(y)\big) f(y)\, dy + \int_\Omega \Gamma(x-y)\varphi(y) f(y)\, dy\\ &= w_I(x) + w_{II}(x).\end{aligned}$$

The first integral is actually over $\Omega \setminus B_{r/2}(x_0)$ since $\varphi \equiv 1$ in $B_{r/2}(x_0)$. Then there is no singularity in the first integral if we restrict x to $B_{r/4}(x_0)$. Hence, w_I is smooth in $B_{r/4}(x_0)$ and $\Delta w_I = 0$ in $B_{r/4}(x_0)$. For the second integral, φf is a C^{k-1}-function of compact support in Ω. We can apply what we just proved in the special case to φf. Then w_{II} is C^k in Ω and $\Delta w_{II} = \varphi f$. Therefore, w is a C^k-function in $B_{r/4}(x_0)$ and

$$\Delta w(x) = \varphi(x) f(x) = f(x),$$

for any $x \in B_{r/4}(x_0)$. Moreover, if f is smooth in Ω, so are w_{II} and w in Ω. □

Lemma 4.4.1 is optimal in the C^∞-category in the sense that the smoothness of f implies the smoothness of w_f. However, it does not seem optimal concerning finite differentiability. For example, Lemma 4.4.1 asserts that w_f is C^2 in Ω if f is C^1 in Ω. Since f is related to second derivatives of w_f, it seems natural to ask whether w_f is C^2 in Ω if f is continuous in Ω. We will explore this issue later.

We now prove a regularity result for general solutions of (4.4.1).

Theorem 4.4.2. *Let Ω be a domain in $\mathbb{R}^n$ and f be continuous in Ω. Suppose $u \in C^2(\Omega)$ satisfies $\Delta u = f$ in Ω. If $f \in C^{k-1}(\Omega)$ for some integer $k \geq 3$, then $u \in C^k(\Omega)$. Moreover, if f is smooth in Ω, then u is smooth in Ω.*

Proof. We take an arbitrary bounded subdomain Ω' in Ω and let $w_{f,\Omega'}$ be the Newtonian potential of f in Ω'. By Lemma 4.4.1, if $f \in C^{k-1}(\Omega)$ for some integer $k \geq 3$, then $w_{f,\Omega'}$ is C^k in Ω' and $\Delta w_{f,\Omega'} = f$ in Ω'. Now we set $v = u - w_{f,\Omega'}$. Since u is C^2 in Ω', so is v. Then, $\Delta v = \Delta u - \Delta w_{f,\Omega'} = 0$ in Ω'. In other words, v is harmonic in Ω', and hence is smooth in Ω' by Theorem 4.1.10. Therefore, $u = v + w_{f,\Omega'}$ is C^k in Ω'. It is obvious that if f is smooth in Ω, then $w_{f,\Omega'}$ and hence u are smooth in Ω'. □

Theorem 4.4.2 is an optimal result concerning the smoothness. Even though Δu is just one particular combination of second derivatives of u, the smoothness of Δu implies the smoothness of all second derivatives.

Next, we solve the Dirichlet problem for the Poisson equation.

Theorem 4.4.3. *Let Ω be a bounded domain in $\mathbb{R}^n$ satisfying the exterior sphere condition at every boundary point, f be a bounded C^1-function in Ω and φ be a continuous function on $\partial\Omega$. Then there exists a solution $u \in C^2(\Omega) \cap C(\bar{\Omega})$ of the Dirichlet problem*

$$\begin{aligned} \Delta u &= f \quad \text{in } \Omega, \\ u &= \varphi \quad \text{on } \partial\Omega. \end{aligned}$$

Moreover, if f is smooth in Ω, then u is smooth in Ω.

Proof. Let w be the Newtonian potential of f in Ω. By Lemma 4.4.1 for $k = 2$, $w \in C^2(\Omega) \cap C(\bar{\Omega})$ and $\Delta w = f$ in Ω. Now consider the Dirichlet problem

$$\begin{aligned} \Delta v &= 0 \quad \text{in } \Omega, \\ v &= \varphi - w \quad \text{on } \partial\Omega. \end{aligned}$$

Theorem 4.3.23 implies the existence of a solution $v \in C^\infty(\Omega) \cap C(\bar{\Omega})$. (The exterior sphere condition is needed in order to apply Theorem 4.3.23.) Then $u = v + w$ is the desired solution of the Dirichlet problem in Theorem 4.4.3. If f is smooth in Ω, then u is smooth there by Theorem 4.4.2. □

Now we raise a question concerning regularity of the lowest order in the classical sense. What is the optimal assumption on f to yield a C^2-solution u of (4.4.1)? We note that the Laplace operator Δ acts on C^2-functions and Δu is continuous for any C^2-function u. It is natural to ask whether the equation (4.4.1) admits any C^2-solutions if f is continuous. The answer turns out to be negative. There exists a continuous function f such that (4.4.1) does not admit any C^2-solutions.

Example 4.4.4. Let f be a function in $B_1 \subset \mathbb{R}^2$ defined by $f(0) = 0$ and

$$f(x) = \frac{x_2^2 - x_1^2}{|x|^2} \left\{ \frac{2}{(-\log|x|)^{1/2}} + \frac{1}{4(-\log|x|)^{3/2}} \right\},$$

for any $x \in B_1 \setminus \{0\}$. Then f is continuous in B_1. Consider

$$\Delta u = f \quad \text{in } B_1. \tag{4.4.3}$$

Define u in B_1 by $u(0) = 0$ and

$$u(x) = (x_1^2 - x_2^2)(-\log |x|)^{1/2},$$

for any $x \in B_1 \setminus \{0\}$. Then $u \in C(B_1) \cap C^\infty(B_1 \setminus \{0\})$. A straightforward calculation shows that u satisfies (4.4.3) in $B_1 \setminus \{0\}$ and

$$\lim_{x \to 0} u_{x_1 x_1}(x) = \infty.$$

Hence, u is not in $C^2(B_1)$. Next, we prove that (4.4.3) has no C^2-solutions. The proof is based on Theorem 4.3.16 concerning removable singularities of harmonic functions. Suppose, to the contrary, that there exists a C^2-solution v of (4.4.3) in B_1. For a fixed $R \in (0, 1)$, the function $w = u - v$ is harmonic in $B_R \setminus \{0\}$. Now $u \in C(\bar{B}_R)$ and $v \in C^2(B_R)$, so $w \in C(B_R)$. Thus w is continuous at the origin. By Theorem 4.3.16, w is harmonic in B_R and therefore belongs to $C^2(B_R)$. In particular, u is C^2 at the origin, which is a contradiction.

Example 4.4.4 illustrates that the C^2-spaces, or any C^k-spaces, are not adapted to the Poisson equation. A further investigation reveals that solutions in this example fail to be C^2 because the modulus of continuity of f does not decay to zero fast enough. If there is a better assumption than the continuity of f, then the modulus of continuity of $\nabla^2 u$ can be estimated. Better adapted to the Poisson equation, or more generally, the elliptic differential equations, are Hölder spaces. The study of the elliptic differential equations in Hölder spaces is known as the *Schauder theory*. In its simplest form, it asserts that all second derivatives of u are Hölder continuous if Δu is. It is beyond the scope of this book to give a presentation of the Schauder theory.

4.4.2. Weak Solutions. In the following, we discuss briefly how to extend the notion of classical solutions of the Poisson equation to less regularized solutions, the so-called weak solutions. These functions have derivatives only in an integral sense and satisfy the Poisson equation also in an integral sense. The same process can be applied to general linear elliptic equations, or even nonlinear elliptic equations, of divergence form.

To introduce weak solutions, we make use of the divergence structure or variation structure of the Laplace operator. Namely, we write the Laplace operator as

$$\Delta u = \operatorname{div}(\nabla u).$$

In fact, we already employed such a structure when we derived energy estimates of solutions of the Dirichlet problem for the Poisson equation in Section 3.2.

Let Ω be a bounded domain in $\mathbb{R}^n$ and f be a bounded continuous function in Ω. Consider

$$-\Delta u = f \quad \text{in } \Omega. \tag{4.4.4}$$

We intentionally put a negative sign in front of Δu. Let $u \in C^2(\Omega)$ be a solution of (4.4.4). Take an arbitrary $\varphi \in C_0^1(\Omega)$. By multiplying (4.4.4) by φ and then integrating by parts, we obtain

$$\int_\Omega \nabla u \cdot \nabla \varphi \, dx = \int_\Omega f\varphi \, dx. \tag{4.4.5}$$

In (4.4.5), φ is referred to as a *test function*. We note that upon integrating by parts, we transfer one derivative for u to test functions. Hence we only need to require u to be C^1 in (4.4.5). This is the advantage in formulating weak solutions.

If $u \in C^2(\Omega)$ satisfies (4.4.5) for any $\varphi \in C_0^1(\Omega)$, we obtain from (4.4.5), upon a simple integration by parts,

$$-\int_\Omega \varphi \Delta u \, dx = \int_\Omega f\varphi \, dx \quad \text{for any } \varphi \in C_0^1(\Omega).$$

This implies easily

$$-\Delta u = f \quad \text{in } \Omega.$$

In conclusion, a C^2-function u satisfying (4.4.5) for any $\varphi \in C_0^1(\Omega)$ is a classical solution of (4.4.4).

We now raise the question whether less regular functions u are allowed in (4.4.5). For any $\varphi \in C_0^1(\Omega)$, it is obvious that the integral in the left-hand side of (4.4.5) makes sense if each component of ∇u is an integrable function in Ω. This suggests that we should introduce derivatives in the integral sense.

Definition 4.4.5. For $i = 1, \cdots, n$, an integrable function u in Ω is said to have a weak x_i-derivative if there exists an integrable function v_i such that

$$\int_\Omega u\varphi_{x_i} \, dx = -\int_\Omega v_i \varphi \, dx \quad \text{for any } \varphi \in C_0^1(\Omega). \tag{4.4.6}$$

Here v_i is called a *weak* x_i*-derivative* of u and is denoted by u_{x_i}, the same way as for classical derivatives.

It is easy to see that weak derivatives are unique if they exist. We also point out that classical derivatives of C^1-functions are weak derivatives upon a simple integration by parts.

Definition 4.4.6. The *Sobolev space* $H^1(\Omega)$ is the collection of L^2-functions in Ω with L^2-weak derivatives in Ω.

The superscript 1 in the notation $H^1(\Omega)$ indicates the order of differentiation. In general, functions in $H^1(\Omega)$ may not have classical derivatives. In fact, they may not be continuous.

We are ready to introduce weak solutions.

Definition 4.4.7. Let $f \in L^2(\Omega)$ and $u \in H^1(\Omega)$. Then u is a *weak solution* of $-\Delta u = f$ in Ω if (4.4.5) holds for any $\varphi \in C_0^1(\Omega)$, where the components of ∇u are given by weak derivatives of u.

We now consider the Dirichlet problem for the Poisson equation with the homogeneous boundary value,

$$\begin{aligned} -\Delta u &= f \quad \text{in } \Omega, \\ u &= 0 \quad \text{on } \partial\Omega. \end{aligned} \tag{4.4.7}$$

We attempt to solve (4.4.7) by methods of functional analysis. It is natural to start with the set

$$\mathcal{C} = \{u \in C^1(\Omega) \cap C(\bar{\Omega}) : \ u = 0 \text{ on } \partial\Omega\}.$$

We note that the left-hand side of (4.4.5) provides an inner product in $\mathcal{C}$. To be specific, we define the H_0^1-*inner product* by

$$(u, v)_{H_0^1(\Omega)} = \int_\Omega \nabla u \cdot \nabla v \, dx,$$

for any $u, v \in \mathcal{C}$. It induces a norm defined by

$$\|u\|_{H_0^1(\Omega)} = \left(\int_\Omega |\nabla u|^2 \, dx \right)^{\frac{1}{2}}.$$

This is simply the L^2-norm of the gradient of u, and it is referred to as the H_0^1-*norm*. Here in the notation H_0^1, the supercript 1 indicates the order of differentiation and the subscript 0 refers to functions vanishing on $\partial\Omega$. The Poincaré inequality in Lemma 3.2.2 has the form

$$\|u\|_{L^2(\Omega)} \le C \|u\|_{H_0^1(\Omega)}, \tag{4.4.8}$$

for any $u \in \mathcal{C}$.

For $f \in L^2(\Omega)$, we define a linear functional F on $\mathcal{C}$ by

$$F(\varphi) = \int_\Omega f\varphi \, dx, \tag{4.4.9}$$

for any $\varphi \in \mathcal{C}$. By the Cauchy inequality and (4.4.8), we have

$$|F(\varphi)| \le \|f\|_{L^2(\Omega)} \|\varphi\|_{L^2(\Omega)} \le C \|f\|_{L^2(\Omega)} \|\varphi\|_{H_0^1(\Omega)}.$$

This means that F is a bounded linear functional on $\mathcal{C}$. If $\mathcal{C}$ were a Hilbert space with respect to the H_0^1-inner product, we would conclude by the Riesz representation theorem that there exists a $u \in \mathcal{C}$ such that

$$(u, \varphi)_{H_0^1(\Omega)} = F(\varphi),$$

for any $\varphi \in \mathcal{C}$. Hence, u satisfies (4.4.5). With $u = 0$ on $\partial\Omega$, u is interpreted as a weak solution of (4.4.7).

However, $\mathcal{C}$ is not complete with respect to the H_0^1-norm, for the same reason that $C(\bar{\Omega})$ is not complete with respect to the L^2-norm. For the remedy, we complete $\mathcal{C}$ under the H_0^1-norm.

Definition 4.4.8. The *Sobolev space* $H_0^1(\Omega)$ is the completion of $C_0^1(\Omega)$ under the H_0^1-norm.

We point out that we may define $H_0^1(\Omega)$ by completing $\mathcal{C}$ under the H_0^1-norm. It yields the same space.

The space $H_0^1(\Omega)$ defined in Definition 4.4.8 is abstract. So what are the elements in $H_0^1(\Omega)$? The next result provides an answer.

Theorem 4.4.9. *The space $H_0^1(\Omega)$ is a subspace of $H^1(\Omega)$ and is a Hilbert space with respect to the H_0^1-inner product.*

Proof. We take a sequence $\{u_k\}$ in $C_0^1(\Omega)$ which is a Cauchy sequence in the $H_0^1(\Omega)$-norm. In other words, $\{u_{k,x_i}\}$ is a Cauchy sequence in $L^2(\Omega)$, for any $i = 1, \cdots, n$. Then there exists a $v_i \in L^2(\Omega)$, for $i = 1, \cdots, n$, such that

$$u_{k,x_i} \to v_i \quad \text{in } L^2(\Omega) \text{ as } k \to \infty.$$

By (4.4.8), we obtain

$$\|u_k - u_l\|_{L^2(\Omega)} \le C\|u_k - u_l\|_{H_0^1(\Omega)}.$$

This implies that $\{u_k\}$ is a Cauchy sequence in $L^2(\Omega)$. We may assume for some $u \in L^2(\Omega)$ that

$$u_k \to u \quad \text{in } L^2(\Omega) \text{ as } k \to \infty.$$

Such a convergence illustrates that elements in $H_0^1(\Omega)$ can be identified as L^2-functions. Hence we have established the inclusion $H_0^1(\Omega) \subset L^2(\Omega)$. Next, we prove that u has L^2-weak derivatives. Since $u_k \in C_0^1(\Omega)$, upon a simple integration by parts, we have

$$\int_\Omega u_k \varphi_{x_i}\,dx = -\int_\Omega u_{k,x_i}\varphi\,dx \quad \text{for any } \varphi \in C_0^1(\Omega).$$

By taking $k \to \infty$, we obtain easily

$$\int_\Omega u \varphi_{x_i}\,dx = -\int_\Omega v_i\varphi\,dx \quad \text{for any } \varphi \in C_0^1(\Omega).$$

Therefore, v_i is the weak x_i-derivative of u. Then $u \in H^1(\Omega)$ since $v_i \in L^2(\Omega)$. In conclusion, $H_0^1(\Omega) \subset H^1(\Omega)$.

With weak derivatives replacing classical derivatives, the inner product $(\cdot,\cdot)_{H_0^1(\Omega)}$ is well defined for functions in $H_0^1(\Omega)$. We then conclude that $H_0^1(\Omega)$ is complete with respect to its induced norm $\|\cdot\|_{H_0^1(\Omega)}$. □

It is easy to see by approximations that (4.4.8) holds for functions in $H_0^1(\Omega)$.

Now we can prove the existence of weak solutions of the Dirichlet problem for the Poisson equation with homogeneous boundary value.

Theorem 4.4.10. *Let Ω be a bounded domain in $\mathbb{R}^n$ and $f \in L^2(\Omega)$. Then the Poisson equation $-\Delta u = f$ admits a weak solution $u \in H_0^1(\Omega)$.*

The proof is based on the Riesz representation theorem, and major steps are already given earlier.

Proof. We define a linear functional F on $H_0^1(\Omega)$ by

$$F(\varphi) = \int_\Omega f\varphi\, dx,$$

for any $\varphi \in H_0^1(\Omega)$. By the Cauchy inequality and (4.4.8), we have

$$|F(\varphi)| \le \|f\|_{L^2(\Omega)}\|\varphi\|_{L^2(\Omega)} \le C\|f\|_{L^2(\Omega)}\|\varphi\|_{H_0^1(\Omega)}.$$

Hence, F is a bounded linear functional on $H_0^1(\Omega)$. By the Riesz representation theorem, there exists a $u \in H_0^1(\Omega)$ such that

$$(u,\varphi)_{H_0^1(\Omega)} = F(\varphi),$$

for any $\varphi \in H_0^1(\Omega)$. Therefore, u is the desired function. □

According to Definition 4.4.7, u in Theorem 4.4.10 is a weak solution of $-\Delta u = f$. Concerning the boundary value, we point out that u is not defined on $\partial\Omega$ in the pointwise sense. We cannot conclude that $u = 0$ at each point on $\partial\Omega$. The boundary condition $u = 0$ on $\partial\Omega$ is interpreted precisely by the fact that $u \in H_0^1(\Omega)$, i.e., u is the limit of a sequence of $C_0^1(\Omega)$-functions in the H_0^1-norm. One consequence is that $u|_{\partial\Omega}$ is a well-defined zero function in $L^2(\partial\Omega)$. Hence, u is referred to as a *weak solution* of the Dirichlet problem (4.4.7).

Now we ask whether u possesses better regularity. The answer is affirmative. To see this, we need to introduce more Sobolev spaces. We first point out that weak derivatives as defined in (4.4.6) can be generalized to higher orders. For any $\alpha \in \mathbb{Z}_+^n$ with $|\alpha| = m$, an integrable function u in Ω

is said to have a *weak x^α-derivative* if there exists an integrable function v_α such that

$$\int_\Omega u\partial^\alpha\varphi\,dx = (-1)^m\int_\Omega v_\alpha\varphi\,dx \quad \text{for any } \varphi\in C_0^m(\Omega).$$

Here v_α is called a *weak x^α-derivative* of u and is denoted by $\partial^\alpha u$, the same notation as for classical derivatives. For any positive integer m, we denote by $H^m(\Omega)$ the collection of L^2-functions with L^2-weak derivatives of order up to m in Ω. This is also a *Sobolev space*. The superscript m indicates the order of differentiation.

We now return to Theorem 4.4.10. We assume, in addition, that Ω is a bounded smooth domain. With $f\in L^2(\Omega)$, the solution u in fact is a function in $H^2(\Omega)$. In other words, u has L^2-weak second derivatives $u_{x_ix_j}$, for $i,j=1,\cdots,n$. Moreover,

$$-\sum_{i=1}^n u_{x_ix_i} = f \quad \text{a.e. in } \Omega.$$

In fact, if $f\in H^k(\Omega)$ for any $k\geq 1$, then $u\in H^{k+2}(\Omega)$. This is the *L^2-theory* for the Poisson equation. We again encounter an optimal regularity result. If Δu is in the space $H^k(\Omega)$, then all second derivatives are in the same space. It is beyond the scope of this book to give a complete presentation of the L^2-theory.

An alternative method to prove the existence of weak solutions is to minimize the functional associated with the Poisson equation. Let Ω be a bounded domain in $\mathbb{R}^n$. For any C^1-function u in Ω, we define the *Dirichlet energy* of u in Ω by

$$E(u) = \frac{1}{2}\int_\Omega |\nabla u|^2\,dx.$$

For any $f\in L^2(\Omega)$, we consider

$$J(u) = E(u) - \int_\Omega fu\,dx = \frac{1}{2}\int_\Omega |\nabla u|^2\,dx - \int_\Omega fu\,dx.$$

For any $u\in C^1(\Omega)\cap C(\bar\Omega)$, we consider C^1-perturbations of u which leave the boundary values of u unchanged. We usually write such perturbations in the form of $u+\varphi$ for $\varphi\in C_0^1(\Omega)$. We now compare $J(u+\varphi)$ and $J(u)$. A straightforward calculation yields

$$J(u+\varphi) = J(u) + E(\varphi) + \int_\Omega \nabla u\cdot\nabla\varphi\,dx - \int_\Omega f\varphi\,dx.$$

We note that $E(\varphi)\geq 0$. Hence, if u is a weak solution of $-\Delta u = f$, we have, by (4.4.5),

$$J(u+\varphi)\geq J(u) \quad \text{for any } \varphi\in C_0^1(\Omega).$$

Therefore, u minimizes J among all functions with the same boundary value. Now we assume that u minimizes J among all functions with the same boundary value. Then for any $\varphi \in C_0^1(\Omega)$,

$$J(u + \varepsilon\varphi) \geq J(u) \quad \text{for any } \varepsilon.$$

In other words,

$$j(\varepsilon) \equiv J(u + \varepsilon\varphi)$$

has a minimum at $\varepsilon = 0$. This implies $j'(0) = 0$. A straightforward calculation shows that u satisfies (4.4.5) for any $\varphi \in C_0^1(\Omega)$. Therefore, u is a weak solution of $-\Delta u = f$. In conclusion, u is a weak solution of $-\Delta u = f$ if and only if u minimizes J among all functions with the same boundary value. The above calculation was performed for functions in $C^1(\Omega)$. A similar calculation can be carried out for functions in $H_0^1(\Omega)$. Hence, an alternative way to solve (4.4.7) in the weak sense is to minimize J in $H_0^1(\Omega)$. We will not provide details in this book.

The weak solutions and Sobolev spaces are important topics in PDEs. The brief discussion here serves only as an introduction. A complete presentation will constitute a book much thicker than this one.

4.5. Exercises

Exercise 4.1. Suppose $u(x)$ is harmonic in some domain in $\mathbb{R}^n$. Prove that

$$v(x) = |x|^{2-n} u\left(\frac{x}{|x|^2}\right)$$

is also harmonic in a suitable domain.

Exercise 4.2. For $n = 2$, find the Green's function for the Laplace operator on the first quadrant.

Exercise 4.3. Find the Green's function for the Laplace operator in the upper half-space $\{x_n > 0\}$ and then derive a formal integral representation for a solution of the Dirichlet problem

$$\begin{aligned} \Delta u &= 0 \quad \text{in } \{x_n > 0\}, \\ u &= \varphi \quad \text{on } \{x_n = 0\}. \end{aligned}$$

Exercise 4.4. (1) Suppose u is a nonnegative harmonic function in $B_R(x_0) \subset \mathbb{R}^n$. Prove by the Poisson integral formula the following Harnack inequality:

$$\left(\frac{R}{R+r}\right)^{n-2} \frac{R-r}{R+r} u(x_0) \leq u(x) \leq \left(\frac{R}{R-r}\right)^{n-2} \frac{R+r}{R-r} u(x_0),$$

where $r = |x - x_0| < R$.

(2) Prove by (1) the Liouville theorem: If u is a harmonic function in $\mathbb{R}^n$ and bounded above or below, then u is constant.

Exercise 4.5. Let u be a harmonic function in $\mathbb{R}^n$ with $\int_{\mathbb{R}^n} |u|^p \, dx < \infty$ for some $p \in (1, \infty)$. Prove that $u \equiv 0$.

Exercise 4.6. Let m be a positive integer and u be a harmonic function in $\mathbb{R}^n$ with $u(x) = O(|x|^m)$ as $|x| \to \infty$. Prove that u is a polynomial of degree at most m.

Exercise 4.7. Suppose $u \in C(\bar{B}_1^+)$ is harmonic in $B_1^+ = \{x \in B_1 : x_n > 0\}$ with $u = 0$ on $\{x_n = 0\} \cap B_1$. Prove that the odd extension of u to B_1 is harmonic in B_1.

Exercise 4.8. Let u be a C^2-solution of

$$\begin{aligned} \Delta u &= 0 \quad \text{in } \mathbb{R}^n \setminus B_R, \\ u &= 0 \quad \text{on } \partial B_R. \end{aligned}$$

Prove that $u \equiv 0$ if

$$\begin{aligned} \lim_{|x| \to \infty} \frac{u(x)}{\ln |x|} &= 0 \quad \text{for } n = 2, \\ \lim_{|x| \to \infty} u(x) &= 0 \quad \text{for } n \geq 3. \end{aligned}$$

Exercise 4.9. Let Ω be a bounded C^1-domain in $\mathbb{R}^n$ satisfying the exterior sphere condition at every boundary point and f be a bounded continuous function in Ω. Suppose $u \in C^2(\Omega) \cap C^1(\bar{\Omega})$ is a solution of

$$\begin{aligned} \Delta u &= f \quad \text{in } \Omega, \\ u &= 0 \quad \text{on } \partial\Omega. \end{aligned}$$

Prove that

$$\sup_{\partial\Omega} \left| \frac{\partial u}{\partial \nu} \right| \leq C \sup_{\Omega} |f|,$$

where C is a positive constant depending only on n and Ω.

Exercise 4.10. Let Ω be a smooth bounded domain in $\mathbb{R}^n$, c be a continuous function in Ω with $c \leq 0$ and α be a continuous function on $\partial\Omega$ with $\alpha \geq 0$. Discuss the uniqueness of the problem

$$\begin{aligned} \Delta u + cu &= f \quad \text{in } \Omega, \\ \frac{\partial u}{\partial \nu} + \alpha u &= \varphi \quad \text{on } \partial\Omega. \end{aligned}$$

Exercise 4.11. Let Ω be a bounded C^1-domain and let φ and α be continuous functions on $\partial\Omega$ with $\alpha \geq \alpha_0$ for a positive constant α_0. Suppose

$u \in C^2(\Omega) \cap C^1(\bar{\Omega})$ satisfies

$$-\Delta u + u^3 = 0 \quad \text{in } \Omega,$$
$$\frac{\partial u}{\partial \nu} + \alpha u = \varphi \quad \text{on } \partial\Omega.$$

Prove that

$$\max_{\Omega} |u| \leq \frac{1}{\alpha_0} \max_{\partial\Omega} |\varphi|.$$

Exercise 4.12. Let f be a continuous function in $\bar{B}_R$. Suppose $u \in C^2(B_R) \cap C(\bar{B}_R)$ satisfies

$$\Delta u = f \quad \text{in } B_R.$$

Prove that

$$|\nabla u(0)| \leq \frac{n}{R} \max_{\partial B_R} |u| + \frac{R}{2} \max_{B_R} |f|.$$

Hint: In B_R^+, set

$$v(x', x_n) = \frac{1}{2}\big(u(x', x_n) - u(x', -x_n)\big).$$

Consider an auxiliary function of the form

$$w(x', x_n) = A|x'|^2 + Bx_n + Cx_n^2.$$

Use the comparison principle to estimate v in B_R^+ and then derive a bound for $v_{x_n}(0)$.

Exercise 4.13. Let u be a nonzero harmonic function in $B_1 \subset \mathbb{R}^n$ and set

$$N(r) = \frac{r \int_{B_r} |\nabla u|^2 dx}{\int_{\partial B_r} u^2 dS} \quad \text{for any } r \in (0, 1).$$

(1) Prove that $N(r)$ is a nondecreasing function in $r \in (0, 1)$ and identify

$$\lim_{r \to 0+} N(r).$$

(2) Prove that, for any $0 < r < R < 1$,

$$\frac{1}{R^{n-1}} \int_{\partial B_R} u^2 dS \leq \left(\frac{R}{r}\right)^{2N(R)} \frac{1}{r^{n-1}} \int_{\partial B_r} u^2 dS.$$

Remark: The quantity $N(r)$ is called the *frequency*. The estimate in (2) for $R = 2r$ is referred to as the *doubling condition.*

Exercise 4.14. Let Ω be a bounded domain in $\mathbb{R}^n$ and f be a bounded function in Ω. Suppose w_f is the Newtonian potential defined in (4.4.2).

(1) Prove that $w_f \in C^1(\mathbb{R}^n)$ and

$$\partial_{x_i} w_f(x) = \int_\Omega \partial_{x_i}\Gamma(x-y)f(y)\,dy,$$

for any $x \in \mathbb{R}^n$ and $i = 1, \cdots, n$.

(2) Assume, in addition, that f is C^α in Ω for some $\alpha \in (0,1)$, i.e., for any $x, y \in \Omega$,

$$|f(x) - f(y)| \le C|x-y|^\alpha.$$

Prove that $w_f \in C^2(\Omega)$, $\Delta w_f = f$ in Ω and the second derivatives of w_f are C^α in Ω.

Chapter 5

Heat Equations

The n-dimensional heat equation is given by $u_t - \Delta u = 0$ for functions $u = u(x,t)$, with $x \in \mathbb{R}^n$ and $t \in \mathbb{R}$. Here, x is the space variable and t the time variable. The heat equation models the temperature of a body conducting heat when the density is constant. Solutions of the heat equation share many properties with harmonic functions, solutions of the Laplace equation.

In Section 5.1, we briefly introduce Fourier transforms. The Fourier transform is an important subject and has a close connection with many fields of mathematics, especially with partial differential equations. In the first part of this section, we discuss basic properties of Fourier transforms and prove the important Fourier inversion formula. In the second part, we use Fourier transforms to discuss several differential equations with constant coefficients, including the heat equation, and we derive explicit expressions for their solutions.

In Section 5.2, we discuss the fundamental solution of the heat equation and its applications. We first discuss the initial-value problem for the heat equation. We prove that the explicit expression for its solution obtained formally by Fourier transforms indeed yields a classical solution under appropriate assumptions on initial values. Then we discuss regularity of arbitrary solutions of the heat equation using the fundamental solution and derive interior gradient estimates.

In Section 5.3, we discuss the maximum principle for the heat equation and its applications. We first prove the weak maximum principle and the strong maximum principle for a class of parabolic equations more general than the heat equation. As applications, we derive a priori estimates of solutions of the initial/boundary-value problem and the initial-value problem.

We also derive interior gradient estimates by the maximum principle. In the final part of this section, we study the Harnack inequality for positive solutions of the heat equation. We point out that the Harnack inequality for the heat equation is more complicated than that for the Laplace equation we discussed earlier.

As in Chapter 4, several results in this chapter are proved by multiple methods. For example, interior gradient estimates are proved by two methods: the fundamental solution and the maximum principle.

5.1. Fourier Transforms

The Fourier transform is an important subject and has a close connection with many fields of mathematics. In this section, we will briefly introduce Fourier transforms and illustrate their applications by studying linear differential equations with constant coefficients.

5.1.1. Basic Properties.

We define the *Schwartz class* $\mathcal{S}$ as the collection of all complex-valued functions $u \in C^\infty(\mathbb{R}^n)$ such that $x^\beta \partial_x^\alpha u(x)$ is bounded in $\mathbb{R}^n$ for any $\alpha, \beta \in \mathbb{Z}_+^n$, i.e.,

$$\sup_{x\in\mathbb{R}^n} |x^\beta \partial_x^\alpha u(x)| < \infty.$$

In other words, the Schwartz class consists of smooth functions in $\mathbb{R}^n$ all of whose derivatives decay faster than any polynomial at infinity. It is easy to check that $u(x) = e^{-|x|^2}$ is in the Schwartz class.

Definition 5.1.1. For any $u \in \mathcal{S}$, the *Fourier transform* $\widehat{u}$ of u is defined by

$$\widehat{u}(\xi) = \frac{1}{(2\pi)^{\frac{n}{2}}} \int_{\mathbb{R}^n} e^{-ix\cdot\xi} u(x)\, dx \quad \text{for any } \xi \in \mathbb{R}^n.$$

We note that the integral on the right-hand side makes sense for $u \in \mathcal{S}$. In fact,

$$|\widehat{u}(\xi)| \le \frac{1}{(2\pi)^{\frac{n}{2}}} \int_{\mathbb{R}^n} |u(x)|\, dx \quad \text{for any } \xi \in \mathbb{R}^n,$$

or

$$\sup_{\mathbb{R}^n} |\widehat{u}| \le \frac{1}{(2\pi)^{\frac{n}{2}}} \|u\|_{L^1(\mathbb{R}^n)}.$$

This suggests that Fourier transforms are well defined for L^1-functions. We will not explore this issue in this book.

We now discuss properties of Fourier transforms. First, it is easy to see that the Fourier transformation is linear, i.e., for any $u_1, u_2 \in \mathcal{S}$ and $c_1, c_2 \in \mathbb{C}$,

$$(c_1u_1 + c_2u_2)\widehat{\ } = c_1\widehat{u}_1 + c_2\widehat{u}_2.$$

The following result illustrates an important property of Fourier transforms.

Lemma 5.1.2. *Let $u \in \mathcal{S}$. Then $\widehat{u} \in \mathcal{S}$ and for any multi-indices $\alpha, \beta \in \mathbb{Z}^n_+$,*

$$\widehat{\partial_x^\alpha u}(\xi) = (i\xi)^\alpha \widehat{u}(\xi)$$

and

$$\partial_\xi^\beta \widehat{u}(\xi) = (-i)^{|\beta|} \widehat{x^\beta u}(\xi).$$

Proof. Upon integrating by parts, we have

$$\begin{aligned}\widehat{\partial_x^\alpha u}(\xi) &= \frac{1}{(2\pi)^{\frac{n}{2}}} \int_{\mathbb{R}^n} e^{-ix\cdot\xi} \partial^\alpha u(x)\, dx \\ &= \frac{1}{(2\pi)^{\frac{n}{2}}} \int_{\mathbb{R}^n} (i\xi)^\alpha e^{-ix\cdot\xi} u(x)\, dx = (i\xi)^\alpha \widehat{u}(\xi).\end{aligned}$$

Next, it follows easily from the definition of $\widehat{u}$ that $\widehat{u} \in C^\infty(\mathbb{R}^n)$. Then we have

$$\begin{aligned}\partial_\xi^\beta \widehat{u}(\xi) &= \frac{1}{(2\pi)^{\frac{n}{2}}} \partial_\xi^\beta \int_{\mathbb{R}^n} e^{-ix\cdot\xi} u(x)\, dx \\ &= \frac{1}{(2\pi)^{\frac{n}{2}}} \int_{\mathbb{R}^n} (-ix)^\beta e^{-ix\cdot\xi} u(x)\, dx = (-i)^{|\beta|} \widehat{x^\beta u}(\xi).\end{aligned}$$

The interchange of the order of differentiation and integration is valid because $x^\beta u \in \mathcal{S}$. To prove $\widehat{u} \in \mathcal{S}$, we take any two multi-indices α and β. It suffices to prove that $\xi^\alpha \partial_\xi^\beta \widehat{u}(\xi)$ is bounded in $\mathbb{R}^n$. For this, we first note that

$$\begin{aligned}\xi^\alpha \partial_\xi^\beta \widehat{u}(\xi) =& (-i)^{|\beta|} \xi^\alpha \widehat{x^\beta u}(\xi) = (-i)^{|\alpha|+|\beta|} (i\xi)^\alpha \widehat{x^\beta u}(\xi) \\ =& (-i)^{|\alpha|+|\beta|} \widehat{\partial_x^\alpha (x^\beta u)}(\xi) \\ =& \frac{1}{(2\pi)^{\frac{n}{2}}} (-i)^{|\alpha|+|\beta|} \int_{\mathbb{R}^n} e^{-ix\cdot\xi} \partial_x^\alpha \left(x^\beta u(x)\right) dx.\end{aligned}$$

Hence

$$\sup_{\xi\in\mathbb{R}^n} |\xi^\alpha \partial_\xi^\beta \widehat{u}(\xi)| \le \frac{1}{(2\pi)^{\frac{n}{2}}} \int_{\mathbb{R}^n} |\partial_x^\alpha \left(x^\beta u(x)\right)|\, dx < \infty,$$

since each term in the integrand decays faster than any polynomial because $x^\beta u \in \mathcal{S}$. □

The next result relates Fourier transforms to translations and dilations.

Lemma 5.1.3. *Let $u \in \mathcal{S}$, $a \in \mathbb{R}^n$, and $k \in \mathbb{R} \setminus \{0\}$. Then*

$$\widehat{u(\cdot - a)}(\xi) = e^{-i\xi\cdot a} \widehat{u}(\xi),$$

and

$$\widehat{u(k\cdot)}(\xi) = \frac{1}{|k|^n}\widehat{u}\left(\frac{\xi}{k}\right).$$

Proof. By a simple change of variables, we have

$$\begin{aligned}\widehat{u(\cdot - a)}(\xi) &= \frac{1}{(2\pi)^{\frac{n}{2}}}\int_{\mathbb{R}^n} e^{-ix\cdot\xi}u(x-a)\,dx \\ &= \frac{1}{(2\pi)^{\frac{n}{2}}}\int_{\mathbb{R}^n} e^{-i(x+a)\cdot\xi}u(x)\,dx = e^{-i\xi\cdot a}\widehat{u}(\xi).\end{aligned}$$

By another change of variables, we have

$$\begin{aligned}\widehat{u(k\cdot)}(\xi) &= \frac{1}{(2\pi)^{\frac{n}{2}}}\int_{\mathbb{R}^n} e^{-ix\cdot\xi}u(kx)\,dx \\ &= \frac{1}{(2\pi)^{\frac{n}{2}}}\int_{\mathbb{R}^n} e^{-i\frac{x}{k}\cdot\xi}u(x)|k|^{-n}\,dx = \frac{1}{|k|^n}\widehat{u}\left(\frac{\xi}{k}\right).\end{aligned}$$

We then obtain the desired results. □

For any $u, v \in \mathcal{S}$, it is easy to check that $u * v \in \mathcal{S}$, where $u * v$ is the *convolution* of u and v defined by

$$(u * v)(x) = \int_{\mathbb{R}^n} u(x-y)v(y)\,dy.$$

Lemma 5.1.4. *Let $u, v \in \mathcal{S}$. Then*

$$\widehat{u * v}(\xi) = (2\pi)^{\frac{n}{2}}\widehat{u}(\xi)\widehat{v}(\xi).$$

Proof. By the definition of the Fourier transform, we have

$$\begin{aligned}\widehat{u * v}(\xi) &= \frac{1}{(2\pi)^{\frac{n}{2}}}\int_{\mathbb{R}^n} e^{-ix\cdot\xi}u * v(x)\,dx \\ &= \frac{1}{(2\pi)^{\frac{n}{2}}}\int_{\mathbb{R}^n} e^{-ix\cdot\xi}\left(\int_{\mathbb{R}^n} u(x-y)v(y)\,dy\right)dx \\ &= \frac{1}{(2\pi)^{\frac{n}{2}}}\int_{\mathbb{R}^n}\int_{\mathbb{R}^n} e^{-i(x-y)\cdot\xi}u(x-y)e^{-iy\cdot\xi}v(y)\,dydx \\ &= \frac{1}{(2\pi)^{\frac{n}{2}}}\int_{\mathbb{R}^n} e^{-iy\cdot\xi}v(y)\left(\int_{\mathbb{R}^n} e^{-i(x-y)\cdot\xi}u(x-y)\,dx\right)dy \\ &= \widehat{u}(\xi)\int_{\mathbb{R}^n} e^{-iy\cdot\xi}v(y)\,dy = (2\pi)^{\frac{n}{2}}\widehat{u}(\xi)\widehat{v}(\xi).\end{aligned}$$

The interchange of the order of integrations can be justified by Fubini's theorem. □

To proceed, we note that

$$\int_{-\infty}^{\infty} e^{-x^2}\,dx = \sqrt{\pi}.$$

The next result will be useful in the following discussions.

Proposition 5.1.5. *Let A be a positive constant and u be the function defined in $\mathbb{R}^n$ by*

$$u(x) = e^{-A|x|^2}.$$

Then

$$\widehat{u}(\xi) = \frac{1}{(2A)^{\frac{n}{2}}} e^{-\frac{|\xi|^2}{4A}}.$$

Proof. By the definition of Fourier transforms, we have

$$\widehat{u}(\xi) = \frac{1}{(2\pi)^{\frac{n}{2}}} \int_{\mathbb{R}^n} e^{-ix\cdot\xi - A|x|^2}\,dx = \prod_{k=1}^{n} \frac{1}{(2\pi)^{\frac{1}{2}}} \int_{-\infty}^{\infty} e^{-ix_k\xi_k - Ax_k^2}\,dx_k.$$

Hence it suffices to compute, for any $\eta \in \mathbb{R}$,

$$\frac{1}{(2\pi)^{\frac{1}{2}}} \int_{-\infty}^{\infty} e^{-it\eta - At^2}\,dt.$$

After the change of variables $s = t\sqrt{A}$, we have

$$\begin{aligned}\int_{-\infty}^{\infty} e^{-it\eta - At^2}\,dt &= e^{-\frac{\eta^2}{4A}} \int_{-\infty}^{\infty} e^{-(t\sqrt{A} + i\frac{\eta}{2\sqrt{A}})^2}\,dt \\ &= \frac{1}{\sqrt{A}} e^{-\frac{\eta^2}{4A}} \int_{-\infty}^{\infty} e^{-(s+i\frac{\eta}{2\sqrt{A}})^2}\,ds = \frac{1}{\sqrt{A}} e^{-\frac{\eta^2}{4A}} \int_L e^{-z^2}\,dz,\end{aligned}$$

where L is the straight line $\operatorname{Im} z = \eta/2\sqrt{A}$ in the complex z-plane. By the Cauchy's integral theorem and the fact that the integrand decays at the exponential rate as $\operatorname{Re} z \to \infty$, we have

$$\int_L e^{-z^2}\,dz = \int_{-\infty}^{\infty} e^{-t^2}\,dt = \sqrt{\pi}.$$

Hence

$$\frac{1}{(2\pi)^{\frac{1}{2}}} \int_{-\infty}^{\infty} e^{-it\eta - At^2}\,dt = \frac{1}{(2A)^{\frac{1}{2}}} e^{-\frac{\eta^2}{4A}}.$$

Therefore,

$$\frac{1}{(2\pi)^{\frac{n}{2}}} \int_{\mathbb{R}^n} e^{-ix\cdot\xi - A|x|^2}\,dx = \frac{1}{(2A)^{\frac{n}{2}}} e^{-\frac{1}{4A}|\xi|^2}.$$

This yields the desired result. □

We now prove the *Fourier inversion formula*, one of the most important results in the theory of Fourier transforms.

Theorem 5.1.6. *Suppose $u \in \mathcal{S}$. Then*

$$u(x) = \frac{1}{(2\pi)^{\frac{n}{2}}} \int_{\mathbb{R}^n} e^{ix\cdot\xi} \widehat{u}(\xi)\, d\xi. \tag{5.1.1}$$

The right-hand side of (5.1.1) is the Fourier transform of $\widehat{u}$ evaluated at $-x$. Hence, $u(x) = (\widehat{u})^\frown(-x)$. It follows that the Fourier transformation $u \mapsto \widehat{u}$ is a one-to-one map of $\mathcal{S}$ onto $\mathcal{S}$.

A natural attempt to prove (5.1.1) is to use the definition of Fourier transforms and rewrite the right-hand side as

$$\frac{1}{(2\pi)^n} \int_{\mathbb{R}^n} \left(\int_{\mathbb{R}^n} e^{ix\cdot\xi} e^{-iy\cdot\xi} u(y)\, dy \right) d\xi.$$

However, as an integral in terms of $(y, \xi) \in \mathbb{R}^n \times \mathbb{R}^n$, it is not absolutely convergent.

Proof. Letting $A = 1/2$ in Proposition 5.1.5, we see that if

$$u_0(x) = e^{-\frac{1}{2}|x|^2}, \tag{5.1.2}$$

then

$$\widehat{u}_0(\xi) = e^{-\frac{1}{2}|\xi|^2}.$$

Since $u_0(x) = u_0(-x)$, we conclude (5.1.1) for $u = u_0$. Now we prove (5.1.1) for any $u \in \mathcal{S}$.

We first consider $u \in \mathcal{S}$ with $u(0) = 0$. We claim that there exist $v_1, \cdots, v_n \in \mathcal{S}$ such that

$$u(x) = \sum_{j=1}^n x_j v_j(x) \quad \text{for any } x \in \mathbb{R}^n.$$

To see this, we note that

$$u(x) = \int_0^1 \frac{d}{dt}\big(u(tx)\big)\, dt = \sum_{j=1}^n x_j \int_0^1 u_{x_j}(tx)\, dt = \sum_{j=1}^n x_j w_j(x),$$

for some $w_j \in C^\infty(\mathbb{R}^n)$, $j = 1, \cdots, n$. By taking $\varphi \in C_0^\infty(\mathbb{R}^n)$ with $\varphi = 1$ in B_1, we write

$$\begin{aligned} u(x) &= \varphi(x)u(x) + \big(1 - \varphi(x)\big)u(x) \\ &= \sum_{j=1}^n x_j \left(\varphi(x) w_j(x) + \frac{x_j}{|x|^2}\big(1 - \varphi(x)\big)u(x) \right). \end{aligned}$$

We note that functions in the parentheses are in $\mathcal{S}$, for $j = 1, \cdots, n$. This proves the claim. Lemma 5.1.2 implies

$$\widehat{u}(\xi) = \sum_{j=1}^n i\partial_{\xi_j} \widehat{v}_j(\xi).$$

We note that $\widehat{v}_j \in \mathcal{S}$ by Lemma 5.1.2. Upon evaluating the right-hand side of (5.1.1) at $x = 0$, we obtain

$$\frac{1}{(2\pi)^{\frac{n}{2}}} \int_{\mathbb{R}^n} \widehat{u}(\xi)\, d\xi = \frac{1}{(2\pi)^{\frac{n}{2}}} \int_{\mathbb{R}^n} \sum_{j=1}^{n} i\partial_{\xi_j} \widehat{v}_j(\xi)\, d\xi = 0.$$

We conclude that (5.1.1) holds at $x = 0$ for all $u \in \mathcal{S}$ with $u(0) = 0$.

We now consider an arbitrary $u \in \mathcal{S}$ and decompose

$$u = u(0)u_0 + \big(u - u(0)u_0\big),$$

where u_0 is defined in (5.1.2). First, (5.1.1) holds for u_0 and hence for $u(0)u_0$. Next, since $u - u(0)u_0$ is zero at $x = 0$, we see that (5.1.1) holds for $u - u(0)u_0$ at $x = 0$. We obtain (5.1.1) for u at $x = 0$. Next, for any $x_0 \in \mathbb{R}^n$, we consider $v(x) = u(x + x_0)$. By Lemma 5.1.3,

$$\widehat{v}(\xi) = e^{ix_0 \cdot \xi} \widehat{u}(\xi).$$

Then by (5.1.1) for v at $x = 0$,

$$u(x_0) = v(0) = \frac{1}{(2\pi)^{\frac{n}{2}}} \int_{\mathbb{R}^n} \widehat{v}(\xi)\, d\xi = \frac{1}{(2\pi)^{\frac{n}{2}}} \int_{\mathbb{R}^n} e^{ix_0 \cdot \xi} \widehat{u}(\xi)\, d\xi.$$

This proves (5.1.1) for u at $x = x_0$. □

Motivated by Theorem 5.1.6, we define $\check{v}$ for any $v \in \mathcal{S}$ by

$$\check{v}(x) = \frac{1}{(2\pi)^{\frac{n}{2}}} \int_{\mathbb{R}^n} e^{ix \cdot \xi} v(\xi)\, d\xi \quad \text{for any } x \in \mathbb{R}^n.$$

The function $\check{v}$ is called the *inverse Fourier transform* of v. It is obvious that $\check{u}(x) = \widehat{u}(-x)$. Theorem 5.1.6 asserts that $u = (\widehat{u})^{\vee}$.

Next, we set, for any $u, v \in L^2(\mathbb{R}^n)$,

$$(u, v)_{L^2(\mathbb{R}^n)} = \int_{\mathbb{R}^n} u\bar{v}\, dx.$$

The following result is referred to as the *Parseval formula*.

Theorem 5.1.7. *Suppose $u, v \in \mathcal{S}$. Then*

$$(u, v)_{L^2(\mathbb{R}^n)} = (\widehat{u}, \widehat{v})_{L^2(\mathbb{R}^n)}.$$

Proof. We note

$$\begin{aligned}(\widehat{u}, \widehat{v})_{L^2(\mathbb{R}^n)} &= \int_{\mathbb{R}^n} \widehat{u}(\xi)\overline{\widehat{v}(\xi)}\, d\xi \\ &= \frac{1}{(2\pi)^{\frac{n}{2}}} \int_{\mathbb{R}^n} \overline{\widehat{v}(\xi)} \left(\int_{\mathbb{R}^n} e^{-ix\cdot\xi} u(x)\, dx \right) d\xi \\ &= \frac{1}{(2\pi)^{\frac{n}{2}}} \int_{\mathbb{R}^n} u(x) \left(\int_{\mathbb{R}^n} \overline{\widehat{v}(\xi) e^{ix\cdot\xi}}\, d\xi \right) dx \\ &= \int_{\mathbb{R}^n} u(x)\bar{v}(x)\, dx = (u, v)_{L^2(\mathbb{R}^n)},\end{aligned}$$

where we applied Theorem 5.1.6 to v. The interchange of the order of integrations can be justified by Fubini's theorem. □

As a consequence, we have *Plancherel's theorem.*

Corollary 5.1.8. *Suppose* $u \in \mathcal{S}$. *Then*

$$\|u\|_{L^2(\mathbb{R}^n)} = \|\widehat{u}\|_{L^2(\mathbb{R}^n)}.$$

In other words, the Fourier transformation is an isometry in $\mathcal{S}$ with respect to the L^2-norm.

Based on Corollary 5.1.8, we can extend Fourier transforms to $L^2(\mathbb{R}^n)$. Note that the Fourier transformation is a linear operator from $\mathcal{S}$ to $\mathcal{S}$ and that $\mathcal{S}$ is dense in $L^2(\mathbb{R}^n)$. For any $u \in L^2(\mathbb{R}^n)$, we can take a sequence $\{u_k\} \subset \mathcal{S}$ such that

$$u_k \to u \quad \text{in } L^2(\mathbb{R}^n) \text{ as } k \to \infty.$$

Corollary 5.1.8 implies

$$\|\widehat{u}_k - \widehat{u}_l\|_{L^2(\mathbb{R}^n)} = \|\widehat{u_k - u_l}\|_{L^2(\mathbb{R}^n)} = \|u_k - u_l\|_{L^2(\mathbb{R}^n)}.$$

Then, $\{\widehat{u}_k\}$ is a Cauchy sequence in $L^2(\mathbb{R}^n)$ and hence converges to a limit in $L^2(\mathbb{R}^n)$. This limit is defined as $\widehat{u}$, i.e.,

$$\widehat{u}_k \to \widehat{u} \quad \text{in } L^2(\mathbb{R}^n) \text{ as } k \to \infty.$$

It is straightforward to check that $\widehat{u}$ is well defined, independently of the choice of sequence $\{\widehat{u}_k\}$.

5.1.2. Examples. The Fourier transform is an important tool in studying linear partial differential equations with constant coefficients. We illustrate this by two examples.

Example 5.1.9. Let f be a function defined in $\mathbb{R}^n$. We consider

$$-\Delta u + u = f \quad \text{in } \mathbb{R}^n. \tag{5.1.3}$$

Obviously, this is an elliptic equation. We obtained an energy identity in Section 3.2 for solutions decaying sufficiently fast at infinity. Now we attempt to solve (5.1.3).

We first seek a formal expression of its solution u by Fourier transforms. In doing so, we will employ properties of Fourier transforms without justifications. By taking the Fourier transform of both sides in (5.1.3), we obtain, by Lemma 5.1.2,

$$(1+|\xi|^2)\widehat{u}(\xi) = \widehat{f}(\xi). \tag{5.1.4}$$

Then

$$\widehat{u}(\xi) = \frac{\widehat{f}(\xi)}{1+|\xi|^2}.$$

By Theorem 5.1.6,

$$u(x) = \frac{1}{(2\pi)^{\frac{n}{2}}}\int_{\mathbb{R}^n} e^{ix\cdot\xi}\frac{\widehat{f}(\xi)}{1+|\xi|^2}\,d\xi. \tag{5.1.5}$$

It remains to verify that this indeed yields a classical solution under appropriate assumptions on f. Before doing so, we summarize the simple process we just carried out. First, we apply Fourier transforms to equation (5.1.3). Basic properties of Fourier transforms allow us to transfer the differential equation (5.1.3) for u to an algebraic equation (5.1.4) for $\widehat{u}$. By solving this algebraic equation, we have an expression for $\widehat{u}$ in terms of $\widehat{f}$. Then, by applying the Fourier inversion formula, we obtain u in terms of $\widehat{f}$. We should point out that it is not necessary to rewrite u in an explicit form in terms of f.

Proposition 5.1.10. *Let $f \in \mathcal{S}$ and u be defined by* (5.1.5). *Then u is a smooth solution of* (5.1.3) *in $\mathcal{S}$. Moreover,*

$$\int_{\mathbb{R}^n}(|u|^2 + 2|\nabla u|^2 + |\nabla^2 u|^2)\,dx = \int_{\mathbb{R}^n}|f|^2\,dx.$$

Proof. We note that the process described above in solving (5.1.3) by Fourier transforms is rigorous if $f \in \mathcal{S}$. In the following, we prove directly from (5.1.5) that u is a smooth solution. By Lemma 5.1.2, $\widehat{f} \in \mathcal{S}$ for $f \in \mathcal{S}$. Then $\widehat{f}/(1+|\xi|^2) \in \mathcal{S}$. Therefore, u defined by (5.1.5) is in $\mathcal{S}$ by Lemma 5.1.2. For any multi-index $\alpha \in \mathbb{Z}_+^n$, we have

$$\partial^\alpha u(x) = \frac{1}{(2\pi)^{\frac{n}{2}}}\int_{\mathbb{R}^n} e^{ix\cdot\xi}\frac{(i\xi)^\alpha\widehat{f}(\xi)}{1+|\xi|^2}\,d\xi.$$

In particular,

$$\Delta u(x) = \sum_{k=1}^n u_{x_kx_k}(x) = -\frac{1}{(2\pi)^{\frac{n}{2}}}\int_{\mathbb{R}^n} e^{ix\cdot\xi}\frac{|\xi|^2\widehat{f}(\xi)}{1+|\xi|^2}\,d\xi,$$

and hence

$$-\Delta u(x)+u(x)=\frac{1}{(2\pi)^{\frac{n}{2}}}\int_{\mathbb{R}^n}e^{ix\cdot\xi}\widehat{f}(\xi)\,d\xi.$$

By Theorem 5.1.6, the right-hand side is $f(x)$.

To prove the integral identity, we obtain from (5.1.4) that

$$|\widehat{u}|^2+2|\xi|^2|\widehat{u}|^2+|\xi|^4|\widehat{u}|^2=|\widehat{f}|^2.$$

By writing it in the form

$$|\widehat{u}|^2+2\sum_{k=1}^n\xi_k^2|\widehat{u}|^2+\sum_{k,l=1}^n\xi_k^2\xi_l^2|\widehat{u}|^2=|\widehat{f}|^2,$$

we have, by Lemma 5.1.2,

$$|\widehat{u}|^2+2\sum_{k=1}^n|\widehat{\partial_{x_k}u}|^2+\sum_{k,l=1}^n|\widehat{\partial_{x_kx_l}u}|^2=|\widehat{f}|^2.$$

A simple integration yields

$$\int_{\mathbb{R}^n}\left(|\widehat{u}|^2+2\sum_{k=1}^n|\widehat{\partial_{x_k}u}|^2+\sum_{k,l=1}^n|\widehat{\partial_{x_kx_l}u}|^2\right)d\xi=\int_{\mathbb{R}^n}|\widehat{f}|^2\,d\xi.$$

By Corollary 5.1.8, we obtain

$$\int_{\mathbb{R}^n}\left(|u|^2+2\sum_{k=1}^n|u_{x_k}|^2+\sum_{k,l=1}^n|u_{x_kx_l}|^2\right)dx=\int_{\mathbb{R}^n}|f|^2\,dx.$$

This is the desired identity. □

Example 5.1.11. Now we discuss the initial-value problem for the nonhomogeneous heat equation and derive an explicit expression for its solution. Let f be a continuous function in $\mathbb{R}^n\times(0,\infty)$ and u_0 a continuous function in $\mathbb{R}^n$. We consider

$$\begin{aligned}u_t-\Delta u&=f\quad\text{in }\mathbb{R}^n\times(0,\infty),\\u(\cdot,0)&=u_0\quad\text{on }\mathbb{R}^n.\end{aligned}\tag{5.1.6}$$

Although called an initial-value problem, (5.1.6) is not the type of initial-value problem we discussed in Section 3.1. The heat equation is of the second order, while only one condition is prescribed on the initial hypersurface $\{t=0\}$, which is characteristic.

Suppose u is a solution of (5.1.6) in $C^2(\mathbb{R}^n\times(0,\infty))\cap C(\mathbb{R}^n\times[0,\infty))$. We now derive formally an expression of u in terms of Fourier transforms. In the following, we employ Fourier transforms with respect to space variables only. With an obvious abuse of notation, we write

$$\widehat{u}(\xi,t)=\frac{1}{(2\pi)^{\frac{n}{2}}}\int_{\mathbb{R}^n}e^{-ix\cdot\xi}u(x,t)\,dx.$$

We take Fourier transforms of both sides of the equation and the initial condition in (5.1.6) and obtain, by Lemma 5.1.2,

$$\begin{aligned}\widehat{u}_t+|\xi|^2\widehat{u}&=\widehat{f}\quad\text{in }\mathbb{R}^n\times(0,\infty),\\ \widehat{u}(\cdot,0)&=\widehat{u}_0\quad\text{on }\mathbb{R}^n.\end{aligned}$$

This is an initial-value problem for an ODE with $\xi\in\mathbb{R}^n$ as a parameter. Its solution is given by

$$\widehat{u}(\xi,t)=\widehat{u}_0(\xi)e^{-|\xi|^2t}+\int_0^t e^{-|\xi|^2(t-s)}\widehat{f}(\xi,s)\,ds.$$

Now we treat t as a parameter instead. For any $t>0$, let $K(x,t)$ satisfy

$$\widehat{K}(\xi,t)=\frac{1}{(2\pi)^{\frac{n}{2}}}e^{-|\xi|^2t}.$$

Then

$$\widehat{u}(\xi,t)=(2\pi)^{\frac{n}{2}}\widehat{K}(\xi,t)\widehat{u}_0(\xi)+(2\pi)^{\frac{n}{2}}\int_0^t\widehat{K}(\xi,t-s)\widehat{f}(\xi,s)\,ds.$$

Therefore Theorem 5.1.6 and Lemma 5.1.4 imply

$$\begin{aligned}u(x,t)=&\int_{\mathbb{R}^n}K(x-y,t)u_0(y)\,dy\\ &+\int_0^t\int_{\mathbb{R}^n}K(x-y,t-s)f(y,s)\,dyds,\end{aligned}\tag{5.1.7}$$

for any $(x,t)\in\mathbb{R}^n\times(0,\infty)$. By Theorem 5.1.6 and Proposition 5.1.5, we have

$$K(x,t)=\frac{1}{(2\pi)^n}\int_{\mathbb{R}^n}e^{ix\cdot\xi}e^{-|\xi|^2t}\,d\xi,$$

or

$$K(x,t)=\frac{1}{(4\pi t)^{\frac{n}{2}}}e^{-\frac{|x|^2}{4t}},\tag{5.1.8}$$

for any $(x,t)\in\mathbb{R}^n\times(0,\infty)$. The function K is called the *fundamental solution* of the heat equation.

The derivation of (5.1.7) is formal. Having derived the integral formula for u, we will prove directly that it indeed defines a solution of the initial-value problem for the heat equation under appropriate assumptions on the initial value u_0 and the nonhomogeneous term f. We will pursue this in the next section.

5.2. Fundamental Solutions

In this section, we discuss the heat equation using the fundamental solution. We first discuss the initial-value problem for the heat equation. We prove that the explicit expression for its solution obtained formally by Fourier transforms indeed yields a classical solution under appropriate assumptions on initial values. Then we discuss regularity of solutions of the heat equation. Finally we discuss solutions of the initial-value problem for nonhomogeneous heat equations.

The n-dimensional heat equation is given by

$$u_t - \Delta u = 0, \tag{5.2.1}$$

for $u = u(x,t)$ with $x \in \mathbb{R}^n$ and $t \in \mathbb{R}$. We note that (5.2.1) is not preserved by the change $t \mapsto -t$. This indicates that the heat equation describes an irreversible process and distinguishes between past and future. This fact will be well illustrated by the Harnack inequality, which we will derive later in the next section. Next, (5.2.1) is preserved under linear transforms $x' = \lambda x$ and $t' = \lambda^2 t$ for any nonzero constant λ, which leave the quotient $|x|^2/t$ invariant. Due to this fact, the expression $|x|^2/t$ appears frequently in connection with the heat equation (5.2.1). In fact, the fundamental solution has such an expression.

If u is a solution of (5.2.1) in a domain in $\mathbb{R}^n \times \mathbb{R}$, then for any (x_0, t_0) in this domain and appropriate $r > 0$,

$$u_{x_0,r}(x,t) = u(x_0 + rx, t_0 + r^2 t)$$

is a solution of (5.2.1) in an appropriate domain in $\mathbb{R}^n \times \mathbb{R}$.

In the following, we denote by $C^{2,1}$ the collection of functions which are C^2 in x and C^1 in t. These are the functions for which the heat equation is well defined classically.

5.2.1. Initial-Value Problems.

We first discuss the initial-value problem for the heat equation. Let u_0 be a continuous function in $\mathbb{R}^n$. We consider

$$\begin{aligned} u_t - \Delta u &= 0 \quad \text{in } \mathbb{R}^n \times (0,\infty), \\ u(\cdot,0) &= u_0 \quad \text{on } \mathbb{R}^n. \end{aligned} \tag{5.2.2}$$

We will seek a solution $u \in C^{2,1}(\mathbb{R}^n \times (0,\infty)) \cap C(\mathbb{R}^n \times [0,\infty))$.

We first consider a special case where u_0 is given by a homogeneous polynomial P of degree d in $\mathbb{R}^n$. We now seek a solution u in $\mathbb{R}^n \times (0,\infty)$ which is a p-homogeneous polynomial of degree d, i.e.,

$$u(\lambda x, \lambda^2 t) = \lambda^d u(x,t),$$

for any $(x, t) \in \mathbb{R}^n \times (0, \infty)$ and $\lambda > 0$. To do this, we expand u as a power series of t with coefficients given by functions of x, i.e.,

$$u(x, t) = \sum_{k=0}^{\infty} a_k(x) t^k.$$

Then a straightforward calculation yields

$$a_0 = P, \quad a_k = \frac{1}{k}\Delta a_{k-1} \quad \text{for any } k \geq 1.$$

Therefore for any $k \geq 0$,

$$a_k = \frac{1}{k!}\Delta^k P.$$

Since P is a polynomial of degree d, it follows that $\Delta^{[d/2]+1}P = 0$, where $[d/2]$ is the integral part of $d/2$, i.e., $[d/2] = d/2$ if d is an even integer and $[d/2] = (d-1)/2$ if d is an odd integer. Hence

$$u(x, t) = \sum_{k=0}^{[\frac{d}{2}]} \frac{1}{k!}\Delta^k P(x) t^k.$$

We note that u in fact exists in $\mathbb{R}^n \times \mathbb{R}$. For $n = 1$, let u_d be a p-homogeneous polynomial of degree d in $\mathbb{R} \times \mathbb{R}$ satisfying the heat equation and $u_d(x, 0) = x^d$. The first five such polynomials are given by

$$\begin{aligned}
u_1(x, t) &= x, \\
u_2(x, t) &= x^2 + 2t, \\
u_3(x, t) &= x^3 + 6xt, \\
u_4(x, t) &= x^4 + 12x^2 t + 12t^2, \\
u_5(x, t) &= x^5 + 20x^3 t + 60xt^2.
\end{aligned}$$

We now return to (5.2.2) for general u_0. In view of Example 5.1.11, we set, for any $(x, t) \in \mathbb{R}^n \times (0, \infty)$,

$$K(x, t) = \frac{1}{(4\pi t)^{\frac{n}{2}}} e^{-\frac{|x|^2}{4t}}, \tag{5.2.3}$$

and

$$u(x, t) = \int_{\mathbb{R}^n} K(x - y, t) u_0(y)\, dy. \tag{5.2.4}$$

In Example 5.1.11, we derived formally by using Fourier transforms that any solution of (5.2.2) is given by (5.2.4). Having derived the integral formula for u, we will prove directly that it indeed defines a solution of (5.2.2) under appropriate assumptions on the initial value u_0.

Definition 5.2.1. The function K defined in $\mathbb{R}^n \times (0, \infty)$ by (5.2.3) is called the *fundamental solution* of the heat equation.

We have the following result concerning properties of the fundamental solution.

Lemma 5.2.2. *Let K be the fundamental solution of the heat equation defined by* (5.2.3). *Then*

(1) *$K(x,t)$ is smooth for any $x \in \mathbb{R}^n$ and $t > 0$;*

(2) *$K(x,t) > 0$ for any $x \in \mathbb{R}^n$ and $t > 0$;*

(3) *$(\partial_t - \Delta)K(x,t) = 0$ for any $x \in \mathbb{R}^n$ and $t > 0$;*

(4) *$\int_{\mathbb{R}^n} K(x,t)dx = 1$ for any $t > 0$;*

(5) *for any $\delta > 0$,*

$$\lim_{t\to 0+} \int_{\mathbb{R}^n \setminus B_\delta} K(x,t)\,dx = 0.$$

Proof. Here (1) and (2) are obvious from the explicit expression of K in (5.2.3). We may also get (3) from (5.2.3) by a straightforward calculation. For (4) and (5), we simply note that

$$\int_{|x|>\delta} K(x,t)dx = \frac{1}{\pi^{\frac{n}{2}}} \int_{|\eta|>\frac{\delta}{2\sqrt{t}}} e^{-|\eta|^2}\,d\eta.$$

This implies (4) for $\delta = 0$ and (5) for $\delta > 0$. □

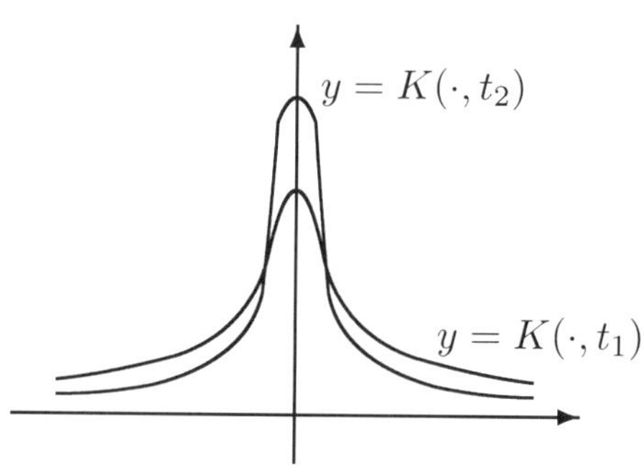

Figure 5.2.1. Graphs of fundamental solutions for $t_2 > t_1 > 0$.

Now we are ready to prove that the integral formula derived by using Fourier transforms indeed yields a classical solution of the initial-value problem for the heat equation under appropriate assumptions on u_0.

Theorem 5.2.3. *Let u_0 be a bounded continuous function in $\mathbb{R}^n$ and u be defined by* (5.2.4). *Then u is smooth in $\mathbb{R}^n \times (0,\infty)$ and satisfies*

$$u_t - \Delta u = 0 \quad \text{in } \mathbb{R}^n \times (0,\infty).$$

Moreover, for any $x_0 \in \mathbb{R}^n$,

$$\lim_{(x,t)\to(x_0,0)} u(x,t) = u_0(x_0).$$

We note that the function u in (5.2.4) is defined only for $t > 0$. We can extend u to $\{t = 0\}$ by setting $u(\cdot, 0) = u_0$ on $\mathbb{R}^n$. Then u is continuous up to $\{t = 0\}$ by Theorem 5.2.3. Therefore, u is a classical solution of the initial-value problem (5.2.2).

The proof of Theorem 5.2.3 proceeds as that of the Poisson integral formula for the Laplace equation in Theorem 4.1.9.

Proof. *Step 1.* We first prove that u is smooth in $\mathbb{R}^n \times (0, \infty)$. For any multi-index $\alpha \in \mathbb{Z}^n_+$ and any nonnegative integer k, we have formally

$$\partial_x^\alpha \partial_t^k u(x,t) = \int_{\mathbb{R}^n} \partial_x^\alpha \partial_t^k K(x-y,t) u_0(y)\, dy.$$

In order to justify the interchange of the order of differentiation and integration, we need to check that, for any nonnegative integer m and any $t > 0$,

$$\int_{\mathbb{R}^n} |x-y|^m e^{-\frac{|x-y|^2}{4t}} |u_0(y)|\, dy < \infty.$$

This follows easily from the exponential decay of the integrand if $t > 0$. Hence u is a smooth function in $\mathbb{R}^n \times (0, \infty)$. Then by Lemma 5.2.2(3),

$$(u_t - \Delta u)(x,t) = \int_{\mathbb{R}^n} (K_t - \Delta_x K)(x-y,t) u_0(y)\, dy = 0.$$

We point out for future references that we used only the boundedness of u_0.

Step 2. We now prove the convergence of $u(x,t)$ to $u_0(x_0)$ as $(x,t) \to (x_0, 0)$. By Lemma 5.2.2(4), we have

$$u_0(x_0) = \int_{\mathbb{R}^n} K(x-y,t) u_0(x_0)\, dy.$$

Then

$$u(x,t) - u_0(x_0) = \int_{\mathbb{R}^n} K(x-y,t)\big(u_0(y) - u_0(x_0)\big)\, dy = I_1 + I_2,$$

where

$$I_1 = \int_{B_\delta(x_0)} \cdots, \quad I_2 = \int_{\mathbb{R}^n \setminus B_\delta(x_0)} \cdots,$$

for a positive constant δ to be determined. For any given $\varepsilon > 0$, we can choose $\delta = \delta(\varepsilon) > 0$ small so that

$$|u_0(y) - u_0(x_0)| < \varepsilon,$$

for any $y \in B_\delta(x_0)$, by the continuity of u_0. Then by Lemma 5.2.2(2) and (4),

$$|I_1| \le \int_{B_\delta(x_0)} K(x-y,t)|u_0(y)-u_0(x_0)|\,dy \le \varepsilon.$$

Since u_0 is bounded, we assume that $|u_0| \le M$ for some positive constant M. We note that $|x-y| \ge \delta/2$ for any $y \in \mathbb{R}^n \setminus B_\delta(x_0)$ and $x \in B_{\delta/2}(x_0)$. By Lemma 5.2.2(5), we can find a $\delta' > 0$ such that

$$\int_{\mathbb{R}^n \setminus B_\delta(x_0)} K(x-y,t)\,dy \le \frac{\varepsilon}{2M},$$

for any $x \in B_{\delta/2}(x_0)$ and $t \in (0,\delta')$, where δ' depends on ε and $\delta = \delta(\varepsilon)$, and hence only on ε. Then

$$|I_2| \le \int_{\mathbb{R}^n \setminus B_\delta(x_0)} K(x-y,t)\big(|u_0(y)| + |u_0(x_0)|\big)\,dy \le \varepsilon.$$

Therefore,

$$|u(x,t) - u_0(0)| \le 2\varepsilon,$$

for any $x \in B_{\delta/2}(x_0)$ and $t \in (0,\delta')$. We then have the desired result. □

Under appropriate assumptions, solutions defined by (5.2.4) decay as time goes to infinity.

Proposition 5.2.4. *Let $u_0 \in L^1(\mathbb{R}^n)$ and u be defined by* (5.2.4). *Then for any $t > 0$,*

$$\sup_{\mathbb{R}^n} |u(\cdot,t)| \le \frac{1}{(4\pi t)^{\frac{n}{2}}} \int_{\mathbb{R}^n} |u_0|\,dx.$$

The proof follows easily from (5.2.4) and the explicit expression for the fundamental solution K in (5.2.3).

Now we discuss a result more general than Theorem 5.2.3 by relaxing the boundedness assumption on u_0. To seek a reasonably more general assumption on initial values, we examine the expression for the fundamental solution K. We note that K in (5.2.3) decays exponentially in space variables with a large decay rate for small time. This suggests that we can allow an exponential growth for initial values. In the convolution formula (5.2.4), a fixed exponential growth from initial values can be offset by the fast exponential decay in the fundamental solution, at least for a short period of time. To see this clearly, we consider an example. For any $\alpha > 0$, set

$$G(x,t) = \frac{1}{(1-4\alpha t)^{\frac{n}{2}}}\, e^{\frac{\alpha}{1-4\alpha t}|x|^2},$$

for any $x \in \mathbb{R}^n$ and $t < 1/4\alpha$. It is straightforward to check that

$$G_t - \Delta G = 0.$$

Note that

$$G(x,0) = e^{\alpha|x|^2} \quad \text{for any } x \in \mathbb{R}^n.$$

Hence, viewed as a function in $\mathbb{R}^n \times [0, 1/4\alpha)$, G has an exponential growth initially for $t = 0$, and in fact for any $t < 1/4\alpha$. The growth rate becomes arbitrarily large as t approaches $1/4\alpha$ and G does not exist beyond $t = 1/4\alpha$.

Now we formulate a general result. If u_0 is continuous and has an exponential growth, then (5.2.4) still defines a solution of the initial-value problem in a short period of time.

Theorem 5.2.5. *Suppose $u_0 \in C(\mathbb{R}^n)$ satisfies*

$$|u_0(x)| \le Me^{A|x|^2} \quad \textit{for any } x \in \mathbb{R}^n,$$

for some constants $M, A \ge 0$. Then u defined by (5.2.4) is smooth in $\mathbb{R}^n \times (0, \frac{1}{4A})$ and satisfies

$$u_t - \Delta u = 0 \quad \textit{in } \mathbb{R}^n \times \left(0, \frac{1}{4A}\right).$$

Moreover, for any $x_0 \in \mathbb{R}^n$,

$$\lim_{(x,t)\to(x_0,0)} u(x,t) = u_0(x_0).$$

The proof is similar to that of Theorem 5.2.3.

Proof. The case $A = 0$ is covered by Theorem 5.2.3. We consider only $A > 0$. First, by the explicit expression for K in (5.2.3) and the assumption on u_0, we have

$$|u(x,t)| \le \frac{M}{(4\pi t)^{\frac{n}{2}}} \int_{\mathbb{R}^n} e^{-\frac{1}{4t}|x-y|^2 + A|y|^2}\, dy.$$

A simple calculation shows that

$$-\frac{1}{4t}|x-y|^2 + A|y|^2 = -\frac{1-4At}{4t}\left|y - \frac{1}{1-4At}x\right|^2 + \frac{A}{1-4At}|x|^2.$$

Hence for any $(x,t) \in \mathbb{R}^n \times (0, 1/(4A))$, we obtain

$$\begin{aligned} |u(x,t)| &\le \frac{M}{(4\pi t)^{\frac{n}{2}}} e^{\frac{A}{1-4At}|x|^2} \int_{\mathbb{R}^n} e^{-\frac{1-4At}{4t}\left|y-\frac{1}{1-4At}x\right|^2}\, dy \\ &\le \frac{M}{(1-4At)^{\frac{n}{2}}} e^{\frac{A}{1-4At}|x|^2}. \end{aligned}$$

The integral defining u in (5.2.4) is convergent absolutely and uniformly for $(x,t) \in \mathbb{R}^n \times [\varepsilon, -\varepsilon + 1/(4A)]$, for any $\varepsilon > 0$ small. Hence, u is continuous in $\mathbb{R}^n \times (0, 1/(4A))$. To show that u has continuous derivatives of arbitrary order in $\mathbb{R}^n \times (0, 1/(4A))$, we need only verify

$$\int_{\mathbb{R}^n} |x-y|^m e^{-\frac{1}{4t}|x-y|^2 + A|y|^2}\, dy < \infty,$$

for any $m \geq 0$. The proof for $m \geq 1$ is similar to that for $m = 0$ and we omit the details.

Next, we need to prove the convergence of $u(x,t)$ to $u_0(x_0)$ as $(x,t) \to (x_0, 0)$. We leave the proof as an exercise. □

Now we discuss properties of the solution u given by (5.2.4) of the initial-value problem (5.2.2). First for any fixed $x \in \mathbb{R}^n$ and $t > 0$, the value of $u(x,t)$ depends on the values of u_0 at all points. Equivalently, the values of u_0 near a point $x_0 \in \mathbb{R}^n$ affect the value of $u(x,t)$ at all x as long as $t > 0$. We interpret this by saying that the effects travel at an *infinite speed*. If the initial value u_0 is nonnegative everywhere and positive somewhere, then the solution u in (5.2.4) at any later time is positive everywhere. We will see later that this is related to the strong maximum principle.

Next, the function $u(x,t)$ in (5.2.4) becomes smooth for $t > 0$, even if the initial value u_0 is simply bounded. This is well illustrated in Step 1 in the proof of Theorem 5.2.3. We did not use any regularity assumption on u_0 there. Compare this with Theorem 3.3.5. Later on, we will prove a general result that any solutions of the heat equation in a domain in $\mathbb{R}^n \times (0, \infty)$ are smooth away from the boundary. Refer to a similar remark at the end of Subsection 4.1.2 for harmonic functions defined by the Poisson integral formula.

We need to point out that (5.2.4) represents only one of infinitely many solutions of the initial-value problem (5.2.2). The solutions are not unique without further conditions on u, such as boundedness or exponential growth. In fact, there exists a nontrivial solution $u \in C^\infty(\mathbb{R}^n \times \mathbb{R})$ of $u_t - \Delta u = 0$, with $u \equiv 0$ for $t \leq 0$. In the following, we construct such a solution of the one-dimensional heat equation.

Proposition 5.2.6. *There exists a nonzero smooth function $u \in C^\infty(\mathbb{R} \times [0, \infty))$ satisfying*

$$\begin{aligned} u_t - u_{xx} &= 0 \quad \textit{in } \mathbb{R} \times [0, \infty), \\ u(\cdot, 0) &= 0 \quad \textit{on } \mathbb{R}. \end{aligned}$$

Proof. We construct a smooth function in $\mathbb{R} \times \mathbb{R}$ such that $u_t - u_{xx} = 0$ in $\mathbb{R} \times \mathbb{R}$ and $u \equiv 0$ for $t < 0$. We treat $\{x = 0\}$ as the initial curve and attempt to find a smooth solution of the initial-value problem

$$\begin{aligned} u_t - u_{xx} &= 0 \quad \text{in } \mathbb{R} \times \mathbb{R}, \\ u(0,t) &= a(t), \ u_x(0,t) = 0 \quad \text{for } t \in \mathbb{R}, \end{aligned}$$

for an appropriate function a in $\mathbb{R}$. We write u as a power series in x:

$$u(x,t) = \sum_{k=0}^{\infty} a_k(t) x^k.$$

Making a simple substitution in the equation $u_t = u_{xx}$ and comparing the coefficients of powers of x, we have

$$a'_{k-2} = k(k-1)a_k \quad \text{for any } k \geq 2.$$

Evaluating u and u_x at $x = 0$, we get

$$a_0 = a, \quad a_1 = 0.$$

Hence for any $k \geq 0$,

$$a_{2k}(t) = \frac{1}{(2k)!} a^{(k)}(t),$$

and

$$a_{2k+1}(t) = 0.$$

Therefore, we have a formal solution

$$u(x,t) = \sum_{k=0}^{\infty} \frac{1}{(2k)!} a^{(k)}(t) x^{2k}.$$

We need to choose $a(t)$ appropriately so that $u(x,t)$ defined above is a smooth function and is identically zero for $t < 0$. To this end, we define

$$a(t) = \begin{cases} e^{-\frac{1}{t^2}} & \text{for } t > 0, \\ 0 & \text{for } t \leq 0. \end{cases}$$

Then it is straightforward to verify that the series defining u is absolutely convergent in $\mathbb{R} \times \mathbb{R}$. This implies that u is continuous. In fact, we can prove that series defining arbitrary derivatives of u are also absolutely convergent in $\mathbb{R} \times \mathbb{R}$. We skip the details and leave the rest of the proof as an exercise. □

Next, we discuss briefly terminal-value problems. For a fixed constant $T > 0$, we consider

$$\begin{aligned} u_t - u_{xx} &= 0 \quad \text{in } \mathbb{R} \times (0,T), \\ u(\cdot, T) &= \varphi \quad \text{on } \mathbb{R}. \end{aligned}$$

Here the function φ is prescribed at the terminal time T. This problem is not well posed. Consider the following example. For any positive integer m, let

$$u_m(x,t) = e^{m^2(T-t)} \sin(mx),$$

for any $(x,t) \in \mathbb{R} \times [0,T)$. Then u_m solves this problem with the terminal value

$$\varphi_m(x) = u_m(x,T) = \sin(mx),$$

for any $x \in \mathbb{R}$. We note that

$$\sup_{\mathbb{R}} |\varphi_m| = 1,$$

and for any $t \in [0,T)$,

$$\sup_{\mathbb{R}} |u_m(\cdot,t)| = e^{m^2(T-t)} \to \infty \quad \text{as } m \to \infty.$$

There is no continuous dependence of solutions on the values prescribed at the terminal time T.

5.2.2. Regularity of Solutions. Next, we discuss regularity of solutions of the heat equation with the help of the fundamental solution. We will do this only in special domains.

For any $(x_0,t_0) \in \mathbb{R}^n \times \mathbb{R}$ and any $R > 0$, we define

$$Q_R(x_0,t_0) = B_R(x_0) \times (t_0 - R^2, t_0].$$

We point out that subsets of the form $Q_R(x_0,t_0)$ play the same role for the heat equation as balls for the Laplace equation. If u is a solution of the heat equation $u_t - \Delta u = 0$ in $Q_R(0)$, then

$$u_R(x,t) = u(Rx, R^2 t)$$

is a solution of the heat equation in $Q_1(0)$.

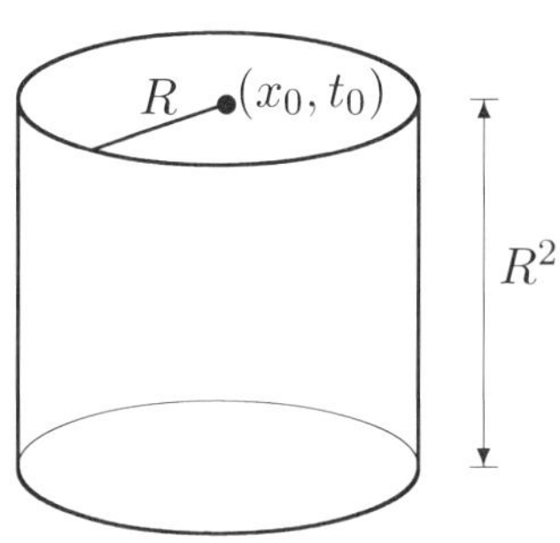

Figure 5.2.2. The region $Q_R(x_0,t_0)$.

For any domain D in $\mathbb{R}^n \times \mathbb{R}$, we denote by $C^{2,1}(D)$ the collection of functions in D which are C^2 in x and C^1 in t.

We first have the following regularity result for solutions of the heat equation.

Theorem 5.2.7. *Let u be a $C^{2,1}$-solution of $u_t - \Delta u = 0$ in $Q_R(x_0, t_0)$ for some $(x_0, t_0) \in \mathbb{R}^n \times \mathbb{R}$ and $R > 0$. Then u is smooth in $Q_R(x_0, t_0)$.*

Proof. For simplicity, we consider the case $(x_0, t_0) = (0, 0)$ and write

$$Q_R = B_R \times (-R^2, 0].$$

Without loss of generality, we assume that u is bounded in $\bar{Q}_R$. Otherwise, we consider u in Q_r for any $r < R$.

We take an arbitrarily fixed point $(x, t) \in Q_R$ and claim that

$$\begin{aligned} u(x,t) = & \int_{B_R} K\big(x - y, t + R^2\big) u(y, -R^2)\, dy \\ & + \int_{-R^2}^{t} \int_{\partial B_R} \bigg[K(x - y, t - s) \frac{\partial u}{\partial \nu_y}(y, s) \\ & \qquad\qquad - u(y, s) \frac{\partial K}{\partial \nu_y}(x - y, t - s) \bigg]\, dS_y ds. \end{aligned}$$

We first assume this identity and prove that it implies the smoothness of u. We note that the integrals in the right-hand side are only over the bottom and the side of the boundary of $B_R \times (-R^2, t]$. The first integral is over $B_R \times \{-R^2\}$. For $(x, t) \in Q_R$, it is obvious that $t + R^2 > 0$ and hence there is no singularity in the first integral. The second integral is over $\partial B_R \times (-R^2, t]$. By the change of variables $\tau = t - s$, we can rewrite it as

$$\int_0^{t+R^2} \int_{\partial B_R} \bigg[K(x - y, \tau) \frac{\partial u}{\partial \nu_y}(y, t - \tau) - u(y, t - \tau) \frac{\partial K}{\partial \nu_y}(x - y, \tau) \bigg]\, dS_y d\tau.$$

There is also no singularity in the integrand since $x \in B_R$, $y \in \partial B_R$, and $\tau > 0$. Hence, we conclude that u is smooth in Q_R.

We now prove the claim. Let K be the fundamental solution of the heat equation as in (5.2.3). Denoting by (y, s) points in Q_R, we set

$$\tilde{K}(y, s) = K(x - y, t - s) = \frac{1}{\big(4\pi(t - s)\big)^{\frac{n}{2}}} e^{-\frac{|x-y|^2}{4(t-s)}} \quad \text{for } s < t.$$

Then

$$\tilde{K}_s + \Delta_y \tilde{K} = 0.$$

Hence,

$$\begin{aligned} 0 = \tilde{K}(u_s - \Delta_y u) &= (u\tilde{K})_s + \sum_{i=1}^{n} (u\tilde{K}_{y_i} - \tilde{K} u_{y_i})_{y_i} - u(\tilde{K}_s + \Delta_y \tilde{K}) \\ &= (u\tilde{K})_s + \sum_{i=1}^{n} (u\tilde{K}_{y_i} - \tilde{K} u_{y_i})_{y_i}. \end{aligned}$$

For any $\varepsilon > 0$ with $t-\varepsilon > -R^2$, we integrate with respect to (y,s) in $B_R \times (-R^2, t-\varepsilon)$. Then

$$\begin{aligned}\int_{B_R} & K(x-y,\varepsilon)u(y,t-\varepsilon)\,dy \\ &= \int_{B_R} K\big(x-y,t-(-R^2)\big)u(y,-R^2)\,dy \\ &\quad + \int_{-R^2}^{t-\varepsilon}\int_{\partial B_R}\bigg[K(x-y,t-s)\frac{\partial u}{\partial \nu_y}(y,s) \\ &\qquad\qquad -u(y,s)\frac{\partial K}{\partial \nu_y}(x-y,t-s)\bigg]\,dS_y ds.\end{aligned}$$

Now it suffices to prove that

$$\lim_{\varepsilon\to 0}\int_{B_R} K(x-y,\varepsilon)u(y,t-\varepsilon)dy = u(x,t).$$

The proof proceeds similarly to that in Step 2 in the proof of Theorem 5.2.3. The integral here over a finite domain introduces few changes here. We omit the details. □

Now we prove interior gradient estimates.

Theorem 5.2.8. *Let u be a bounded $C^{2,1}$-solution of $u_t - \Delta u = 0$ in $Q_R(x_0,t_0)$ for some $(x_0,t_0)\in\mathbb{R}^n\times\mathbb{R}$ and $R>0$. Then*

$$|\nabla_x u(x_0,t_0)| \le \frac{C}{R}\sup_{Q_R(x_0,t_0)}|u|,$$

where C is a positive constant depending only on n.

Proof. We consider the case $(x_0,t_0)=(0,0)$ and $R=1$ only. The general case follows from a simple translation and dilation. (Refer to Lemma 4.1.11 for a similar dilation for harmonic functions.) In the following, we write $Q_r = B_r\times(-r^2,0]$ for any $r\in(0,1]$.

We first modify the proof of Theorem 5.2.7 to express u in terms of the fundamental solution and cutoff functions. We denote points in Q_1 by (y,s). Let K be the fundamental solution of the heat equation given in (5.2.3). As in the proof of Theorem 5.2.7, we set, for any fixed $(x,t)\in Q_{1/4}$,

$$\tilde K(y,s) = K(x-y,t-s) = \frac{1}{\big(4\pi(t-s)\big)^{\frac{n}{2}}}e^{-\frac{|x-y|^2}{4(t-s)}} \quad \text{for } s<t.$$

By choosing a cutoff function $\varphi\in C^\infty(Q_1)$ with $\operatorname{supp}\varphi\subset Q_{3/4}$ and $\varphi\equiv 1$ in $Q_{1/2}$, we set

$$v = \varphi\tilde K.$$

We need to point out that $v(y,s)$ is defined only for $s<t$. For such a function v, we have

$$0=v(u_s-\Delta_y u)=(uv)_s+\sum_{i=1}^{n}(uv_{y_i}-vu_{y_i})_{y_i}-u(v_s+\Delta_y v).$$

For any $\varepsilon>0$, we integrate with respect to (y,s) in $B_1\times(-1,t-\varepsilon)$. We note that there is no boundary integral over $B_1\times\{-1\}$ and $\partial B_1\times(-1,t-\varepsilon)$, since φ vanishes there. Hence

$$\int_{B_1}(\varphi u)(y,t-\varepsilon)K(x-y,\varepsilon)\,dy=\int_{B_1\times(-1,t-\varepsilon)}u(\partial_s+\Delta_y)(\varphi\tilde{K})\,dyds.$$

Then similarly to the proof of Theorem 5.2.3, we have, as $\varepsilon\to 0$,

$$\varphi(x,t)u(x,t)=\int_{B_1\times(-1,t)}u(\partial_s+\Delta_y)(\varphi\tilde{K})\,dyds.$$

In view of

$$\tilde{K}_s+\Delta_y\tilde{K}=0,$$

we obtain for any $(x,t)\in Q_{1/4}$ that

$$u(x,t)=\int_{B_1\times(-1,t)}u\big((\varphi_s+\Delta_y\varphi)\tilde{K}+2\nabla_y\varphi\cdot\nabla_y\tilde{K}\big)\,dyds.$$

We note that each term in the integrand involves a derivative of φ, which is zero in $Q_{1/2}$ since $\varphi\equiv 1$ there. Then the domain of integration D is actually given by

$$D=B_{\frac{3}{4}}\times\left(-(3/4)^2,t\right]\setminus B_{\frac{1}{2}}\times\left(-(1/2)^2,t\right].$$

The distance between any $(y,s)\in D$ and any $(x,t)\in Q_{1/4}$ has a positive lower bound. Therefore, the integrand has no singularity in D. (This gives an alternate proof of the smoothness of u in $Q_{1/4}$.)

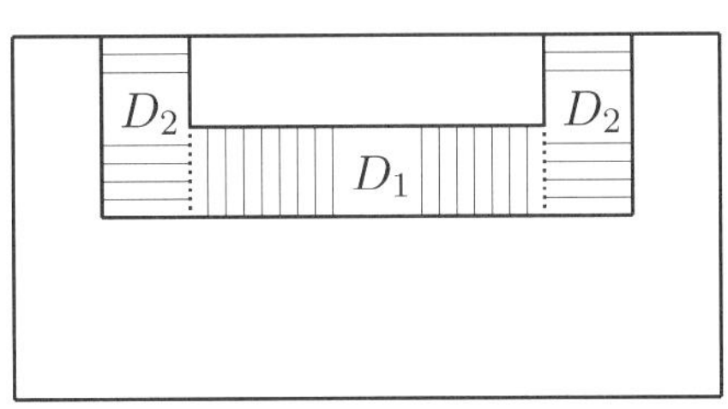

Figure 5.2.3. A decomposition of D for $n=1$.

Next, we have, for any $(x,t)\in Q_{1/4}$,

$$\nabla_x u(x,t)=\int_D u\big((\varphi_s+\Delta_y\varphi)\nabla_x\tilde{K}+2\nabla_y\varphi\cdot\nabla_x\nabla_y\tilde{K}\big)\,dyds.$$

Let C be a positive constant such that

$$2|\nabla_y\varphi| \leq C, \quad |\varphi_s| + |\nabla_y^2\varphi| \leq C.$$

Hence

$$|\nabla_x u(x,t)| \leq C\int_D \left(|\nabla_x\tilde{K}| + |\nabla_x\nabla_y\tilde{K}|\right)|u|\,dyds.$$

By the explicit expression for $\tilde{K}$, we have

$$|\nabla_x\tilde{K}| \leq C\frac{|x-y|}{(t-s)^{\frac{n}{2}+1}}e^{-\frac{|x-y|^2}{4(t-s)}},$$

and

$$|\nabla_x\nabla_y\tilde{K}| \leq C\frac{|x-y|^2+(t-s)}{(t-s)^{\frac{n}{2}+2}}e^{-\frac{|x-y|^2}{4(t-s)}}.$$

Obviously, for any $(x,t) \in Q_{1/4}$ and any $(y,s) \in D$,

$$|x-y| \leq 1, \quad 0 < t-s \leq 1.$$

Therefore, for any $(x,t) \in Q_{1/4}$,

$$|\nabla_x u(x,t)| \leq C\sum_{i=1}^{2}\int_D \frac{1}{(t-s)^{\frac{n}{2}+i}}e^{-\frac{|x-y|^2}{4(t-s)}}|u(y,s)|\,dyds.$$

Now we claim that, for any $(x,t) \in Q_{1/4}$, $(y,s) \in D$ and $i = 1,2$,

$$\frac{1}{(t-s)^{\frac{n}{2}+i}}e^{-\frac{|x-y|^2}{4(t-s)}} \leq C.$$

Then we obtain easily for any $(x,t) \in Q_{1/4}$ that

$$|\nabla_x u(x,t)| \leq C\sup_{Q_1}|u|.$$

To prove the claim, we decompose D into two parts,

$$D_1 = B_{\frac{1}{2}} \times \left(-(3/4)^2, -(1/2)^2\right),$$
$$D_2 = \left(B_{\frac{3}{4}} \setminus B_{\frac{1}{2}}\right) \times \left(-(3/4)^2, t\right).$$

We first consider D_1. For any $(x,t) \in Q_{1/4}$ and $(y,s) \in D_1$, we have

$$t-s \geq \frac{1}{8},$$

and hence

$$\frac{1}{(t-s)^{\frac{n}{2}+i}}e^{-\frac{|x-y|^2}{4(t-s)}} \leq 8^{\frac{n}{2}+i}.$$

Next, we consider D_2. For any $(x,t) \in Q_{1/4}$ and $(y,s) \in D_2$, we have

$$|y-x| \geq \frac{1}{4}, \quad 0 < t-s < \left(\frac{3}{4}\right)^2,$$

and hence, with $\tau = (t-s)^{-1}$,

$$\frac{1}{(t-s)^{\frac{n}{2}+i}} e^{-\frac{|x-y|^2}{4(t-s)}} \leq \frac{1}{(t-s)^{\frac{n}{2}+i}} e^{-\frac{1}{4^3(t-s)}} = \tau^{\frac{n}{2}+i} e^{-\frac{\tau}{4^3}} \leq C,$$

for any $\tau > (4/3)^2$. This finishes the proof of the claim. □

Next, we estimate derivatives of arbitrary order.

Theorem 5.2.9. *Let u be a bounded $C^{2,1}$-solution of $u_t - \Delta u = 0$ in $Q_R(x_0, t_0)$ for some $(x_0, t_0) \in \mathbb{R}^n \times \mathbb{R}$ and $R > 0$. Then for any nonnegative integers m and k,*

$$|\partial_t^k \nabla_x^m u(x_0, t_0)| \leq \frac{C^{m+2k}}{R^{m+2k}} n^k e^{m+2k-1}(m+2k)! \sup_{Q_R(x_0,t_0)} |u|,$$

where C is a positive constant depending only on n.

Proof. For x-derivatives, we proceed as in the proof of Theorem 4.1.12 and obtain that, for any $\alpha \in \mathbb{Z}_+^n$ with $|\alpha| = m$,

$$|\partial_x^\alpha u(x_0, t_0)| \leq \frac{C^m e^{m-1} m!}{R^m} \sup_{Q_R(x_0,t_0)} |u|.$$

For t-derivatives, we have $u_t = \Delta u$ and hence $\partial_t^k u = \Delta^k u$ for any positive integer k. We note that there are n^k terms of x-derivatives of u of order $2k$ in $\Delta^k u$. Hence

$$|\partial_t^k \nabla_x^m u(x_0, t_0)| \leq n^k \max_{|\beta|=m+2k} |\partial_x^\beta u(x_0, t_0)|.$$

This implies the desired result easily. □

The next result concerns the analyticity of solutions of the heat equation on any time slice.

Theorem 5.2.10. *Let u be a $C^{2,1}$-solution of $u_t - \Delta u = 0$ in $Q_R(x_0, t_0)$ for some $(x_0, t_0) \in \mathbb{R}^n \times \mathbb{R}$ and $R > 0$. Then $u(\cdot, t)$ is analytic in $B_R(x_0)$ for any $t \in (t_0 - R^2, t_0]$. Moreover, for any nonnegative integer k, $\partial_t^k u(\cdot, t)$ is analytic in $B_R(x_0)$ for any $t \in (t_0 - R^2, t_0]$.*

The proof is identical to that of Theorem 4.1.14 and is omitted. In general, solutions of $u_t - \Delta u = 0$ are not analytic in t. This is illustrated by Proposition 5.2.6.

5.2.3. Nonhomogeneous Problems. Now we discuss the initial-value problem for the nonhomogeneous equation. Let f be continuous in $\mathbb{R}^n \times (0, \infty)$. Consider

$$\begin{aligned} u_t - \Delta u &= f \quad \text{in } \mathbb{R}^n \times (0, \infty), \\ u(\cdot, 0) &= 0 \quad \text{on } \mathbb{R}^n. \end{aligned}$$

Let K be the fundamental solution of the heat equation as in (5.2.3), i.e.,

$$K(x,t) = \frac{1}{(4\pi t)^{\frac{n}{2}}} e^{-\frac{|x|^2}{4t}},$$

for any $(x,t) \in \mathbb{R}^n \times (0, \infty)$. Define

$$u(x,t) = \int_0^t \int_{\mathbb{R}^n} K(x-y, t-s) f(y,s)\, dyds, \tag{5.2.5}$$

for any $(x,t) \in \mathbb{R}^n \times (0, \infty)$. If f is bounded in $\mathbb{R}^n \times (0, \infty)$, it is straightforward to check that the integral in the right-hand side of (5.2.5) is well defined and continuous in $(x,t) \in \mathbb{R}^n \times (0, \infty)$. By Lemma 5.2.2(4), we have

$$|u(x,t)| \le \sup_{\mathbb{R}^n \times (0,t)} |f| \int_0^t \int_{\mathbb{R}^n} K(y,s)\, dyds = t \sup_{\mathbb{R}^n \times (0,t)} |f|.$$

Hence

$$\sup_{\mathbb{R}^n} |u(\cdot, t)| \to 0 \quad \text{as } t \to 0.$$

To discuss whether u is differentiable, we note that

$$\begin{aligned} K_{x_i}(x,t) &= -\frac{1}{(4\pi t)^{\frac{n}{2}}} \frac{x_i}{2t} e^{-\frac{|x|^2}{4t}}, \\ K_{x_i x_j}(x,t) &= \frac{1}{(4\pi t)^{\frac{n}{2}}} \left(\frac{x_i x_j}{4t^2} - \frac{\delta_{ij}}{2t} \right) e^{-\frac{|x|^2}{4t}}. \end{aligned}$$

For any $t > 0$, by the change of variables $x = 2z\sqrt{t}$, we have

$$\int_{\mathbb{R}^n} |K_{x_i}(x,t)|\, dx = \frac{1}{\pi^{\frac{n}{2}}} \int_{\mathbb{R}^n} e^{-|z|^2} \frac{|z_i|}{\sqrt{t}}\, dz = \frac{1}{\sqrt{\pi t}},$$

and

$$\int_{\mathbb{R}^n} |K_{x_i x_j}(x,t)|\, dx = \frac{1}{\pi^{\frac{n}{2}} t} \int_{\mathbb{R}^n} \left| z_i z_j - \frac{\delta_{ij}}{2} \right| e^{-|z|^2}\, dz.$$

Hence $K_{x_i} \in L^1(\mathbb{R}^n \times (0,T))$ and $K_{x_i x_j} \notin L^1(\mathbb{R}^n \times (0,T))$ for any $T > 0$. A formal differentiation of (5.2.5) yields

$$u_{x_i}(x,t) = \int_0^t \int_{\mathbb{R}^n} K_{x_i}(x-y, t-s) f(y,s)\, dyds. \tag{5.2.6}$$

We denote by I the integral in the right-hand side. If f is bounded in $\mathbb{R}^n \times (0, \infty)$, then

$$\begin{aligned} |I| &\leq \sup_{\mathbb{R}^n \times (0,t)} |f| \int_0^t \int_{\mathbb{R}^n} |K_{x_i}(x-y, t-s)| \, dyds \\ &\leq \sup_{\mathbb{R}^n \times (0,t)} |f| \int_0^t \frac{1}{\sqrt{\pi(t-s)}} \, ds = \frac{2\sqrt{t}}{\sqrt{\pi}} \sup_{\mathbb{R}^n \times (0,t)} |f|. \end{aligned}$$

Hence, the integral in the right-hand side of (5.2.6) is well defined and continuous in $(x, t) \in \mathbb{R}^n \times (0, \infty)$. We will justify (5.2.6) later in the proof of Theorem 5.2.11 under extra assumptions. Even assuming the validity of (5.2.6), we cannot continue differentiating (5.2.6) to get the second x-derivatives of u if f is merely bounded, since $K_{x_i x_j} \notin L^1(\mathbb{R}^n \times (0, T))$ for any $T > 0$. In order to get the second x-derivatives of u, we need extra assumptions on f.

Theorem 5.2.11. *Let f be a bounded continuous function in $\mathbb{R}^n \times (0, \infty)$ with bounded and continuous $\nabla_x f$ in $\mathbb{R}^n \times (0, \infty)$ and u be defined by (5.2.5) for $(x, t) \in \mathbb{R}^n \times (0, \infty)$. Then u is $C^{2,1}$ in $\mathbb{R}^n \times (0, \infty)$ and satisfies*

$$u_t - \Delta u = f \quad \text{in } \mathbb{R}^n \times (0, \infty),$$

and for any $x_0 \in \mathbb{R}^n$,

$$\lim_{(x,t) \to (x_0, 0)} u(x, t) = 0.$$

Moreover, if f is smooth with bounded derivatives of arbitrary order in $\mathbb{R}^n \times (0, \infty)$, then u is smooth in $\mathbb{R}^n \times (0, \infty)$.

Proof. We first assume that f and $\nabla_x f$ are continuous and bounded in $\mathbb{R}^n \times (0, \infty)$. By the explicit expression for K and the change of variables $y = x + 2z\sqrt{t-s}$, we obtain from (5.2.5) that

$$u(x, t) = \frac{1}{\pi^{\frac{n}{2}}} \int_0^t \int_{\mathbb{R}^n} e^{-|z|^2} f(x + 2z\sqrt{t-s}, s) \, dzds, \tag{5.2.7}$$

for any $(x, t) \in \mathbb{R}^n \times (0, \infty)$. It follows easily that the limit of $u(x, t)$ is zero as $t \to 0$.

A simple differentiation yields

$$\begin{aligned} u_{x_i}(x, t) &= \frac{1}{\pi^{\frac{n}{2}}} \int_0^t \int_{\mathbb{R}^n} e^{-|z|^2} \partial_{x_i} f(x + 2z\sqrt{t-s}, s) \, dzds \\ &= \frac{1}{\pi^{\frac{n}{2}}} \int_0^t \int_{\mathbb{R}^n} e^{-|z|^2} \frac{1}{2\sqrt{t-s}} \partial_{z_i} f(x + 2z\sqrt{t-s}, s) \, dzds. \end{aligned}$$

Upon integrating by parts, we have

$$u_{x_i}(x, t) = \frac{1}{\pi^{\frac{n}{2}}} \int_0^t \int_{\mathbb{R}^n} e^{-|z|^2} \frac{z_i}{\sqrt{t-s}} f(x + 2z\sqrt{t-s}, s) \, dzds.$$

(We note that this is (5.2.6) by the change of variables $y = x + 2z\sqrt{t-s}$.) A differentiation under the integral signs yields

$$u_{x_ix_j}(x,t) = \frac{1}{\pi^{\frac{n}{2}}}\int_0^t\int_{\mathbb{R}^n} e^{-|z|^2}\frac{z_i}{\sqrt{t-s}} f_{x_j}(x+2z\sqrt{t-s},s)\,dzds.$$

A similar differentiation of (5.2.7) yields

$$\begin{aligned} u_t(x,t) &= \frac{1}{\pi^{\frac{n}{2}}}\int_{\mathbb{R}^n} e^{-|z|^2} f(x,t)\,dz \\ &\quad + \frac{1}{\pi^{\frac{n}{2}}}\int_0^t\int_{\mathbb{R}^n}\sum_{i=1}^n e^{-|z|^2}\frac{z_i}{\sqrt{t-s}} f_{x_i}(x+2z\sqrt{t-s},s)\,dzds. \end{aligned}$$

In view of the boundedness of $\nabla_x f$, we conclude that u_t and $u_{x_ix_j}$ are continuous in $(x,t)\in\mathbb{R}^n\times(0,\infty)$. We note that the first term in the right-hand side of $u_t(x,t)$ is simply $f(x,t)$. Hence,

$$u_t(x,t) - \Delta u(x,t) = u_t(x,t) - \sum_{i=1}^n u_{x_ix_i}(x,t) = f(x,t),$$

for any $(x,t)\in\mathbb{R}^n\times(0,\infty)$.

If f has bounded x-derivatives of arbitrary order in $\mathbb{R}^n\times(0,\infty)$, by (5.2.7) we conclude that x-derivatives of u of arbitrary order exist and are continuous in $\mathbb{R}^n\times(0,\infty)$. By the equation $u_t = \Delta u + f$, we then conclude that u_t and all its x-derivatives exist and are continuous in $\mathbb{R}^n\times(0,\infty)$. Next,

$$u_{tt} = \Delta u_t + f_t = \Delta_x(\Delta_x u + f) + f_t.$$

Hence u_{tt} and all its x-derivatives exist and are continuous in $\mathbb{R}^n\times(0,\infty)$. Continuing this process, all derivatives of u exist and are continuous in $\mathbb{R}^n\times(0,\infty)$. □

By combining Theorem 5.2.3 and Theorem 5.2.11, we conclude that, under the assumptions on u_0 and f as above, the function u given by

$$\begin{aligned} u(x,t) &= \int_{\mathbb{R}^n} K(x-y,t)u_0(y)\,dy \\ &\quad + \int_0^t\int_{\mathbb{R}^n} K(x-y,t-s)f(y,s)\,dyds \end{aligned}$$

is a solution of

$$\begin{aligned} u_t - \Delta u &= f \quad \text{in } \mathbb{R}^n\times(0,\infty), \\ u(\cdot,0) &= u_0 \quad \text{on } \mathbb{R}^n. \end{aligned}$$

Theorem 5.2.11 is optimal in the C^∞-category in the sense that the smoothness of f implies the smoothness of u. However, it is not optimal concerning finite differentiability. In the equation $u_t - \Delta u = f$, f is related to the second x-derivatives and the first t-derivative of u. Theorem 5.2.11 asserts that the continuity of f and its first x-derivatives implies the continuity of $\nabla_x^2 u$ and u_t. It is natural to ask whether the continuity of f itself is sufficient. This question has a negative answer, and an example can be constructed by modifying Example 4.4.4. Hence, spaces of functions with continuous derivatives are not adequate for optimal regularity. What is needed is the Hölder spaces adapted to the heat equation, referred to as the *parabolic Hölder spaces.* The study of the nonhomogeneous heat equation, or more generally, nonhomogeneous parabolic differential equations, in parabolic Hölder spaces is known as the parabolic version of the Schauder theory. It is beyond the scope of this book to give a presentation of the Schauder theory. Refer to Subsection 4.4.1 for discussions of the Poisson equation.

5.3. The Maximum Principle

In this section, we discuss the maximum principle for a class of parabolic differential equations slightly more general than the heat equation. As applications of the maximum principle, we derive a priori estimates for mixed problems and initial-value problems, interior gradient estimates and the Harnack inequality.

5.3.1. The Weak Maximum Principle. Let D be a domain in $\mathbb{R}^n \times \mathbb{R}$. The *parabolic boundary* $\partial_p D$ of D consists of points $(x_0, t_0) \in \partial D$ such that $B_r(x_0) \times (t_0 - r^2, t_0]$ contains points not in D, for any $r > 0$. We denote by $C^{2,1}(D)$ the collection of functions in D which are C^2 in x and C^1 in t.

We often discuss the heat equation or general parabolic equations in cylinders of the following form. Suppose $\Omega \subset \mathbb{R}^n$ is a bounded domain. For any $T > 0$, set

$$\Omega_T = \Omega \times (0, T] = \{(x, t) : \ x \in \Omega,\ 0 < t \leq T\}.$$

We note that Ω_T includes the top portion of its geometric boundary. The parabolic boundary $\partial_p \Omega_T$ of Ω_T is given by

$$\partial_p \Omega_T = (\Omega \times \{t = 0\}) \cup (\partial\Omega \times (0, T]) \cup \big(\partial\Omega \times \{0\}\big).$$

In other words, parabolic boundary consists of the bottom, the side and the bottom corner of the geometric boundary.

For simplicity of presentation, we will prove the weak maximum principle only in domains of the form Ω_T. We should point out that the results in this subsection hold for general domains in $\mathbb{R}^n \times \mathbb{R}$.

We first prove the *weak maximum principle* for the heat equation, which asserts that any subsolution of the heat equation attains its maximum on the parabolic boundary. Here, a $C^{2,1}(\Omega_T)$-function u is a subsolution of the heat equation if $u_t - \Delta u \le 0$ in Ω_T.

Theorem 5.3.1. *Suppose $u \in C^{2,1}(\Omega_T) \cap C(\bar{\Omega}_T)$ satisfies*

$$u_t - \Delta u \le 0 \quad \text{in } \Omega_T.$$

Then u attains on $\partial_p\Omega_T$ its maximum in $\bar{\Omega}_T$, i.e.,

$$\max_{\bar{\Omega}_T} u = \max_{\partial_p\Omega_T} u.$$

Proof. We first consider a special case where $u_t - \Delta u < 0$ and prove that u cannot attain in Ω_T its maximum in $\bar{\Omega}_T$. Suppose, to the contrary, that there exists a point $P_0 = (x_0, t_0) \in \Omega_T$ such that

$$u(P_0) = \max_{\bar{\Omega}_T} u.$$

Then $\nabla_x u(P_0) = 0$ and the Hessian matrix $\nabla_x^2 u(P_0)$ is nonpositive definite. Moreover, $u_t(P_0) = 0$ if $t_0 \in (0, T)$, and $u_t(P_0) \ge 0$ if $t_0 = T$. Hence $u_t - \Delta u \ge 0$ at P_0, which is a contradiction.

We now consider the general case. For any $\varepsilon > 0$, let

$$u_\varepsilon(x, t) = u(x, t) - \varepsilon t.$$

Then

$$(\partial_t - \Delta)u_\varepsilon = u_t - \Delta u - \epsilon < 0.$$

By the special case we just discussed, u_ε cannot attain in Ω_T its maximum. Hence

$$\max_{\bar{\Omega}_T} u_\varepsilon = \max_{\partial_p\Omega_T} u_\varepsilon.$$

Then

$$\begin{aligned}\max_{\bar{\Omega}_T} u(x,t) &= \max_{\bar{\Omega}_T}(u_\varepsilon(x,t) + \varepsilon t)) \le \max_{\bar{\Omega}_T} u_\varepsilon(x,t) + \varepsilon T \\ &= \max_{\partial_p\Omega_T} u_\varepsilon(x,t) + \varepsilon T \le \max_{\partial_p\Omega_T} u(x,t) + \varepsilon T.\end{aligned}$$

Letting $\varepsilon \to 0$, we obtain the desired result. □

Next, we consider a class of parabolic equations slightly more general than the heat equation. Let c be a continuous function in Ω_T. Consider

$$Lu = u_t - \Delta u + cu \quad \text{in } \Omega_T.$$

We prove the following *weak maximum principle* for subsolutions of L. Here, a $C^{2,1}(\Omega_T)$-function u is a *subsolution* of L if $Lu \le 0$ in Ω_T. Similarly, a $C^{2,1}(\Omega_T)$-function u is a *supersolution* of L if $Lu \ge 0$ in Ω_T.

Theorem 5.3.2. *Let c be a continuous function in Ω_T with $c \geq 0$. Suppose $u \in C^{2,1}(\Omega_T) \cap C(\bar{\Omega}_T)$ satisfies*

$$u_t - \Delta u + cu \leq 0 \quad \text{in } \Omega_T.$$

Then u attains on $\partial_p \Omega_T$ its nonnegative maximum in $\bar{\Omega}_T$, i.e.,

$$\max_{\bar{\Omega}_T} u \leq \max_{\partial_p \Omega_T} u^+.$$

We note that u^+ is the nonnegative part of u given by $u^+ = \max\{0, u\}$. The proof of Theorem 5.3.2 is a simple modification of that of Theorem 5.3.1 and is omitted.

Now, we consider a more general case.

Theorem 5.3.3. *Let c be a continuous function in Ω_T with $c \geq -c_0$ for a nonnegative constant c_0. Suppose $u \in C^{2,1}(\Omega_T) \cap C(\bar{\Omega}_T)$ satisfies*

$$\begin{aligned} u_t - \Delta u + cu &\leq 0 \quad \text{in } \Omega_T, \\ u &\leq 0 \quad \text{on } \partial_p \Omega_T. \end{aligned}$$

Then $u \leq 0$ in Ω_T.

Continuous functions in $\bar{\Omega}_T$ always have global minima. Therefore, $c \geq -c_0$ in Ω_T for some nonnegative constant c_0 if c is continuous in $\bar{\Omega}_T$. Such a condition is introduced to emphasize the role of the minimum of c.

Proof. Let $v(x,t) = e^{-c_0 t} u(x,t)$. Then $u = c^{c_0 t} v$ and

$$u_t - \Delta u + cu = e^{c_0 t}\big(v_t - \Delta v + (c + c_0)v\big).$$

Hence

$$v_t - \Delta v + (c + c_0)v \leq 0.$$

With $c + c_0 \geq 0$, we obtain, by Theorem 5.3.2, that

$$\max_{\bar{\Omega}_T} v \leq \max_{\partial_p \Omega_T} v^+ = \max_{\partial_p \Omega_T} e^{-c_0 t} u^+ = 0.$$

Hence $u \leq 0$ in Ω_T. □

The following result is referred to as the *comparison principle.*

Corollary 5.3.4. *Let c be a continuous function in Ω_T with $c \geq -c_0$ for a nonnegative constant c_0. Suppose $u, v \in C^{2,1}(\Omega_T) \cap C(\bar{\Omega}_T)$ satisfy*

$$\begin{aligned} u_t - \Delta u + cu &\leq v_t - \Delta v + cv \quad \text{in } \Omega_T, \\ u &\leq v \quad \text{on } \partial_p \Omega_T. \end{aligned}$$

Then $u \leq v$ in Ω_T.

In the following, we simply say *by the maximum principle* when we apply Theorem 5.3.2, Theorem 5.3.3 or Corollary 5.3.4.

Before we discuss applications of maximum principles, we compare maximum principles for elliptic equations and parabolic equations. Consider

$$L_e u = -\Delta u + c(x)u \quad \text{in } \Omega$$

and

$$L_p u = u_t - \Delta u + c(x,t)u \quad \text{in } \Omega_T \equiv \Omega \times (0,T).$$

We note that the elliptic operator L_e here has a form different from those in Section 4.3.1, where we used the form $\Delta + c$. Hence, we should change the assumption on the sign of c accordingly. If $c \geq 0$, then

$$\begin{aligned} L_e u \leq 0 &\Rightarrow u \text{ attains its nonnegative maximum on } \partial\Omega, \\ L_p u \leq 0 &\Rightarrow u \text{ attains its nonnegative maximum on } \partial_p\Omega_T. \end{aligned}$$

If $c \equiv 0$, the nonnegativity condition can be removed. For $c \geq 0$, comparison principles can be stated as follows:

$$\begin{aligned} L_e u \leq L_e v \text{ in } \Omega,\ u \leq v \text{ on } \partial\Omega &\Rightarrow u \leq v \text{ in } \Omega, \\ L_p u \leq L_p v \text{ in } \Omega_T,\ u \leq v \text{ on } \partial_p\Omega_T &\Rightarrow u \leq v \text{ in } \Omega_T. \end{aligned}$$

In fact, the comparison principle for parabolic equations holds for $c \geq -c_0$, for a nonnegative constant c_0.

In applications, we need to construct auxiliary functions for comparisons. Usually, we take $|x|^2$ or $e^{\pm\alpha|x|^2}$ for elliptic equations and $Kt + |x|^2$ for parabolic equations. Sometimes, auxiliary functions are constructed with the help of the fundamental solutions for the Laplace equation and the heat equation.

5.3.2. The Strong Maximum Principle. The weak maximum principle asserts that subsolutions of parabolic equations attain on the parabolic boundary their nonnegative maximum if the coefficient of the zeroth-order term is nonnegative. In fact, these subsolutions can attain their nonnegative maximum only on the parabolic boundary, unless they are constant on suitable subsets. This is the strong maximum principle. We shall point out that the weak maximum principle suffices for most applications to the initial/boundary-value problem with values of the solutions prescribed on the parabolic boundary of the domain.

We first prove the following result.

Lemma 5.3.5. *Let (x_0, t_0) be a point in $\mathbb{R}^n \times \mathbb{R}$, R and T be positive constants and Q be the set defined by*

$$Q = B_R(x_0) \times (t_0 - T, t_0].$$

Suppose c is a continuous function in $\bar{Q}$ and $u \in C^{2,1}(Q) \cap C(\bar{Q})$ satisfies

$$u_t - \Delta u + cu \geq 0 \quad \text{in } Q.$$

If $u \geq 0$ in Q and

$$u(x_0, t_0 - T) > 0,$$

then

$$u(x,t) > 0 \quad \text{for any } (x,t) \in Q.$$

Lemma 5.3.5 asserts that a nonnegative supersolution, if positive *somewhere* initially, becomes positive *everywhere* at all later times. This can be interpreted as *infinite-speed propagation.*

Proof. Take an arbitrary $t_* \in (t_0 - T, t_0]$. We will prove that

$$u(x, t_*) > 0 \quad \text{for any } x \in B_R(x_0).$$

Without loss of generality, we assume that $x_0 = 0$ and $t_* = 0$. We take $\alpha > 0$ such that $t_0 - T = -\alpha R^2$ and set

$$D = B_R \times (-\alpha R^2, 0].$$

By the assumption $u(0, -\alpha R^2) > 0$ and the continuity of u, we can assume that

$$u(x, -\alpha R^2) \geq m \quad \text{for any } x \in \bar{B}_{\varepsilon R},$$

for some constants $m > 0$ and $\varepsilon \in (0,1)$. Here, m can be taken as the (positive) minimum of $u(\cdot, -\alpha R^2)$ on $\bar{B}_{\varepsilon R}$.

Now we set

$$D_0 = \left\{ (x,t) \in B_R \times (-\alpha R^2, 0] : \ |x|^2 - \frac{1-\varepsilon^2}{\alpha} t < R^2 \right\} \subset D.$$

It is easy to see that

$$D_0 \cap \{t = 0\} = B_R, \quad \bar{D}_0 \cap \{t = -\alpha R^2\} = \bar{B}_{\varepsilon R}.$$

Set

$$w_1(t) = \frac{1-\varepsilon^2}{\alpha} t + R^2,$$

$$w_2(x,t) = w_1(t) - |x|^2 = \frac{1-\varepsilon^2}{\alpha} t + R^2 - |x|^2,$$

and for some β to be determined,

$$w = w_1^{-\beta} w_2^2.$$

We will consider w_1, w_2 and w in D_0.

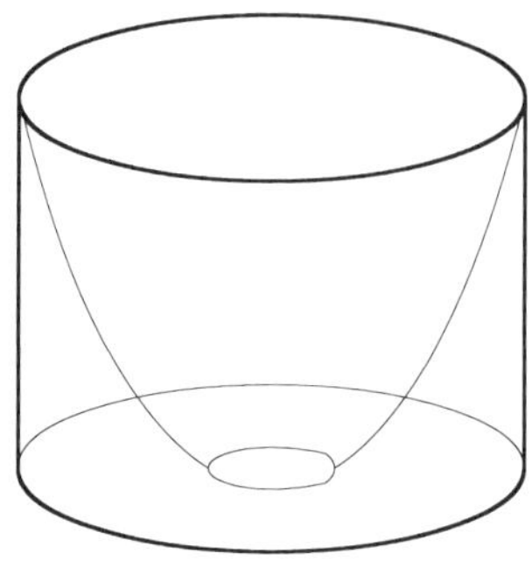

Figure 5.3.1. The domain D_0.

We first note that $\varepsilon^2 R^2 \le w_1 \le R^2$ and $w_2 \ge 0$ in D_0. A straightforward calculation yields

$$\begin{aligned} w_t &= -\beta w_1^{-\beta-1}\partial_t w_1 w_2^2 + 2w_1^{-\beta} w_2 \partial_t w_2 \\ &= w_1^{-\beta-1}\left(-\frac{\beta(1-\varepsilon^2)}{\alpha} w_2^2 + \frac{2(1-\varepsilon^2)}{\alpha} w_1 w_2\right), \end{aligned}$$

and

$$\Delta w = w_1^{-\beta}(2w_2 \Delta w_2 + 2|\nabla w_2|^2) = w_1^{-\beta}(-4nw_2 + 8|x|^2).$$

Since $|x|^2 = w_1 - w_2$, we have

$$\Delta w = w_1^{-\beta}\big(8w_1 - (4n+8)w_2\big) = w_1^{-\beta-1}\big(8w_1^2 - (4n+8)w_1 w_2\big).$$

Therefore,

$$\begin{aligned} w_t - \Delta w + cw = w_1^{-\beta-1} &\left(\left(-\frac{\beta(1-\varepsilon^2)}{\alpha} + cw_1\right) w_2^2\right. \\ &\left. + \left(\frac{2(1-\varepsilon^2)}{\alpha} + 4n + 8\right) w_1 w_2 - 8w_1^2\right). \end{aligned}$$

Hence

$$\begin{aligned} w_t - \Delta w + cw \le -w_1^{-\beta-1} &\left(\left(\frac{\beta(1-\varepsilon^2)}{\alpha} - R^2|c|\right) w_2^2\right. \\ &\left. - \left(\frac{2(1-\varepsilon^2)}{\alpha} + 4n + 8\right) w_1 w_2 + 8w_1^2\right). \end{aligned}$$

The expression in the parentheses is a quadratic form in w_1 and w_2 with a positive coefficient of w_1^2. Hence, we can make this quadratic form nonnegative by choosing β sufficiently large, depending only on ε, α, R and $\sup|c|$. Hence,

$$w_t - \Delta w + cw \le 0 \quad \text{in } D_0.$$

Note that the parabolic boundary $\partial_p D_0$ consists of two parts Σ_1 and Σ_2 given by

$$\Sigma_1 = \{(x,t) : \ |x| < \varepsilon R, \ t = -\alpha R^2\},$$
$$\Sigma_2 = \left\{(x,t) : \ |x|^2 - \frac{1-\varepsilon^2}{\alpha} t = R^2, \ -\alpha R^2 \le t \le 0\right\}.$$

For $(x,t) \in \Sigma_1$, we have $t = -\alpha R^2$ and $|x| < \varepsilon R$, and hence

$$w(x, -\alpha R) = (\varepsilon^2 R^2)^{-\beta}(\varepsilon^2 R^2 - |x|^2)^2 \le (\varepsilon R)^{-2\beta+4}.$$

Next, on Σ_2, we have $w = 0$. In the following, we set

$$v = m(\varepsilon R)^{2\beta-4} w \quad \text{in } D_0,$$

where m is the minimum of u over $\bar{\Sigma}_1$ defined earlier. Then

$$v_t - \Delta v + cv \le 0 \quad \text{in } D_0,$$

and

$$v \le u \quad \text{on } \partial_p D_0,$$

since $u \ge m$ on Σ_1 and $u \ge 0$ on Σ_2. In conclusion,

$$v_t - \Delta v + cv \le u_t - \Delta u + cu \quad \text{in } D_0,$$
$$v \le u \quad \text{on } \partial_p D_0.$$

By the maximum principle, we have

$$v \le u \quad \text{in } D_0.$$

This holds in particular at $t = 0$. By evaluating v at $t = 0$, we obtain

$$u(x,0) \ge m\varepsilon^{2\beta-4}\left(1 - \frac{|x|^2}{R^2}\right)^2 \quad \text{for any } x \in B_R.$$

This implies the desired result. □

We point out that the final estimate in the proof yields a lower bound of u over $B_R \times \{0\}$ in terms of the lower bound of u over $B_{\varepsilon R} \times \{-\alpha R^2\}$. This is an important estimate.

Now, we are ready to prove the *strong maximum principle*.

Theorem 5.3.6. *Let Ω be a bounded domain in $\mathbb{R}^n$ and T be a positive constant. Suppose c is a continuous function in $\Omega \times (0,T]$ with $c \ge 0$, and $u \in C^{2,1}(\Omega \times (0,T])$ satisfies*

$$u_t - \Delta u + cu \le 0 \quad \text{in } \Omega \times (0,T].$$

If for some $(x_*, t_*) \in \Omega \times (0,T]$,

$$u(x_*, t_*) = \sup_{\Omega \times (0,T]} u \ge 0,$$

then

$$u(x,t) = u(x_*, t_*) \quad \textit{for any } (x,t) \in \Omega \times (0, t_*).$$

Proof. Set

$$M = \sup_{\Omega\times(0,T]} u \geq 0,$$

and

$$v = M - u \quad \text{in } \Omega \times (0,T].$$

Then $v(x_*, t_*) = 0$, $v \geq 0$ in $\Omega \times (0,T]$ and

$$v_t - \Delta v + cv \geq 0 \quad \text{in } \Omega \times (0,T].$$

We will prove that $v(x_0, t_0) = 0$ for any $(x_0, t_0) \in \Omega \times (0, t_*)$.

To this end, we connect (x_0, t_0) and (x_*, t_*) by a smooth curve $\gamma \subset \Omega\times(0,T]$ along which the t-component is increasing. In fact, we first connect x_0 and x_* by a smooth curve $\gamma_0 = \gamma_0(s) \subset \Omega$, for $s \in [0,1]$, with $\gamma_0(0) = x_0$ and $\gamma_0(1) = x_*$. Then we may take γ to be the curve given by $(\gamma_0(s), st_* + (1-s)t_0)$. With such a γ, there exist a positive constant R and finitely

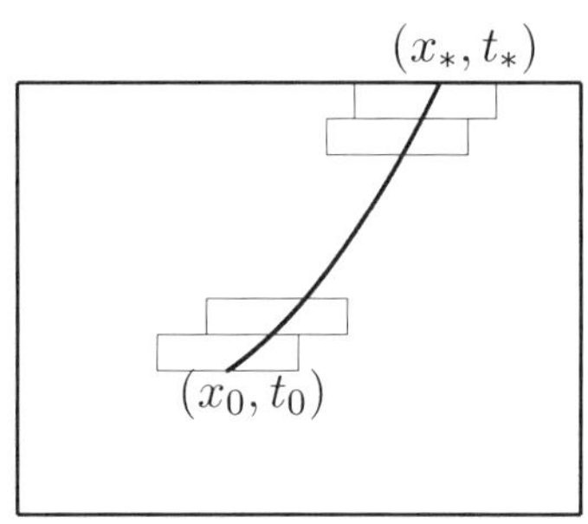

Figure 5.3.2. γ and the corresponding covering.

many points (x_k, t_k) on γ, for $k = 1, \cdots, N$, with $(x_N, t_N) = (x_*, t_*)$, such that

$$\gamma \subset \bigcup_{k=0}^{N-1} B_R(x_k) \times [t_k, t_k + R^2] \subset \Omega \times (0,T].$$

We may require that $t_k = t_{k-1} + R^2$ for $k = 0, \cdots, N-1$.

If $v(x_0, t_0) > 0$, then, applying Lemma 5.3.5 in $B_R(x_0) \times [t_0, t_0 + R^2]$, we conclude that

$$v(x,t) > 0 \quad \text{in } B_R(x_0) \times (t_0, t_0 + R^2],$$

and in particular, $v(x_1, t_1) > 0$. We may continue this process finitely many times to obtain $v(x_*, t_*) = v(x_N, t_N) > 0$. This contradicts the assumption. Therefore, $v(x_0, t_0) = 0$ and hence $u(x_0, t_0) = M$. □

Related to the strong maximum principle is the following Hopf lemma in the parabolic version.

Lemma 5.3.7. *Let (x_0, t_0) be a point in $\mathbb{R}^n \times \mathbb{R}$, R and η be two positive constants and D be the set defined by*

$$D = \{(x,t) \in \mathbb{R}^n \times \mathbb{R} : \ |x - x_0|^2 + \eta(t_0 - t) < R^2, \ t \le t_0\}.$$

Suppose c is a continuous function in $\bar{D}$ with $c \ge 0$, and $u \in C^{2,1}(D) \cap C(\bar{D})$ satisfies

$$u_t - \Delta u + cu \le 0 \quad \text{in } D.$$

Assume, in addition, for some $\tilde{x} \in \mathbb{R}^n$ with $|\tilde{x} - x_0| = R$, that

$$\begin{aligned} &u(x,t) \le u(\tilde{x}, t_0) \quad \text{for any } (x,t) \in D \text{ and } u(\tilde{x}, t_0) \ge 0, \\ &u(x,t) < u(\tilde{x}, t_0) \quad \text{for any } (x,t) \in \bar{D} \text{ with } |x - x_0| \le \frac{1}{2}R. \end{aligned}$$

If $\nabla_x u$ is continuous up to $(\tilde{x}, t_0)$, then

$$\nu \cdot \nabla_x u(\tilde{x}, t_0) > 0,$$

where ν is the unit vector given by $\nu = (\tilde{x} - x_0)/|\tilde{x} - x_0|$.

Proof. Without loss of generality, we assume that $(x_0, t_0) = (0, 0)$. Then

$$D = \{(x,t) \in \mathbb{R}^n \times \mathbb{R} : \ |x|^2 - \eta t < R^2, \ t \le 0\}.$$

By the continuity of u up to ∂D, we have

$$u(x,t) \le u(\tilde{x}, 0) \quad \text{for any } (x,t) \in \bar{D}.$$

For positive constants α and ε to be determined, we set

$$w(x,t) = e^{-\alpha(|x|^2 - \eta t)} - e^{-\alpha R^2}$$

and

$$v(x,t) = u(x,t) - u(\tilde{x}, 0) + \varepsilon w(x,t).$$

We consider w and v in

$$D_0 = \left\{ (x,t) \in D : \ |x| > \frac{1}{2}R \right\}.$$

A direct calculation yields

$$\begin{aligned} w_t - \Delta w + cw &= -e^{-\alpha(|x|^2 - \eta t)} \big(4\alpha^2 |x|^2 - 2n\alpha - \eta\alpha - c \big) - ce^{-\alpha R^2} \\ &\le -e^{-\alpha(|x|^2 - \eta t)} \big(4\alpha^2 |x|^2 - 2n\alpha - \eta\alpha - c \big), \end{aligned}$$

where we used $c \ge 0$ in D. By taking into account that $R/2 \le |x| \le R$ in D_0 and choosing α sufficiently large, we have

$$4\alpha^2 |x|^2 - 2n\alpha - \eta\alpha - c > 0 \quad \text{in } D_0,$$

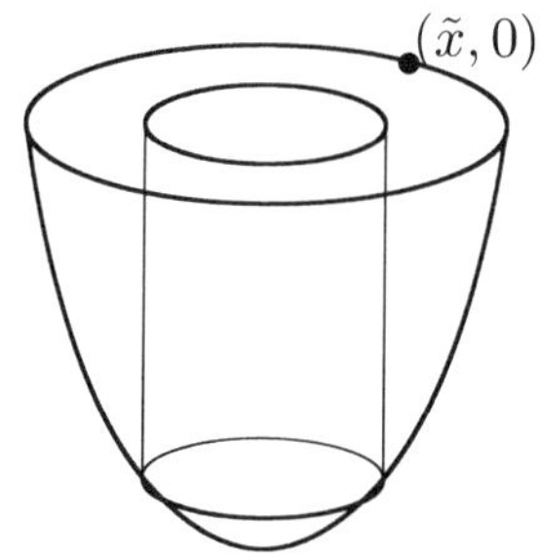

Figure 5.3.3. The domain D_0.

and hence

$$w_t - \Delta w + cw \le 0 \quad \text{in } D_0.$$

Since $c \ge 0$ and $u(\tilde{x}, 0) \ge 0$, we obtain for any $\varepsilon > 0$ that

$$v_t - \Delta v + cv = u_t - \Delta u + cu + \varepsilon(w_t - \Delta w + cw) - cu(\tilde{x}, 0) \le 0 \quad \text{in } D_0.$$

The parabolic boundary $\partial_p D_0$ consists of two parts Σ_1 and Σ_2 given by

$$\Sigma_1 = \left\{(x, t):\ |x|^2 - \eta t < R^2,\ t \le 0,\ |x| = \frac{1}{2}R\right\},$$

$$\Sigma_2 = \left\{(x, t):\ |x|^2 - \eta t = R^2,\ t \le 0,\ |x| \ge \frac{1}{2}R\right\}.$$

First, on $\bar{\Sigma}_1$, we have $u - u(\tilde{x}, 0) < 0$, and hence $u - u(\tilde{x}, 0) < -\varepsilon$ for some $\varepsilon > 0$. Note that $w \le 1$ on Σ_1. Then for such an ε, we obtain $v < 0$ on Σ_1. Second, for $(x, t) \in \Sigma_2$, we have $w(x, t) = 0$ and $u(x, t) \le u(\tilde{x}, 0)$. Hence $v(x, t) \le 0$ for any $(x, t) \in \Sigma_2$ and $v(\tilde{x}, 0) = 0$. Therefore, $v \le 0$ on Σ_2. In conclusion,

$$\begin{aligned} v_t - \Delta v + cv &\le 0 \quad \text{in } D_0, \\ v &\le 0 \quad \text{on } \partial_p D_0. \end{aligned}$$

By the maximum principle, we have

$$v \le 0 \quad \text{in } D_0.$$

Then, by $v(\tilde{x}, 0) = 0$, v attains at $(\tilde{x}, 0)$ its maximum in $\bar{D}_0$. In particular,

$$v(x, 0) \le v(\tilde{x}, 0) \quad \text{for any } x \in B_R \setminus B_{\frac{1}{2}R}.$$

Hence, we obtain

$$\frac{\partial v}{\partial \nu}(\tilde{x}, 0) \ge 0,$$

and then

$$\frac{\partial u}{\partial \nu}(\tilde{x}, 0) \ge -\varepsilon \frac{\partial w}{\partial \nu}(\tilde{x}, 0) = 2\varepsilon\alpha R e^{-\alpha R^2} > 0.$$

This is the desired result. □

To conclude our discussion of the strong maximum principle, we briefly compare our approaches for elliptic equations and parabolic equations. For elliptic equations, we first prove the Hopf lemma and then prove the strong maximum principle as its consequence. See Subsection 4.3.2 for details. For parabolic equations, we first prove infinite speed of propagation and then obtain the strong maximum principle as a consequence. It is natural to ask whether we can prove the strong maximum principle by Lemma 5.3.7, the parabolic Hopf lemma. By an argument similar to the proof of Theorem 4.3.9, we can conclude that, if a subsolution u attains its nonnegative maximum at an interior point $(x_0, t_0) \in \Omega \times (0, T]$, then u is constant on $\Omega \times \{t_0\}$. In order to conclude that u is constant in $\Omega \times (0, t_0)$ as asserted by Theorem 5.3.6, we need a result concerning the t-derivative at the interior maximum point, similar to that concerning the x-derivative in the Hopf lemma. We will not pursue this issue in this book.

5.3.3. A Priori Estimates. In the rest of this section, we discuss applications of the maximum principle. We point out that only the weak maximum principle is needed.

As the first application, we derive an estimate of the sup-norms of solutions of initial/boundary-value problems with Dirichlet boundary values. Compare this with the estimate in integral norms in Theorem 3.2.4.

As before, for a bounded domain $\Omega \subset \mathbb{R}^n$ and a positive constant T, we set

$$\Omega_T = \Omega \times (0, T] = \{(x, t) : \ x \in \Omega, \, 0 < t \leq T\}.$$

Theorem 5.3.8. *Let c be continuous in Ω_T with $c \geq -c_0$ for a nonnegative constant c_0. Suppose $u \in C^{2,1}(\Omega_T) \cap C(\bar{\Omega}_T)$ is a solution of*

$$\begin{aligned} u_t - \Delta u + cu &= f \quad \text{in } \Omega_T, \\ u(\cdot, 0) &= u_0 \quad \text{on } \Omega, \\ u &= \varphi \quad \text{on } \partial\Omega \times (0, T), \end{aligned}$$

for some $f \in C(\bar{\Omega}_T)$, $u_0 \in C(\bar{\Omega})$ and $\varphi \in C(\partial\Omega \times [0, T])$. Then

$$\sup_{\Omega_T} |u| \leq e^{c_0 T} \left(\max\left\{ \sup_{\Omega} |u_0|, \ \sup_{\partial\Omega \times (0,T)} |\varphi| \right\} + T \sup_{\Omega_T} |f| \right).$$

Proof. Set $Lu = u_t - \Delta u + cu$ and

$$B = \max\left\{ \sup_{\Omega} |u_0|, \ \sup_{\partial\Omega \times (0,T)} |\varphi| \right\}, \quad F = \sup_{\Omega_T} |f|.$$

Then

$$L(\pm u) \le F \quad \text{in } \Omega_T,$$
$$\pm u \le B \quad \text{on } \partial_p\Omega_T.$$

Set

$$v(x,t) = e^{c_0 t}(B + Ft).$$

Since $c + c_0 \ge 0$ and $e^{c_0 t} \ge 1$ in Ω_T, we have

$$Lv = (c_0 + c)e^{c_0 t}(B + Ft) + e^{c_0 t}F \ge F \quad \text{in } \Omega_T$$

and

$$v \ge B \quad \text{on } \partial_p\Omega_T.$$

Hence,

$$L(\pm u) \le Lv \quad \text{in } \Omega_T,$$
$$\pm u \le v \quad \text{on } \partial_p\Omega_T.$$

By the maximum principle, we obtain

$$\pm u \le v \quad \text{in } \Omega_T.$$

Therefore,

$$|u(x,t)| \le e^{c_0 t}(B + Ft) \quad \text{for any } (x,t) \in \Omega_T.$$

This implies the desired estimate. □

Next, we derive a priori estimates of solutions of initial-value problems.

Theorem 5.3.9. *Let c be continuous in $\mathbb{R}^n \times (0,T]$ with $c \ge -c_0$ for a nonnegative constant c_0. Suppose $u \in C^{2,1}(\mathbb{R}^n \times (0,T]) \cap C(\mathbb{R}^n \times [0,T])$ is a bounded solution of*

$$u_t - \Delta u + cu = f \quad \text{in } \mathbb{R}^n \times (0,T],$$
$$u(\cdot,0) = u_0 \quad \text{on } \mathbb{R}^n,$$

for some bounded $f \in C(\mathbb{R}^n \times (0,T])$ and $u_0 \in C(\mathbb{R}^n)$. Then

$$\sup_{\mathbb{R}^n \times (0,T)} |u| \le e^{c_0 T}\left(\sup_{\mathbb{R}^n} |u_0| + T \sup_{\mathbb{R}^n \times (0,T)} |f|\right).$$

We note that the maximum principle is established in bounded domains such as $\Omega \times (0,T]$. In studying solutions of the initial-value problem where solutions are defined in $\mathbb{R}^n \times (0,T]$, we should first derive suitable estimates of solutions in $B_R \times (0,T]$ and then let $R \to \infty$. For this purpose, we need to impose extra assumptions on u as $x \to \infty$. For example, u is assumed to be bounded in Theorem 5.3.9 and to be of the exponential growth in Theorem 5.3.10.

Proof. Set $Lu = u_t - \Delta u + cu$ and

$$F = \sup_{\mathbb{R}^n \times (0,T]} |f|, \quad B = \sup_{\mathbb{R}^n} |u_0|.$$

Then

$$\begin{aligned} L(\pm u) &\le F \quad \text{in } \mathbb{R}^n \times (0,T], \\ \pm u &\le B \quad \text{on } \mathbb{R}^n. \end{aligned}$$

Since u is bounded, we assume that $|u| \le M$ in $\mathbb{R}^n \times (0,T]$ for a positive constant M. For any $R > 0$, consider

$$w(x,t) = e^{c_0 t}(B + Ft) + v_R(x,t) \quad \text{in } B_R \times (0,T],$$

where v_R is a function to be chosen. By $c + c_0 \ge 0$ and $e^{c_0 t} \ge 1$, we have

$$Lw = (c + c_0)e^{c_0 t}(B + Ft) + e^{c_0 t}F + Lv_R \ge F + Lv_R \quad \text{in } B_R \times (0,T].$$

Moreover,

$$w(\cdot,0) = B + v_R(\cdot,0) \quad \text{in } B_R,$$

and

$$w \ge v_R \quad \text{on } \partial B_R \times (0,T].$$

We will choose v_R such that

$$\begin{aligned} Lv_R &\ge 0 \quad \text{in } B_R \times (0,T], \\ v_R(\cdot,0) &\ge 0 \quad \text{in } B_R, \\ v_R &\ge \pm u \quad \text{on } \partial B_R \times [0,T]. \end{aligned}$$

To construct such a v_R, we consider

$$v_R(x,t) = \frac{M}{R^2} e^{c_0 t}(2nt + |x|^2).$$

Obviously, $v_R \ge 0$ for $t = 0$ and $v_R \ge M$ on $|x| = R$. Next,

$$Lv_R = \frac{M}{R^2} e^{c_0 t}(c + c_0)(2nt + |x|^2) \ge 0 \quad \text{in } B_R \times (0,T].$$

With such a v_R, we have

$$\begin{aligned} L(\pm u) &\le Lw \quad \text{in } B_R \times (0,T], \\ \pm u &\le w \quad \text{on } \partial_p (B_R \times (0,T]). \end{aligned}$$

Then the maximum principle yields $\pm u \le w$ in $B_R \times (0,T]$. Hence for any $(x,t) \in B_R \times (0,T]$,

$$|u(x,t)| \le e^{c_0 t}(B + Ft) + \frac{M}{R^2} e^{c_0 t}(2nt + |x|^2).$$

Now we fix an arbitrary $(x,t) \in \mathbb{R}^n \times (0,T]$. By choosing $R > |x|$ and then letting $R \to \infty$, we have

$$|u(x,t)| \le e^{c_0 t}(B + Ft).$$

This yields the desired estimate. □

Next, we prove the uniqueness of solutions of initial-value problems for the heat equation under the assumption of exponential growth.

Theorem 5.3.10. *Let $u \in C^{2,1}(\mathbb{R}^n \times (0,T]) \cap C(\mathbb{R}^n \times [0,T])$ satisfy*

$$\begin{aligned} u_t - \Delta u &= 0 \quad \text{in } \mathbb{R}^n \times (0,T], \\ u(\cdot,0) &= 0 \quad \text{on } \mathbb{R}^n. \end{aligned}$$

Suppose, for some positive constants M and A,

$$|u(x,t)| \leq Me^{A|x|^2},$$

for any $(x,t) \in \mathbb{R}^n \times (0,T]$. Then $u \equiv 0$ in $\mathbb{R}^n \times [0,T]$.

Proof. For any constant $\alpha > A$, we prove that

$$u = 0 \quad \text{in } \mathbb{R}^n \times \left[0, \frac{1}{4\alpha}\right].$$

We then extend $u = 0$ in the t-direction successively to $[\frac{1}{4\alpha}, \frac{2}{4\alpha}]$, $[\frac{2}{4\alpha}, \frac{3}{4\alpha}]$, $\cdots$, until $t = T$.

For any constant $R > 0$, consider

$$v_R(x,t) = \frac{Me^{(A-\alpha)R^2}}{(1-4\alpha t)^{\frac{n}{2}}} e^{\frac{\alpha|x|^2}{1-4\alpha t}},$$

for any $(x,t) \in B_R \times (0, 1/4\alpha)$. We note that v_R is modified from the example we discussed preceding Theorem 5.2.5. Then

$$\partial_t v_R - \Delta v_R = 0 \quad \text{in } B_R \times \left(0, \frac{1}{4\alpha}\right).$$

Obviously,

$$v_R(\cdot,0) \geq 0 = \pm u(\cdot,0) \quad \text{in } B_R.$$

Next, for any $(x,t) \in \partial B_R \times (0, 1/4\alpha)$,

$$v_R(x,t) \geq Me^{(A-\alpha)R^2} e^{\alpha R^2} = Me^{AR^2} \geq \pm u(x,t).$$

In conclusion,

$$\pm u \leq v_R \quad \text{on } \partial_p \left(B_R \times \left(0, \frac{1}{4\alpha}\right)\right).$$

By the maximum principle, we have

$$\pm u \leq v_R \quad \text{in } B_R \times \left(0, \frac{1}{4\alpha}\right).$$

Therefore,

$$|u(x,t)| \leq v_R(x,t) \quad \text{for any } (x,t) \in B_R \times \left(0, \frac{1}{4\alpha}\right).$$

Now we fix an arbitrary $(x,t) \in \mathbb{R}^n \times (0, 1/4\alpha)$ and then choose $R > |x|$. We note that $v_R(x,t) \to 0$ as $R \to \infty$, since $\alpha > A$. We therefore obtain $u(x,t) = 0$. □

5.3.4. Interior Gradient Estimates. We now give an alternative proof, based on the maximum principle, of the interior gradient estimate. We do this only for solutions of the heat equation. Recall that for any $r > 0$,

$$Q_r = B_r \times (-r^2, 0].$$

Theorem 5.3.11. *Suppose $u \in C^{2,1}(Q_1) \cap C(\bar{Q}_1)$ satisfies*

$$u_t - \Delta u = 0 \quad \text{in } Q_1.$$

Then

$$\sup_{Q_{\frac{1}{2}}} |\nabla_x u| \leq C \sup_{\partial_p Q_1} |u|,$$

where C is a positive constant depending only on n.

The proof is similar to that of Theorem 4.3.13, the interior gradient estimate for harmonic functions.

Proof. We first note that u is smooth in Q_1 by Theorem 5.2.7. A straightforward calculation yields

$$\begin{aligned}(\partial_t - \Delta)|\nabla_x u|^2 &= -2 \sum_{i,j=1}^n u_{x_i x_j}^2 + 2 \sum_{i=1}^n u_{x_i}(u_t - \Delta u)_{x_i} \\ &= -2 \sum_{i,j=1}^n u_{x_i x_j}^2.\end{aligned}$$

To get interior estimates, we need to introduce a cutoff function. For any smooth function φ in $C^\infty(Q_1)$ with $\operatorname{supp}\varphi \subset Q_{3/4}$, we have

$$\begin{aligned}(\partial_t - \Delta)(\varphi|\nabla_x u|^2) = (\varphi_t - \Delta\varphi)|\nabla_x u|^2 \\ - 4 \sum_{i,j=1}^n \varphi_{x_i} u_{x_j} u_{x_i x_j} - 2\varphi \sum_{i,j=1}^n u_{x_i x_j}^2.\end{aligned}$$

Now we take $\varphi = \eta^2$ for some $\eta \in C^\infty(Q_1)$ with $\eta \equiv 1$ in $Q_{1/2}$ and $\operatorname{supp}\eta \subset Q_{3/4}$. Then

$$\begin{aligned}(\partial_t - \Delta)(\eta^2|\nabla_x u|^2) = (2\eta\eta_t - 2\eta\Delta\eta - 2|\nabla_x \eta|^2)|\nabla_x u|^2 \\ - 8\eta \sum_{i,j=1}^n \eta_{x_i} u_{x_j} u_{x_i x_j} - 2\eta^2 \sum_{i,j=1}^n u_{x_i x_j}^2.\end{aligned}$$

By the Cauchy inequality, we obtain

$$8|\eta\eta_{x_i} u_{x_j} u_{x_i x_j}| \leq 8\eta_{x_i}^2 u_{x_j}^2 + 2\eta^2 u_{x_i x_j}^2.$$

Hence,

$$(\partial_t - \Delta)(\eta^2|\nabla_x u|^2) \le \Big(2\eta\eta_t - 2\eta\Delta\eta + 6|\nabla_x \eta|^2\Big)|\nabla_x u|^2$$
$$\le C|\nabla_x u|^2,$$

where C is a positive constant depending only on η and n. Note that

$$(\partial_t - \Delta)(u^2) = -2|\nabla_x u|^2 + 2u(u_t - \Delta u) = -2|\nabla_x u|^2.$$

By taking a constant α large enough, we get

$$(\partial_t - \Delta)(\eta^2|\nabla_x u|^2 + \alpha u^2) \le (C - 2\alpha)|\nabla_x u|^2 \le 0.$$

By the maximum principle, we have

$$\sup_{Q_1}(\eta^2|\nabla_x u|^2 + \alpha u^2) \le \sup_{\partial_p Q_1}(\eta^2|\nabla_x u|^2 + \alpha u^2).$$

This implies the desired result since $\eta = 0$ on $\partial_p Q_1$ and $\eta = 1$ in $Q_{1/2}$. □

5.3.5. Harnack Inequalities. For positive harmonic functions, the Harnack inequality asserts that their values in compact subsets are comparable. In this section, we study the Harnack inequality for positive solutions of the heat equation. In seeking a proper form of the Harnack inequality for solutions of the heat equation, we begin our discussion with the fundamental solution.

We fix an arbitrary $\xi \in \mathbb{R}^n$ and consider for any $(x, t) \in \mathbb{R}^n \times (0, \infty)$,

$$u(x,t) = \frac{1}{(4\pi t)^{\frac{n}{2}}} e^{-\frac{|x-\xi|^2}{4t}}.$$

Then u satisfies the heat equation $u_t - \Delta u = 0$ in $\mathbb{R}^n \times (0, \infty)$. For any (x_1, t_1) and $(x_2, t_2) \in \mathbb{R}^n \times (0, \infty)$,

$$\frac{u(x_1,t_1)}{u(x_2,t_2)} = \left(\frac{t_2}{t_1}\right)^{\frac{n}{2}} e^{\frac{|x_2-\xi|^2}{4t_2} - \frac{|x_1-\xi|^2}{4t_1}}.$$

Recall that

$$\frac{(p+q)^2}{a+b} \le \frac{p^2}{a} + \frac{q^2}{b},$$

for any $a, b > 0$ and any $p, q \in \mathbb{R}$, and the equality holds if and only if $bp = aq$. This implies, for any $t_2 > t_1 > 0$,

$$\frac{|x_2 - \xi|^2}{t_2} \le \frac{|x_2 - x_1|^2}{t_2 - t_1} + \frac{|x_1 - \xi|^2}{t_1},$$

and the equality holds if and only if

$$\xi = \frac{t_2 x_1 - t_1 x_2}{t_2 - t_1}.$$

Therefore,

$$u(x_1,t_1) \le \left(\frac{t_2}{t_1}\right)^{\frac{n}{2}} e^{\frac{|x_2-x_1|^2}{4(t_2-t_1)}} u(x_2,t_2),$$

for any $x_1, x_2 \in \mathbb{R}^n$ and any $t_2 > t_1 > 0$, and the equality holds if ξ is chosen as above. This simple calculation suggests that the Harnack inequality for the heat equation has an "evolution" feature: the value of a positive solution at a certain time is controlled from above by the value at a later time. Hence, if we attempt to establish the estimate

$$u(x_1,t_1) \le Cu(x_2,t_2),$$

the constant C should depend on t_2/t_1, $|x_2 - x_1|$, and most importantly $(t_2 - t_1)^{-1} (> 0)$.

Suppose u is a positive solution of the heat equation and set $v = \log u$. In order to derive an estimate for the quotient

$$\frac{u(x_1,t_1)}{u(x_2,t_2)},$$

it suffices to get an estimate for the difference

$$v(x_1,t_1) - v(x_2,t_2).$$

To this end, we need an estimate of v_t and $|\nabla v|$. For a hint of proper forms, we again turn our attention to the fundamental solution of the heat equation.

Consider for any $(x,t) \in \mathbb{R}^n \times (0,\infty)$,

$$u(x,t) = \frac{1}{(4\pi t)^{\frac{n}{2}}} e^{-\frac{|x|^2}{4t}}.$$

Then

$$v(x,t) = \log u(x,t) = -\frac{n}{2}\log(4\pi t) - \frac{|x|^2}{4t},$$

and hence

$$v_t = -\frac{n}{2t} + \frac{|x|^2}{4t^2}, \quad \nabla v = -\frac{x}{2t}.$$

Therefore,

$$v_t = -\frac{n}{2t} + |\nabla v|^2.$$

We have the following *differential Harnack inequality* for arbitrary positive solutions of the heat equation.

Theorem 5.3.12. *Suppose $u \in C^{2,1}(\mathbb{R}^n \times (0,T])$ satisfies*

$$u_t = \Delta u, \quad u > 0 \quad \text{in } \mathbb{R}^n \times (0,T].$$

Then $v = \log u$ satisfies

$$v_t + \frac{n}{2t} \ge |\nabla v|^2 \quad \text{in } \mathbb{R}^n \times (0,T].$$

The differential Harnack inequality implies the Harnack inequality by a simple integration.

Corollary 5.3.13. *Suppose $u \in C^{2,1}(\mathbb{R}^n \times (0,T])$ satisfies*

$$u_t = \Delta u, \quad u > 0 \quad \text{in } \mathbb{R}^n \times (0,T].$$

Then for any $(x_1,t_1), (x_2,t_2) \in \mathbb{R}^n \times (0,T]$ with $t_2 > t_1 > 0$,

$$\frac{u(x_1,t_1)}{u(x_2,t_2)} \leq \left(\frac{t_2}{t_1}\right)^{\frac{n}{2}} \exp\left\{\frac{|x_2-x_1|^2}{4(t_2-t_1)}\right\}.$$

Proof. Let $v = \log u$ be as in Theorem 5.3.12 and take an arbitrary path $x = x(t)$ for $t \in [t_1,t_2]$ with $x(t_i) = x_i$, $i = 1,2$. By Theorem 5.3.12, we have

$$\frac{d}{dt}v(x(t),t) = v_t + \nabla v \cdot \frac{dx}{dt} \geq |\nabla v|^2 + \nabla v \cdot \frac{dx}{dt} - \frac{n}{2t}.$$

By completing the square, we obtain

$$\frac{d}{dt}v(x(t),t) \geq -\frac{1}{4}\left|\frac{dx}{dt}\right|^2 - \frac{n}{2t}.$$

Then a simple integration yields

$$v(x_1,t_1) \leq v(x_2,t_2) + \frac{n}{2}\log\frac{t_2}{t_1} + \frac{1}{4}\int_{t_1}^{t_2}\left|\frac{dx}{dt}\right|^2 dt.$$

To seek an optimal path which makes the last integral minimal, we require

$$\frac{d^2x}{dt^2} = 0$$

along the path. Hence we set, for some $a, b \in \mathbb{R}^n$,

$$x(t) = at + b.$$

Since $x_i = at_i + b$, $i = 1,2$, we take

$$a = \frac{x_2 - x_1}{t_2 - t_1}, \qquad b = \frac{t_2x_1 - t_1x_2}{t_2 - t_1}.$$

Then,

$$\int_{t_1}^{t_2}\left|\frac{dx}{dt}\right|^2 dt = \frac{|x_2-x_1|^2}{t_2-t_1}.$$

Therefore, we obtain

$$v(x_1,t_1) \leq v(x_2,t_2) + \frac{n}{2}\log\frac{t_2}{t_1} + \frac{1}{4}\frac{|x_2-x_1|^2}{t_2-t_1},$$

or

$$\frac{u(x_1,t_1)}{u(x_2,t_2)} \leq \left(\frac{t_2}{t_1}\right)^{\frac{n}{2}} \exp\left\{\frac{|x_2-x_1|^2}{4(t_2-t_1)}\right\}.$$

This is the desired estimate. □

Now we begin to prove the differential Harnack inequality. The basic idea is to apply the maximum principle to an appropriate combination of derivatives of v. In our case, we consider $|\nabla v|^2 - v_t$ and intend to derive an upper bound. First, we derive a parabolic equation satisfied by $|\nabla v|^2 - v_t$. A careful analysis shows that some terms in this equation cannot be controlled. So we introduce a parameter $\alpha \in (0,1)$ and consider $\alpha|\nabla v|^2 - v_t$ instead. After we apply the maximum principle, we let $\alpha \to 1$. The proof below is probably among the most difficult ones in this book.

Proof of Theorem 5.3.12. Without loss of generality, we assume that u is continuous up to $\{t = 0\}$. Otherwise, we consider u in $\mathbb{R}^n \times [\varepsilon, T]$ for any constant $\varepsilon \in (0, T)$ and then let $\varepsilon \to 0$. We divide the proof into several steps. In the following, we avoid notions of summations if possible.

Step 1. We first derive some equations involving derivatives of $v = \log u$. A simple calculation yields

$$v_t = \Delta v + |\nabla v|^2.$$

Consider $w = \Delta v$. Then

$$w_t = \Delta v_t = \Delta(\Delta v + |\nabla v|^2) = \Delta w + \Delta|\nabla v|^2.$$

Since

$$\Delta|\nabla v|^2 = 2|\nabla^2 v|^2 + 2\nabla v \cdot \nabla(\Delta v) = 2|\nabla^2 v|^2 + 2\nabla v \cdot \nabla w,$$

we have

$$w_t - \Delta w - 2\nabla v \cdot \nabla w = 2|\nabla^2 v|^2. \tag{5.3.1}$$

Note that ∇v is to be controlled and appears as a coefficient in the equation (5.3.1). So it is convenient to derive an equation for ∇v. Set $\tilde{w} = |\nabla v|^2$. Then,

$$\begin{aligned}\tilde{w}_t &= 2\nabla v \cdot \nabla v_t = 2\nabla v \cdot \nabla(\Delta v + |\nabla v|^2)\\ &= 2\nabla v \cdot \nabla(\Delta v) + 2\nabla v \cdot \nabla \tilde{w}\\ &= \Delta|\nabla v|^2 - 2|\nabla^2 v|^2 + 2\nabla v \cdot \nabla \tilde{w}\\ &= \Delta \tilde{w} + 2\nabla v \cdot \nabla \tilde{w} - 2|\nabla^2 v|^2.\end{aligned}$$

Therefore,

$$\tilde{w}_t - \Delta \tilde{w} - 2\nabla v \cdot \nabla \tilde{w} = -2|\nabla^2 v|^2. \tag{5.3.2}$$

Note that, by the Cauchy inequality,

$$|\nabla^2 v|^2 = \sum_{i,j=1}^n v_{x_i x_j}^2 \geq \sum_{i=1}^n v_{x_i x_i}^2 \geq \frac{1}{n}\left(\sum_{i=1}^n v_{x_i x_i}\right)^2 = \frac{1}{n}(\Delta v)^2.$$

Hence, (5.3.1) implies

$$w_t - \Delta w - 2\nabla v \cdot \nabla w \geq \frac{2}{n} w^2.$$

Step 2. For a constant $\alpha \in (0,1)$, set

$$f = \alpha |\nabla v|^2 - v_t.$$

Then

$$\begin{aligned} f &= \alpha |\nabla v|^2 - \Delta v - |\nabla v|^2 = -\Delta v - (1-\alpha)|\nabla v|^2 \\ &= -w - (1-\alpha)\tilde{w}, \end{aligned}$$

and hence by (5.3.1) and (5.3.2),

$$f_t - \Delta f - 2\nabla v \cdot \nabla f = -2\alpha |\nabla^2 v|^2.$$

Next, we estimate $|\nabla^2 v|^2$ by f. Note that

$$\begin{aligned} |\nabla^2 v|^2 &\geq \frac{1}{n}(\Delta v)^2 = \frac{1}{n}\big(|\nabla v|^2 - v_t\big)^2 = \frac{1}{n}\big((1-\alpha)|\nabla v|^2 + f\big)^2 \\ &= \frac{1}{n}\big(f^2 + 2(1-\alpha)|\nabla v|^2 f + (1-\alpha)^2 |\nabla v|^4\big) \\ &\geq \frac{1}{n}\big(f^2 + 2(1-\alpha)|\nabla v|^2 f\big). \end{aligned}$$

We obtain

$$f_t - \Delta f - 2\nabla v \cdot \nabla f \leq -\frac{2\alpha}{n}\big(f^2 + 2(1-\alpha)|\nabla v|^2 f\big). \tag{5.3.3}$$

We should point out that $|\nabla v|^2$ in the right-hand side plays an important role later on.

Step 3. Now we introduce a cutoff function $\varphi \in C_0^\infty(\mathbb{R}^n)$ with $\varphi \geq 0$ and set

$$g = t\varphi f.$$

We derive a differential inequality for g. Note that

$$\begin{aligned} g_t &= \varphi f + t\varphi f_t, \\ \nabla g &= t\varphi \nabla f + t f \nabla \varphi, \\ \Delta g &= t\varphi \Delta f + 2t\nabla \varphi \cdot \nabla f + t f \Delta \varphi. \end{aligned}$$

Then,

$$\begin{aligned}
t\varphi f_t &= g_t - \frac{g}{t},\\
t\varphi\nabla f &= \nabla g - \frac{\nabla\varphi}{\varphi}g,\\
t\varphi\Delta f &= \Delta g - 2\frac{\nabla\varphi}{\varphi}\cdot\left(\nabla g - \frac{\nabla\varphi}{\varphi}g\right) - \frac{\Delta\varphi}{\varphi}g\\
&= \Delta g - 2\frac{\nabla\varphi}{\varphi}\cdot\nabla g + \left(2\frac{|\nabla\varphi|^2}{\varphi^2} - \frac{\Delta\varphi}{\varphi}\right)g.
\end{aligned}$$

Multiplying (5.3.3) by $t^2\varphi^2$ and substituting $f_t, \nabla f$ and Δf by above equalities, we obtain

$$\begin{aligned}
&t\varphi(g_t - \Delta g) + 2t(\nabla\varphi - \varphi\nabla v)\cdot\nabla g\\
&\quad\leq g\left\{\varphi - \frac{2\alpha}{n}g + t\left(2\frac{|\nabla\varphi|^2}{\varphi} - \Delta\varphi - \frac{4\alpha(1-\alpha)}{n}\varphi|\nabla v|^2 - 2\nabla\varphi\cdot\nabla v\right)\right\}.
\end{aligned}$$

To eliminate $|\nabla v|$ from the right-hand side, we complete the square for the last two terms. (Here we need $\alpha < 1$! Otherwise, we cannot control the expression $-2\nabla\varphi\cdot\nabla v$ in the right-hand side.) Hence,

$$\begin{aligned}
&t\varphi(g_t - \Delta g) + 2t(\nabla\varphi - \varphi\nabla v)\cdot\nabla g\\
&\quad\leq g\left\{\varphi - \frac{2\alpha}{n}g + t\left(2\frac{|\nabla\varphi|^2}{\varphi} - \Delta\varphi + \frac{n}{4\alpha(1-\alpha)}\frac{|\nabla\varphi|^2}{\varphi}\right)\right\},
\end{aligned}$$

whenever g is nonnegative. We point out that there are no unknown expressions in the right-hand side except g. By choosing $\varphi = \eta^2$ for some $\eta \in C_0^\infty(\mathbb{R}^n)$ with $\eta \geq 0$, we get

$$\begin{aligned}
&t\eta^2(g_t - \Delta g) + 2t(2\eta\nabla\eta - \eta^2\nabla v)\cdot\nabla g\\
&\quad\leq g\left\{\eta^2 - \frac{2\alpha}{n}g + t\left(6|\nabla\eta|^2 - 2\eta\Delta\eta + \frac{n}{\alpha(1-\alpha)}|\nabla\eta|^2\right)\right\},
\end{aligned}$$

whenever g is nonnegative. Now we fix a cutoff function $\eta_0 \in C_0^\infty(B_1)$, with $0 \leq \eta_0 \leq 1$ in B_1 and $\eta_0 = 1$ in $B_{1/2}$. For any fixed $R \geq 1$, we consider $\eta(x) = \eta_0(x/R)$. Then

$$\begin{aligned}
&\left(6|\nabla\eta|^2 - 2\eta\Delta\eta + \frac{n}{\alpha(1-\alpha)}|\nabla\eta|^2\right)(x)\\
&\quad= \frac{1}{R^2}\left(6|\nabla\eta_0|^2 - 2\eta_0\Delta\eta_0 + \frac{n}{\alpha(1-\alpha)}|\nabla\eta_0|^2\right)\left(\frac{x}{R}\right).
\end{aligned}$$

Therefore, we obtain that in $B_R \times (0,T)$,

$$t\eta^2(g_t - \Delta g) + 2t(2\eta\nabla\eta - \eta^2\nabla v)\cdot\nabla g \leq g\left(1 - \frac{2\alpha}{n}g + \frac{C_\alpha t}{R^2}\right),$$

whenever g is nonnegative. Here, C_α is a positive constant depending only on α and η_0. We point out that the unknown expression ∇v in the left-hand side appears as a coefficient of ∇g and is unharmful.

Step 4. We claim that

$$1 - \frac{2\alpha}{n} g + \frac{C_\alpha t}{R^2} \geq 0 \quad \text{in } B_R \times (0, T]. \tag{5.3.4}$$

Note that g vanishes on the parabolic boundary of $B_R \times (0, T)$ since $g = t\eta^2 f$. Suppose, to the contrary, that

$$h \equiv 1 - \frac{2\alpha}{n} g + \frac{C_\alpha t}{R^2}$$

has a negative minimum at $(x_0, t_0) \in B_R \times (0, T]$. Hence,

$$h(x_0, t_0) < 0,$$

and

$$h_t \leq 0, \quad \nabla h = 0, \quad \Delta h \geq 0 \quad \text{at } (x_0, t_0).$$

Thus,

$$g(x_0, t_0) > 0,$$

and

$$g_t \geq 0, \quad \nabla g = 0, \quad \Delta g \leq 0 \quad \text{at } (x_0, t_0).$$

Then at (x_0, t_0), we get

$$\begin{aligned} 0 &\leq t\eta^2(g_t - \Delta g) + 2t(2\eta\nabla\eta - \eta^2\nabla v) \cdot \nabla g \\ &\leq g\left(1 - \frac{2\alpha}{n} g + \frac{C_\alpha t}{R^2}\right) < 0. \end{aligned}$$

This is a contradiction. Hence (5.3.4) holds in $B_R \times (0, T)$.

Therefore, we obtain

$$1 - \frac{2\alpha}{n} t\eta^2(\alpha|\nabla v|^2 - v_t) + \frac{C_\alpha t}{R^2} \geq 0 \quad \text{in } B_R \times (0, T]. \tag{5.3.5}$$

For any fixed $(x, t) \in \mathbb{R}^n \times (0, T]$, choose $R > |x|$. Recall that $\eta = \eta_0(\cdot/R)$ and $\eta_0 = 1$ in $B_{1/2}$. Letting $R \to \infty$, we obtain

$$1 - \frac{2\alpha}{n} t(\alpha|\nabla v|^2 - v_t) \geq 0.$$

We then let $\alpha \to 1$ and get the desired estimate. □

We also have the following *differential Harnack inequality* for positive solutions in finite regions.

Theorem 5.3.14. *Suppose $u \in C^{2,1}(B_1 \times (0,1])$ satisfies*

$$u_t - \Delta u = 0, \ \ u > 0 \quad \text{in } B_1 \times (0,1].$$

Then for any $\alpha \in (0,1)$, $v = \log u$ satisfies

$$v_t - \alpha|\nabla v|^2 + \frac{n}{2\alpha t} + C \geq 0 \quad \text{in } B_{1/2} \times (0,1],$$

where C is a positive constant depending only on n and α.

Proof. We simply take $R = 1$ in (5.3.5). □

Now we state the *Harnack inequality* in finite regions.

Corollary 5.3.15. *Suppose $u \in C^{2,1}(B_1 \times (0,1])$ satisfies*

$$u_t - \Delta u = 0, \ \ u \geq 0 \quad \text{in } B_1 \times (0,1].$$

Then for any $(x_1, t_1), (x_2, t_2) \in B_{1/2} \times (0,1]$ with $t_2 > t_1$,

$$u(x_1, t_1) \leq Cu(x_2, t_2),$$

where C is a positive constant depending only on n, t_2/t_1 and $(t_2 - t_1)^{-1}$.

The proof is left as an exercise.

We point out that u is assumed to be positive in Theorem 5.3.14 and only nonnegative in Corollary 5.3.15.

The Harnack inequality implies the following form of the strong maximum principle: *Let u be a nonnegative solution of the heat equation $u_t - \Delta u = 0$ in $B_1 \times (0,1]$. If $u(x_0, t_0) = 0$ for some $(x_0, t_0) \in B_1 \times (0,1]$, then $u = 0$ in $B_1 \times (0, t_0]$.* This may be interpreted as *infinite-speed propagation.*

5.4. Exercises

Exercise 5.1. Prove the following statements by straightforward calculations:

(1) $K(x,t) = t^{-\frac{n}{2}} e^{-\frac{|x|^2}{4t}}$ satisfies the heat equation for $t > 0$.

(2) For any $\alpha > 0$, $G(x,t) = (1 - 4\alpha t)^{-\frac{n}{2}} e^{\frac{\alpha |x|^2}{1 - 4\alpha t}}$ satisfies the heat equation for $t < 1/4\alpha$.

Exercise 5.2. Let u_0 be a continuous function in $\mathbb{R}^n$ and u be defined in (5.2.4). Suppose $u_0(x) \to 0$ uniformly as $x \to \infty$. Prove

$$\lim_{t \to \infty} u(x,t) = 0 \quad \text{uniformly in } x.$$

Exercise 5.3. Prove the convergence in Theorem 5.2.5.

Exercise 5.4. Let u_0 be a bounded and continuous function in $[0,\infty)$ with $u_0(0)=0$. Find an integral representation for the solution of the problem

$$\begin{aligned} u_t - u_{xx} &= 0 \quad \text{for } x>0, t>0,\\ u(x,0) &= u_0(x) \quad \text{for } x>0,\\ u(0,t) &= 0 \quad \text{for } t>0. \end{aligned}$$

Exercise 5.5. Let $u \in C^{2,1}(\mathbb{R}^n \times (-\infty,0))$ be a solution of

$$u_t - \Delta u = 0 \quad \text{in } \mathbb{R}^n \times (-\infty,0).$$

Suppose that for some nonnegative integer m,

$$|u(x,t)| \le C(1+|x|+\sqrt{|t|})^m,$$

for any $(x,t) \in \mathbb{R}^n \times (-\infty,0)$. Prove that u is a polynomial of degree at most m.

Exercise 5.6. Prove that u constructed in the proof of Proposition 5.2.6 is smooth in $\mathbb{R} \times \mathbb{R}$.

Exercise 5.7. Let Ω be a bounded domain in $\mathbb{R}^n$ and $u_0 \in C(\bar{\Omega})$. Suppose $u \in C^{2,1}(\Omega \times (0,\infty)) \cap C(\bar{\Omega} \times [0,\infty))$ is a solution of

$$\begin{aligned} u_t - \Delta u &= 0 \quad \text{in } \Omega \times (0,\infty),\\ u(\cdot,0) &= u_0 \quad \text{on } \Omega,\\ u &= 0 \quad \text{on } \partial\Omega \times (0,\infty). \end{aligned}$$

Prove that

$$\sup_\Omega |u(\cdot,t)| \le Ce^{-\mu t} \sup_\Omega |u_0| \quad \text{for any } t>0,$$

where μ and C are positive constants depending only on n and Ω.

Exercise 5.8. Let Ω be a bounded domain in $\mathbb{R}^n$, c be continuous in $\bar{\Omega} \times [0,T]$ with $c \ge -c_0$ for a nonnegative constant c_0, and u_0 be continuous in Ω with $u_0 \ge 0$. Suppose $u \in C^{2,1}(\Omega \times (0,T]) \cap C(\bar{\Omega} \times [0,T])$ is a solution of

$$\begin{aligned} u_t - \Delta u + cu &= -u^2 \quad \text{in } \Omega \times (0,T],\\ u(\cdot,0) &= u_0 \quad \text{on } \Omega,\\ u &= 0 \quad \text{on } \partial\Omega \times (0,T). \end{aligned}$$

Prove that

$$0 \le u \le e^{c_0 T} \sup_\Omega u_0 \quad \text{in } \Omega \times (0,T].$$

Exercise 5.9. Let Ω be a bounded domain in $\mathbb{R}^n$, u_0 and f be continuous in $\bar{\Omega}$, and φ be continuous on $\partial\Omega \times [0,T]$. Suppose $u \in C^{2,1}(\Omega \times (0,T]) \cap C(\bar{\Omega} \times [0,T])$ is a solution of

$$\begin{aligned} u_t - \Delta u &= e^{-u} - f(x) \quad \text{in } \Omega \times (0,T], \\ u(\cdot, 0) &= u_0 \quad \text{on } \Omega, \\ u &= \varphi \quad \text{on } \partial\Omega \times (0,T). \end{aligned}$$

Prove that

$$-M \leq u \leq Te^M + M \quad \text{in } \Omega \times (0,T],$$

where

$$M = T \sup_{\Omega} |f| + \sup \left\{ \sup_{\Omega} |u_0|, \sup_{\partial\Omega \times (0,T)} |\varphi| \right\}.$$

Exercise 5.10. Let $Q = (0,l) \times (0,\infty)$ and $u_0 \in C^1[0,l]$ with $u_0(0) = u_0(l) = 0$. Suppose $u \in C^{3,1}(Q) \cap C^1(\bar{Q})$ is a solution of

$$\begin{aligned} u_t - u_{xx} &= 0 \quad \text{in } Q, \\ u(\cdot, 0) &= u_0 \quad \text{on } (0,l), \\ u(0,\cdot) = u(l,\cdot) &= 0 \quad \text{on } (0,\infty). \end{aligned}$$

Prove that

$$\sup_Q |u_x| \leq \sup_{[0,l]} |u_0'|.$$

Exercise 5.11. Let Ω be a bounded domain in $\mathbb{R}^n$. Suppose $u_1, \cdots, u_m \in C^{2,1}(\Omega \times (0,T]) \cap C(\bar{\Omega} \times [0,T])$ satisfy

$$\partial_t u_i = \Delta u_i \quad \text{in } \Omega \times (0,T],$$

for $i = 1, \cdots, m$. Assume that f is a convex function in $\mathbb{R}^m$. Prove that

$$\sup_{\Omega \times (0,T]} f(u_1, \cdots, u_m) \leq \sup_{\partial_p(\Omega \times (0,T])} f(u_1, \cdots, u_m).$$

Exercise 5.12. Let u_0 be a bounded continuous function in $\mathbb{R}^n$. Suppose $u \in C^{2,1}(\mathbb{R}^n \times (0,T]) \cap C(\mathbb{R}^n \times [0,T])$ satisfies

$$\begin{aligned} u_t - \Delta u &= 0 \quad \text{in } \mathbb{R}^n \times (0,T], \\ u(\cdot, 0) &= u_0 \quad \text{on } \mathbb{R}^n. \end{aligned}$$

Assume that u and ∇u are bounded in $\mathbb{R}^n \times (0,T]$. Prove that

$$\sup_{\mathbb{R}^n} |\nabla u(\cdot, t)| \leq \frac{1}{\sqrt{2t}} \sup_{\mathbb{R}^n} |u_0|,$$

for any $t \in (0,T]$.

Hint: With $|u_0| \leq M$ in $\mathbb{R}^n$, consider

$$w = u^2 + 2t|\nabla u|^2 - M^2.$$

Exercise 5.13. Prove Corollary 5.3.15.

Chapter 6

Wave Equations

The n-dimensional wave equation is given by $u_{tt} - \Delta u = 0$ for functions $u = u(x, t)$, with $x \in \mathbb{R}^n$ and $t \in \mathbb{R}$. Here, x is the space variable and t the time variable. The wave equation represents vibrations of strings or propagation of sound waves in tubes for $n = 1$, waves on the surface of shallow water for $n = 2$, and acoustic or light waves for $n = 3$.

In Section 6.1, we discuss the initial-value problem and mixed problems for the one-dimensional wave equation. We derive explicit expressions for solutions of these problems by various methods and study properties of these solutions. We illustrate that characteristic curves play an important role in studying the one-dimensional wave equation. They determine the domain of dependence and the range of influence.

In Section 6.2, we study the initial-value problem for the wave equation in higher-dimensional spaces. We derive explicit expressions for solutions in odd dimensions by the method of spherical averages and in even dimensions by the method of descent. We study properties of these solutions with the help of these formulas and illustrate the importance of characteristic cones for the higher-dimensional wave equation. Among applications of these explicit expressions, we discuss global behaviors of solutions and prove that solutions decay at certain rates as time goes to infinity. We will also solve the initial-value problem for the nonhomogeneous wave equation by Duhamel's principle.

In Section 6.3, we discuss energy estimates for solutions of the initial-value problem for a class of hyperbolic equations slightly more general than the wave equation. We introduce the important concept of space-like and time-like hypersurfaces. We demonstrate that initial-value problems for hyperbolic equations with initial values prescribed on space-like hypersurfaces are well posed. We point out that energy estimates are fundamental and form the basis for the existence of solutions of general hyperbolic equations.

6.1. One-Dimensional Wave Equations

In this section, we discuss initial-value problems and initial/boundary-value problems for the one-dimensional wave equation. We first study initial-value problems.

6.1.1. Initial-Value Problems. For $f \in C(\mathbb{R} \times (0, \infty))$, $\varphi \in C^2(\mathbb{R})$ and $\psi \in C^1(\mathbb{R})$, we seek a solution $u \in C^2(\mathbb{R} \times [0, \infty))$ of the problem

$$\begin{aligned} u_{tt} - u_{xx} &= f \quad \text{in } \mathbb{R} \times (0, \infty), \\ u(\cdot, 0) = \varphi, \ u_t(\cdot, 0) &= \psi \quad \text{on } \mathbb{R}. \end{aligned} \tag{6.1.1}$$

We will derive expressions for its solutions by several different methods.

Throughout this section, we denote points in $\mathbb{R} \times (0, \infty)$ by (x, t). However, when (x, t) is taken as a fixed point, we denote arbitrary points by (y, s).

The characteristic curves for the one-dimensional wave equation are given by the straight lines $s = \pm y + c$. (Refer to Section 3.1 for the detail.) In particular, for any $(x, t) \in \mathbb{R} \times (0, \infty)$, there are two characteristic curves through (x, t) given by

$$s - y = t - x \quad \text{and} \quad s + y = t + x.$$

These two characteristic curves intercept the x-axis at $(x-t, 0)$ and $(x+t, 0)$, respectively, and form a triangle $C_1(x, t)$ with the x-axis given by

$$C_1(x, t) = \{(y, s) : \ |y - x| < t - s, s > 0\}.$$

This is the cone we introduced in Section 2.3 for $n = 1$. We usually refer to $C_1(x, t)$ as the *characteristic triangle.*

We first consider the homogeneous wave equation

$$u_{tt} - u_{xx} = 0 \quad \text{in } \mathbb{R} \times (0, \infty). \tag{6.1.2}$$

We introduce new coordinates (ξ, η) along characteristic curves by

$$\xi = x - t, \ \eta = x + t.$$

In the new coordinates, the wave equation has the form

$$u_{\xi\eta} = 0.$$

By a simple integration, we obtain

$$u(\xi, \eta) = g(\xi) + h(\eta),$$

for some functions g and h in $\mathbb{R}$. Therefore,

$$u(x, t) = g(x - t) + h(x + t). \tag{6.1.3}$$

This provides a general form for solutions of (6.1.2).

As a consequence of (6.1.3), we derive an important formula for the solution of the wave equation. Let u be a C^2-solution of (6.1.2). Consider a parallelogram bounded by four characteristic curves in $\mathbb{R} \times (0, \infty)$, which is referred to as a *characteristic parallelogram*. (This parallelogram is in fact a rectangle.) Suppose A, B, C, D are its four vertices. Then

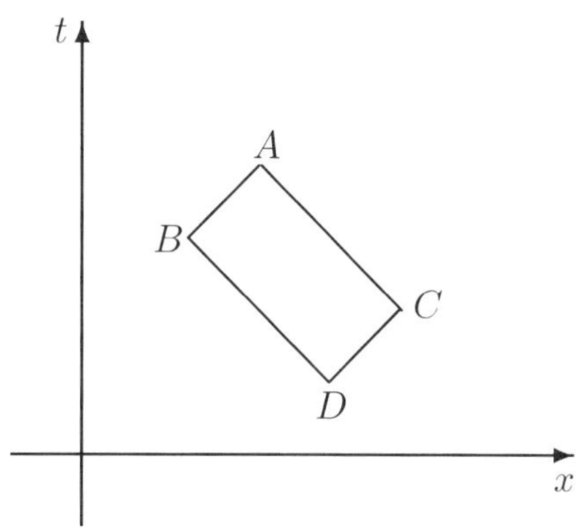

Figure 6.1.1. A characteristic parallelogram.

$$u(A) + u(D) = u(B) + u(C). \tag{6.1.4}$$

In other words, the sums of the values of u at opposite vertices are equal. This follows easily from (6.1.3). In fact, if we set $A = (x_A, t_A)$, $B = (x_B, t_B)$, $C = (x_C, t_C)$ and $D = (x_D, t_D)$, we have

$$x_B - t_B = x_A - t_A, \quad x_B + t_B = x_D + t_D,$$

and

$$x_C - t_C = x_D - t_D, \quad x_C + t_C = x_A + t_A.$$

We then get (6.1.4) by (6.1.3) easily. An alternative method to prove (6.1.4) is to consider it in (ξ, η)-coordinates, where A, B, C, D are the vertices of a rectangle with sides parallel to the axes. Then we simply integrate $u_{\xi\eta}$, which is zero, in this rectangle to get the desired relation.

We now solve (6.1.1) for the case $f \equiv 0$. Let u be a C^2-solution which is given by (6.1.3) for some functions g and h. By evaluating u and u_t at $t = 0$, we have

$$u(x, 0) = g(x) + h(x) = \varphi(x),$$
$$u_t(x, 0) = -g'(x) + h'(x) = \psi(x).$$

Then

$$g'(x) = \frac{1}{2}\varphi'(x) - \frac{1}{2}\psi(x),$$
$$h'(x) = \frac{1}{2}\varphi'(x) + \frac{1}{2}\psi(x).$$

A simple integration yields

$$g(x) = \frac{1}{2}\varphi(x) - \frac{1}{2}\int_0^x \psi(s)\, ds + c,$$

for a constant c. Then a substitution into the expression of $u(x,0)$ implies

$$h(x) = \frac{1}{2}\varphi(x) + \frac{1}{2}\int_0^x \psi(s) ds - c.$$

Therefore,

$$u(x,t) = \frac{1}{2}\big(\varphi(x-t) + \varphi(x+t)\big) + \frac{1}{2}\int_{x-t}^{x+t} \psi(s)\, ds. \tag{6.1.5}$$

This is *d'Alembert's formula.* It clearly shows that regularity of $u(\cdot,t)$ for any $t > 0$ is the same as that of the initial value $u(\cdot,0)$ and is 1-degree better than $u_t(\cdot,0)$. There is no improvement of regularity.

We see from (6.1.5) that $u(x,t)$ is determined uniquely by the initial values in the interval $[x-t, x+t]$ of the x-axis, which is the base of the characteristic triangle $C_1(x,t)$. This interval is the *domain of dependence* for the solution u at the point (x,t). We note that the endpoints of this interval are cut out by the characteristic curves through (x,t). Conversely, the initial values at a point $(x_0,0)$ of the x-axis influence $u(x,t)$ at points (x,t) in the wedge-shaped region bounded by characteristic curves through $(x_0,0)$, i.e., for $x_0 - t < x < x_0 + t$, which is often referred to as the *range of influence.*

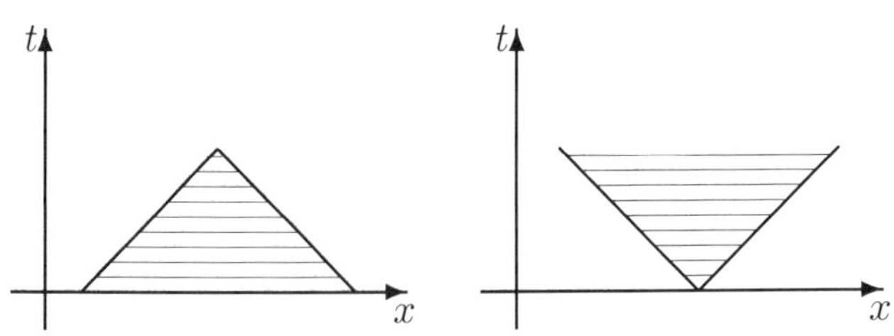

Figure 6.1.2. The domain of dependence and the range of influence.

Next, we consider the case $f \equiv 0$ and $\varphi \equiv 0$ and solve (6.1.1) by the method of characteristics. We write

$$u_{tt} - u_{xx} = (\partial_t + \partial_x)(\partial_t - \partial_x)u.$$

By setting $v = u_t - u_x$, we decompose (6.1.1) into two initial-value problems for first-order PDEs,

$$(6.1.6) \qquad \begin{aligned} u_t - u_x &= v \quad \text{in } \mathbb{R} \times (0, \infty), \\ u(\cdot, 0) &= 0 \quad \text{on } \mathbb{R}, \end{aligned}$$

and

$$(6.1.7) \qquad \begin{aligned} v_t + v_x &= 0 \quad \text{in } \mathbb{R} \times (0, \infty), \\ v(\cdot, 0) &= \psi \quad \text{on } \mathbb{R}. \end{aligned}$$

The initial-value problem (6.1.7) was discussed in Example 2.2.3. Its solution is given by

$$v(x, t) = \psi(x - t).$$

The initial-value problem (6.1.6) was discussed in Example 2.2.4. Its solution is given by

$$u(x, t) = \int_0^t \psi(x + t - 2\tau)\, d\tau.$$

By a change of variables, we obtain

$$u(x, t) = \frac{1}{2} \int_{x-t}^{x+t} \psi(s)\, ds.$$

This is simply a special case of d'Alembert's formula (6.1.5).

Now we derive an expression of solutions in the general case. For any $(x, t) \in \mathbb{R} \times (0, \infty)$, consider the characteristic triangle

$$C_1(x, t) = \{(y, s) : \ |y - x| < t - s, s > 0\}.$$

The boundary of $C_1(x, t)$ consists of three parts,

$$\begin{aligned} L_+ &= \{(y, s) : \ s = -y + x + t, \ 0 < s < t\}, \\ L_- &= \{(y, s) : \ s = y - x + t, \ 0 < s < t\}, \end{aligned}$$

and

$$L_0 = \{(y, 0) : \ x - t < y < x + t\}.$$

We note that L_+ and L_- are parts of the characteristic curves through (x, t). Let $\nu = (\nu_1, \nu_2)$ be the unit exterior normal vector of $\partial C_1(x, t)$. Then

$$\nu = \begin{cases} (1, 1)/\sqrt{2} & \text{on } L_+, \\ (-1, 1)/\sqrt{2} & \text{on } L_-, \\ (0, -1) & \text{on } L_0. \end{cases}$$

Upon integrating by parts, we have

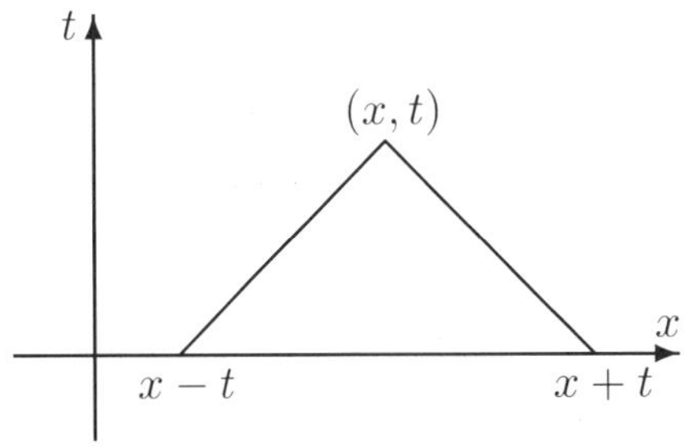

Figure 6.1.3. A characteristic triangle.

$$\begin{aligned}\int_{C_1(x,t)} f dyds &= \int_{C_1(x,t)} (u_{tt} - u_{xx})\, dyds = \int_{\partial C_1(x,t)} (u_t\nu_2 - u_x\nu_1)\, dl \\ &= \int_{L_+} \frac{1}{\sqrt{2}}(u_t - u_x)\, dl + \int_{L_-} \frac{1}{\sqrt{2}}(u_t + u_x)\, dl \\ &\quad - \int_{x-t}^{x+t} u_t(s,0)\, ds,\end{aligned}$$

where the orientation of the integrals over L_+ and L_- is counterclockwise. Note that $(\partial_t - \partial_x)/\sqrt{2}$ is a directional derivative along L_+ with unit length and with direction matching the orientation of the integral over L_+. Hence

$$\int_{L_+} \frac{1}{\sqrt{2}}(u_t - u_x)\, dl = u(x,t) - u(x+t,0).$$

On the other hand, $(\partial_t + \partial_x)/\sqrt{2}$ is a directional derivative along L_- with unit length and with direction opposing the orientation of the integral over L_-. Hence

$$\int_{L_-} \frac{1}{\sqrt{2}}(u_t + u_x)\, dl = -\big(u(x-t,0) - u(x,t)\big).$$

Therefore, a simple substitution yields

$$\begin{aligned}u(x,t) &= \frac{1}{2}\big(\varphi(x+t) + \varphi(x-t)\big) + \frac{1}{2}\int_{x-t}^{x+t} \psi(s)\, ds \\ &\quad + \frac{1}{2}\int_0^t \int_{x-(t-\tau)}^{x+(t-\tau)} f(y,\tau)\, dyd\tau.\end{aligned} \tag{6.1.8}$$

Theorem 6.1.1. *Let $m \geq 2$ be an integer, $\varphi \in C^m(\mathbb{R}), \psi \in C^{m-1}(\mathbb{R})$ and $f \in C^{m-1}(\mathbb{R} \times [0,\infty))$. Suppose u is defined by (6.1.8). Then $u \in C^m(\mathbb{R} \times (0,\infty))$ and*

$$u_{tt} - u_{xx} = f \quad \text{in } \mathbb{R} \times (0,\infty).$$

Moreover, for any $x_0 \in \mathbb{R}$,

$$\lim_{(x,t)\to(x_0,0)} u(x,t) = \varphi(x_0), \qquad \lim_{(x,t)\to(x_0,0)} u_t(x,t) = \psi(x_0).$$

Hence, u defined by (6.1.8) is a solution of (6.1.1). In fact, u is C^m in $\mathbb{R} \times [0,\infty)$.

The proof is a straightforward calculation and is omitted. Obviously, C^2-solutions of (6.1.1) are unique.

Formula (6.1.8) illustrates that the value $u(x,t)$ is determined by f in the triangle $C_1(x,t)$, by ψ on the interval $[x-t, x+t] \times \{0\}$ and by φ at the two points $(x+t, 0)$ and $(x-t, 0)$.

In fact, without using the explicit expression of solutions in (6.1.8), we can derive energy estimates, the estimates for the L^2-norms of solutions of (6.1.1) and their derivatives in terms of the L^2-norms of φ, ψ and f. To obtain energy estimates, we take any constants $0 < T < \bar{t}$ and use the domain

$$\{(x,t) : \ |x| < \bar{t} - t, 0 < t < T\}.$$

We postpone the derivation until the final section of this chapter.

6.1.2. Mixed Problems. In the following, we study mixed problems. For simplicity, we discuss the wave equation only, with no nonhomogeneous terms.

First, we study the *half-space problem*. Let $\varphi \in C^2[0,\infty)$, $\psi \in C^1[0,\infty)$ and $\alpha \in C^2[0,\infty)$. We consider

$$\begin{aligned} &u_{tt} - u_{xx} = 0 \quad \text{in } (0,\infty) \times (0,\infty), \\ &u(\cdot,0) = \varphi, \ u_t(\cdot,0) = \psi \quad \text{on } [0,\infty), \\ &u(0,t) = \alpha(t) \quad \text{for } t > 0. \end{aligned} \tag{6.1.9}$$

We will construct a C^2-solution under appropriate compatibility conditions. We note that the origin is the corner of the region $(0,\infty) \times (0,\infty)$. In order to have a C^2-solution u, the initial values φ and ψ and the boundary value α have to match at the corner to generate the same u and its first-order and second-order derivatives when computed either from φ and ψ or from α. If (6.1.9) admits a solution which is C^2 in $[0,\infty) \times [0,\infty)$, a simple calculation shows that

$$\varphi(0) = \alpha(0), \ \psi(0) = \alpha'(0), \ \varphi''(0) = \alpha''(0). \tag{6.1.10}$$

This is the *compatibility condition* for (6.1.9). It is the necessary condition for the existence of a C^2-solution of (6.1.9). We will show that it is also sufficient.

We first consider the case $\alpha \equiv 0$ and solve (6.1.9) by the *method of reflection.* In this case, the compatibility condition (6.1.10) has the form

$$\varphi(0) = 0, \quad \psi(0) = 0, \quad \varphi''(0) = 0.$$

Now we assume that this holds and proceed to construct a C^2-solution of (6.1.9). We extend φ and ψ to $\mathbb{R}$ by *odd reflection.* In other words, we set

$$\tilde{\varphi}(x) = \begin{cases} \varphi(x) & \text{for } x \geq 0 \\ -\varphi(-x) & \text{for } x < 0, \end{cases}$$

$$\tilde{\psi}(x) = \begin{cases} \psi(x) & \text{for } x \geq 0 \\ -\psi(-x) & \text{for } x < 0. \end{cases}$$

Then $\tilde{\varphi}$ and $\tilde{\psi}$ are C^2 and C^1 in $\mathbb{R}$, respectively. Let $\tilde{u}$ be the unique C^2-solution of the initial-value problem

$$\begin{aligned} &\tilde{u}_{tt} - \tilde{u}_{xx} = 0 \quad \text{in } \mathbb{R} \times (0, \infty), \\ &\tilde{u}(\cdot, 0) = \tilde{\varphi}, \ \tilde{u}_t(\cdot, 0) = \tilde{\psi} \quad \text{in } \mathbb{R}. \end{aligned}$$

We now prove that $\tilde{u}(x,t)$ is the solution of (6.1.9) when we restrict x to $[0, \infty)$. We need only prove that

$$\tilde{u}(0, t) = 0 \quad \text{for any } t > 0.$$

In fact, for $v(x,t) = -\tilde{u}(-x,t)$, a simple calculation yields

$$\begin{aligned} &v_{tt} - v_{xx} = 0 \quad \text{in } \mathbb{R} \times (0, \infty), \\ &v(\cdot, 0) = \tilde{\varphi}, \ v_t(\cdot, 0) = \tilde{\psi} \quad \text{in } \mathbb{R}. \end{aligned}$$

In other words, v is also a C^2-solution of the initial-value problem for the wave equation with the same initial values as $\tilde{u}$. By the uniqueness, $\tilde{u}(x,t) = v(x,t) = -\tilde{u}(-x,t)$ and hence $\tilde{u}(0,t) = 0$. In fact, $\tilde{u}$ is given by d'Alembert's formula (6.1.5), i.e.,

$$\tilde{u}(x,t) = \frac{1}{2}\big(\tilde{\varphi}(x+t) + \tilde{\varphi}(x-t)\big) + \frac{1}{2}\int_{x-t}^{x+t} \tilde{\psi}(s)\, ds.$$

By restricting (x,t) to $[0,\infty) \times [0,\infty)$, we have, for any $x \geq t \geq 0$,

$$u(x,t) = \frac{1}{2}\big(\varphi(x+t) + \varphi(x-t)\big) + \frac{1}{2}\int_{x-t}^{x+t} \psi(s)\, ds,$$

and for any $t \geq x \geq 0$,

$$u(x,t) = \frac{1}{2}\big(\varphi(x+t) - \varphi(t-x)\big) + \frac{1}{2}\int_{t-x}^{x+t} \psi(s)\, ds, \tag{6.1.11}$$

since $\tilde{\varphi}$ and $\tilde{\psi}$ are odd in $\mathbb{R}$. We point out that (6.1.11) will be needed in solving the initial-value problem for the wave equation in higher dimensions.

Now we consider the general case of (6.1.9) and construct a solution in $[0,\infty)\times[0,\infty)$ by an alternative method. We first decompose $[0,\infty)\times[0,\infty)$ into two regions by the straight line $t=x$. We note that $t=x$ is the characteristic curve for the wave equation in the domain $[0,\infty)\times[0,\infty)$ passing through the origin, which is the corner of $[0,\infty)\times[0,\infty)$. We will solve for u in these two regions separately. First, we set

$$\Omega_1=\{(x,t):\ x>t>0\},$$

and

$$\Omega_2=\{(x,t):\ t>x>0\}.$$

We denote by u_1 the solution in Ω_1. Then, u_1 is determined by (6.1.5) from the initial values. In fact,

$$u_1(x,t)=\frac{1}{2}\big(\varphi(x+t)+\varphi(x-t)\big)+\frac{1}{2}\int_{x-t}^{x+t}\psi(s)\,ds,$$

for any $(x,t)\in\Omega_1$. Set for $x>0$,

$$\gamma(x)=u_1(x,x)=\frac{1}{2}\big(\varphi(2x)+\varphi(0)\big)+\frac{1}{2}\int_0^{2x}\psi(s)\,ds.$$

We note that $\gamma(x)$ is the value of the solution u along the straight line $t=x$ for $x>0$. Next, we consider

$$\begin{aligned}&u_{tt}-u_{xx}=0\quad\text{in }\Omega_2,\\&u(0,t)=\alpha(t),\ u(x,x)=\gamma(x).\end{aligned}$$

We denote its solution by u_2. For any $(x,t)\in\Omega_2$, consider the characteristic parallelogram with vertices (x,t), $(0,t-x)$, $(\frac{t-x}{2},\frac{t-x}{2})$ and $(\frac{t+x}{2},\frac{t+x}{2})$. In other words, one vertex is (x,t), one vertex is on the boundary $\{x=0\}$ and the other two vertices are on $\{t=x\}$. By (6.1.4), we have

$$u_2(x,t)+u_2\left(\frac{t-x}{2},\frac{t-x}{2}\right)=u_2(0,t-x)+u_2\left(\frac{t+x}{2},\frac{t+x}{2}\right).$$

Hence

$$\begin{aligned}u_2(x,t)&=\alpha(t-x)-\gamma\left(\frac{t-x}{2}\right)+\gamma\left(\frac{x+t}{2}\right)\\&=\alpha(t-x)+\frac{1}{2}\big(\varphi(x+t)-\varphi(t-x)\big)+\frac{1}{2}\int_{t-x}^{x+t}\psi(s)\,ds,\end{aligned}$$

for any $(x,t)\in\Omega_2$. Set $u=u_1$ in Ω_1 and $u=u_2$ in Ω_2. Now we check that $u,u_t,u_x,u_{tt},u_{xx},u_{tx}$ are continuous along $\{t=x\}$. By a direct calculation,

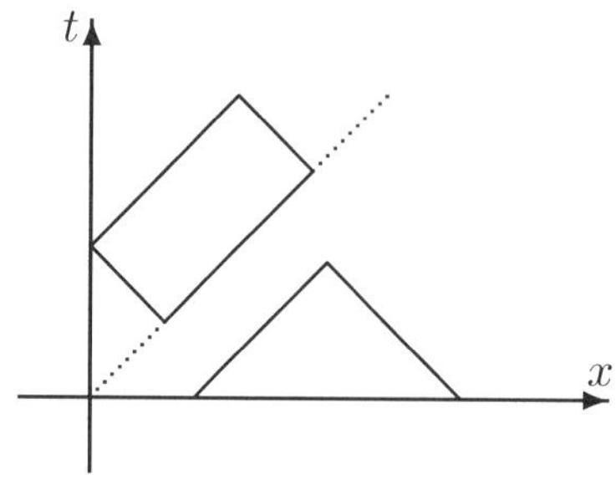

Figure 6.1.4. Division by a characteristic curve.

we have

$$\begin{aligned}
&u_1(x,t)|_{t=x} - u_2(x,t)|_{t=x} = \gamma(0) - \alpha(0) = \varphi(0) - \alpha(0),\\
&\partial_x u_1(x,t)|_{t=x} - \partial_x u_2(x,t)|_{t=x} = -\psi(0) + \alpha'(0),\\
&\partial_x^2 u_1(x,t)|_{t=x} - \partial_x^2 u_2(x,t)|_{t=x} = \varphi''(0) - \alpha''(0).
\end{aligned}$$

Then (6.1.10) implies

$$u_1 = u_2, \quad \partial_x u_1 = \partial_x u_2, \quad \partial_x^2 u_1 = \partial_x^2 u_2 \quad \text{on } \{t = x\}.$$

It is easy to get $\partial_t u_1 = \partial_t u_2$ on $\{t = x\}$ by $u_1 = u_2$ and $\partial_x u_1 = \partial_x u_2$ on $\{t = x\}$. Similarly, we get $\partial_{xt} u_1 = \partial_{xt} u_2$ and $\partial_{tt} u_1 = \partial_{tt} u_2$ on $\{t = x\}$. Therefore, u is C^2 across $t = x$. Hence, we obtain the following result.

Theorem 6.1.2. *Suppose* $\varphi \in C^2[0,\infty), \psi \in C^1[0,\infty), \alpha \in C^2[0,\infty)$ *and the compatibility condition* (6.1.10) *holds. Then there exists a solution* $u \in C^2([0,\infty) \times [0,\infty))$ *of* (6.1.9).

We can also derive a priori energy estimates for solutions of (6.1.9). For any constants $T > 0$ and $x_0 > T$, we use the following domain for energy estimates:

$$\{(x,t) : \ 0 < x < x_0 - t, 0 < t < T\}.$$

Now we consider the initial/boundary-value problem. For a positive constant $l > 0$, assume that $\varphi \in C^2[0,l]$, $\psi \in C^1[0,l]$ and $\alpha, \beta \in C^2[0,\infty)$. Consider

$$\begin{aligned}
&u_{tt} - u_{xx} = 0 \quad \text{in } (0,l) \times (0,\infty),\\
&u(\cdot,0) = \varphi, \ u_t(\cdot,0) = \psi \quad \text{on } [0,l],\\
&u(0,t) = \alpha(t), \ u(l,t) = \beta(t) \quad \text{for } t > 0.
\end{aligned} \tag{6.1.12}$$

The *compatibility condition* is given by

$$\begin{aligned}
&\varphi(0) = \alpha(0), \ \psi(0) = \alpha'(0), \ \varphi''(0) = \alpha''(0),\\
&\varphi(l) = \beta(0), \ \psi(l) = \beta'(0), \ \varphi''(l) = \beta''(0).
\end{aligned} \tag{6.1.13}$$

We first consider the special case $\alpha = \beta \equiv 0$. We discussed this case using separation of variables in Section 3.3 if $l = \pi$. We now construct solutions by the method of reflection. We first extend φ to $[-l, 0]$ by odd reflection. In other words, we define

$$\tilde{\varphi}(x) = \begin{cases} \varphi(x) & \text{for } x \in [0, l], \\ -\varphi(-x) & \text{for } x \in [-l, 0]. \end{cases}$$

We then extend $\tilde{\varphi}$ to $\mathbb{R}$ as a $2l$-periodic function. Then $\tilde{\varphi}$ is odd in $\mathbb{R}$. We extend ψ similarly. The extended functions $\tilde{\varphi}$ and $\tilde{\psi}$ are C^2 and C^1 on $\mathbb{R}$, respectively. Let $\tilde{u}$ be the unique solution of the initial-value problem

$$\begin{aligned} &\tilde{u}_{tt} - \tilde{u}_{xx} = 0 \quad \text{in } \mathbb{R} \times (0, \infty), \\ &\tilde{u}(\cdot, 0) = \tilde{\varphi}, \ \tilde{u}_t(\cdot, 0) = \tilde{\psi} \quad \text{on } \mathbb{R}. \end{aligned}$$

We now prove that $\tilde{u}(x, t)$ is a solution of (6.1.12) when we restrict x to $[0, l]$. We need only prove that

$$\tilde{u}(0, t) = 0, \ \tilde{u}(l, t) = 0 \quad \text{for any } t > 0.$$

The proof is similar to that for the half-space problem. We prove that $\tilde{u}(0, t) = 0$ by introducing $v(x, t) = -\tilde{u}(-x, t)$ and prove $\tilde{u}(l, t) = 0$ by introducing $w(x, t) = -\tilde{u}(2l - x, t)$.

We now discuss the general case and construct a solution of (6.1.12) by an alternative method. We decompose $[0, l] \times [0, \infty)$ into infinitely many regions by the characteristic curves through the corners and through the intersections of the characteristic curves with the boundaries. Specifically, we first consider the characteristic curve $t = x$. It starts from $(0, 0)$, one of the two corners, and intersects the right portion of the boundary $x = l$ at (l, l). Meanwhile, the characteristic curve $x + t = l$ starts from $(l, 0)$, the other corner, and intersects the left portion of the boundary $x = 0$ at $(0, l)$. These two characteristic curves intersect at $(l/2, l/2)$. We then consider the characteristic curve $t - x = l$ from $(0, l)$ and the characteristic curve $t + x = 2l$ from (l, l). They intersect the right portion of the boundary at $(l, 2l)$ and the left portion of the boundary at $(0, 2l)$, respectively. We continue this process. We first solve for u in the characteristic triangle with vertex $(l/2, l/2)$. In this region, u is determined by the initial values. Then we can solve for u by forming characteristic parallelograms in the triangle with vertices $(0, 0)$, $(l/2, l/2)$ and $(0, l)$ and in the triangle with vertices $(l, 0)$, $(l/2, l/2)$ and (l, l). In the next step, we solve for u again by forming characteristic parallelogram in the rectangle with vertices $(0, l)$, $(l/2, l/2)$, (l, l) and $(l/2, 3l/2)$. We note that this rectangle is a characteristic parallelogram. By continuing this process, we can find u in the entire region $[0, l] \times [0, \infty)$.

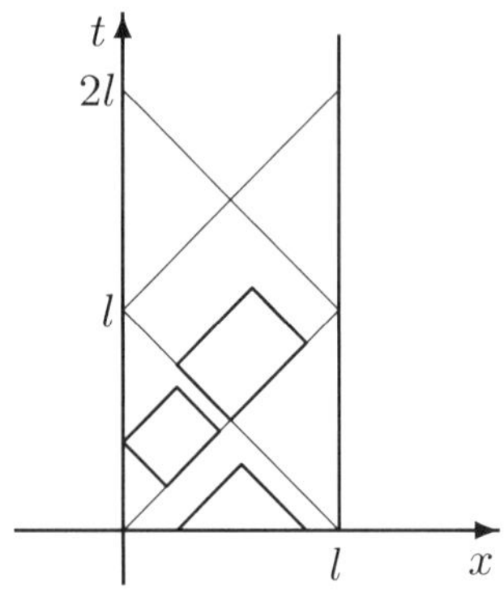

Figure 6.1.5. A decomposition by characteristic curves.

Theorem 6.1.3. *Suppose* $\varphi \in C^2[0,l], \psi \in C^1[0,l], \alpha, \beta \in C^2[0,\infty)$ *and the compatibility condition* (6.1.13) *holds. Then there exists a solution* $u \in C^2([0,l] \times [0,\infty))$ *of* (6.1.12).

Theorem 6.1.3 includes Theorem 3.3.8 in Chapter 3 as a special case.

Now we summarize various problems discussed in this section. We emphasize that characteristic curves play an important role in studying the one-dimensional wave equation.

First, presentations of problems depend on characteristic curves. Let Ω be a piecewise smooth domain in $\mathbb{R}^2$ whose boundary is not characteristic. In the following, we shall treat the initial curve as a part of the boundary and treat initial values as a part of boundary values. We intend to prescribe appropriate values on the boundary to ensure the well-posedness for the wave equation. To do this, we take an arbitrary point on the boundary and examine characteristic curves through this point. We then count how many characteristic curves enter the domain Ω in the positive t-direction. In this section, we discussed cases where Ω is given by the upper half-space $\mathbb{R} \times (0,\infty)$, the first quadrant $(0,\infty) \times (0,\infty)$ and $I \times (0,\infty)$ for a finite interval I. We note that the number of boundary values is the same as the number of characteristic curves entering the domain in the positive t-direction. In summary, we have

$$u|_{t=0} = \varphi,\ u_t|_{t=0} = \psi \text{ for initial-value problems;}$$
$$u|_{t=0} = \varphi,\ u_t|_{t=0} = \psi,\ u|_{x=0} = \alpha \text{ for half-space problems;}$$
$$u|_{t=0} = \varphi,\ u_t|_{t=0} = \psi,\ u|_{x=0} = \alpha,\ u|_{x=l} = \beta$$
$$\text{for initial/boundary-value problems.}$$

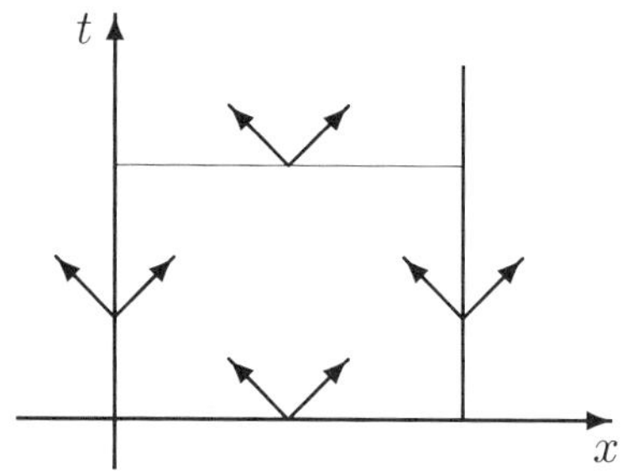

Figure 6.1.6. Characteristic directions.

Second, characteristic curves determine the domain of dependence and the range of influence. In fact, as illustrated by (6.1.5), initial values propagate along characteristic curves.

Last, characteristic curves also determine domains for energy estimates. We indicated domains of integration for initial-value problems and for half-space problems. We will explore energy estimates in detail in Section 6.3.

6.2. Higher-Dimensional Wave Equations

In this section, we discuss the initial-value problem for the wave equation in higher dimensions. Our main task is to derive an expression for its solutions and discuss their properties.

6.2.1. The Method of Spherical Averages.

Let $\varphi \in C^2(\mathbb{R}^n)$ and $\psi \in C^1(\mathbb{R}^n)$. Consider

$$\begin{aligned} &u_{tt} - \Delta u = 0 \quad \text{in } \mathbb{R}^n \times (0, \infty), \\ &u(\cdot, 0) = \varphi, \ u_t(\cdot, 0) = \psi \quad \text{on } \mathbb{R}^n. \end{aligned} \tag{6.2.1}$$

We will solve this initial-value problem by the *method of spherical averages*.

We first discuss briefly spherical averages. Let w be a continuous function in $\mathbb{R}^n$. For any $x \in \mathbb{R}^n$ and $r > 0$, set

$$W(x; r) = \frac{1}{\omega_n r^{n-1}} \int_{\partial B_r(x)} w(y)\, dS_y,$$

where ω_n is the surface area of the unit sphere in $\mathbb{R}^n$. Then $W(x; r)$ is the average of w over the sphere $\partial B_r(x)$. Now, w can be recovered from W by

$$\lim_{r \to 0} W(x; r) = w(x) \quad \text{for any } x \in \mathbb{R}^n.$$

Next, we suppose u is a C^2-solution of (6.2.1). For any $x \in \mathbb{R}^n$, $t > 0$ and $r > 0$, set

$$U(x;r,t) = \frac{1}{\omega_n r^{n-1}} \int_{\partial B_r(x)} u(y,t)\, dS_y \tag{6.2.2}$$

and

$$\begin{aligned} \Phi(x;r) &= \frac{1}{\omega_n r^{n-1}} \int_{\partial B_r(x)} \varphi(y)\, dS_y, \\ \Psi(x;r) &= \frac{1}{\omega_n r^{n-1}} \int_{\partial B_r(x)} \psi(y)\, dS_y. \end{aligned} \tag{6.2.3}$$

In other words, $U(x;r,t)$, $\Phi(x,r)$ and $\Psi(x,r)$ are the averages of $u(\cdot,t)$, φ and ψ over the sphere $\partial B_r(x)$, respectively. Then U determines u by

$$\lim_{r \to 0} U(x;r,t) = u(x,t).$$

Now we transform the differential equation for u to a differential equation for U. We claim that, for each fixed $x \in \mathbb{R}^n$, $U(x;r,t)$ satisfies the *Euler-Poisson-Darboux equation*

$$U_{tt} = U_{rr} + \frac{n-1}{r} U_r \quad \text{for } r > 0 \text{ and } t > 0, \tag{6.2.4}$$

with initial values

$$U(x;r,0) = \Phi(x;r), \; U_t(x;r,0) = \Psi(x;r) \quad \text{for } r > 0.$$

It is worth pointing out that we treat x as a parameter in forming the equation (6.2.4) and its initial values. To verify (6.2.4), we first write

$$U(x;r,t) = \frac{1}{\omega_n} \int_{|\omega|=1} u(x + r\omega, t)\, dS_\omega.$$

By differentiating under the integral sign and then integrating by parts, we have

$$\begin{aligned} U_r &= \frac{1}{\omega_n} \int_{|\omega|=1} \frac{\partial u}{\partial \nu}(x + r\omega, t)\, dS_\omega = \frac{1}{\omega_n r^{n-1}} \int_{\partial B_r(x)} \frac{\partial u}{\partial \nu}(y,t)\, dS_y \\ &= \frac{1}{\omega_n r^{n-1}} \int_{B_r(x)} \Delta u(y,t)\, dy. \end{aligned}$$

Then by the equation in (6.2.1),

$$r^{n-1} U_r = \frac{1}{\omega_n} \int_{B_r(x)} \Delta u(y,t)\, dy = \frac{1}{\omega_n} \int_{B_r(x)} u_{tt}(y,t)\, dy.$$

Hence

$$\begin{aligned}(r^{n-1}U_r)_r &= \frac{1}{\omega_n}\int_{\partial B_r(x)} u_{tt}(y,t)\,dS_y\\ &= \frac{1}{\omega_n}\partial_{tt}\int_{\partial B_r(x)} u(y,t)\,dS_y = r^{n-1}U_{tt}.\end{aligned}$$

For the initial values, we simply have for any $r>0$,

$$\begin{aligned}U(x;r,0) &= \frac{1}{\omega_n r^{n-1}}\int_{\partial B_r(x)} \varphi(y)\,dS_y,\\ U_t(x;r,0) &= \frac{1}{\omega_n r^{n-1}}\int_{\partial B_r(x)} \psi(y)\,dS_y.\end{aligned}$$

6.2.2. Dimension Three. We note that the Euler-Poisson-Darboux equation is a one-dimensional hyperbolic equation. In general, it is a tedious process to solve the corresponding initial-value problems for general n. However, this process is relatively easy for $n=3$. If $n=3$, we have

$$U_{tt} = U_{rr} + \frac{2}{r}U_r.$$

Hence for $r>0$ and $t>0$,

$$(rU)_{tt} = (rU)_{rr}.$$

We note that rU satisfies the *one-dimensional* wave equation. Set

$$\tilde U(x;r,t) = rU(x;r,t)$$

and

$$\tilde\Phi(x;r) = r\Phi(x;r), \quad \tilde\Psi(x;r) = r\Psi(x;r).$$

Then for each fixed $x\in\mathbb{R}^3$,

$$\begin{aligned}&\tilde U_{tt} = \tilde U_{rr} \quad \text{for } r>0 \text{ and } t>0,\\ &\tilde U(x;r,0) = \tilde\Phi(x;r),\ \tilde U_t(x;r,0) = \tilde\Psi(x;r) \quad \text{for } r>0,\\ &\tilde U(x;0,t) = 0 \quad \text{for } t>0.\end{aligned}$$

This is a half-space problem for $\tilde U$ studied in Section 6.1. By (6.1.11), we obtain formally for any $t\ge r>0$,

$$\tilde U(x;r,t) = \frac12\big(\tilde\Phi(x;r+t) - \tilde\Phi(x;t-r)\big) + \frac12\int_{t-r}^{r+t}\tilde\Psi(x;s)\,ds.$$

Hence,

$$\begin{aligned}U(x;r,t) = &\frac{1}{2r}\big((t+r)\Phi(x;t+r) - (t-r)\Phi(x;t-r)\big)\\ &+ \frac{1}{2r}\int_{t-r}^{t+r} s\Psi(x;s)\,ds.\end{aligned}$$

Letting $r \to 0$, we obtain

$$u(x,t) = \lim_{r\to 0} U(x;r,t) = \partial_t\big(t\Phi(x;t)\big) + t\Psi(x;t).$$

Note that the area of the unit sphere in $\mathbb{R}^3$ is 4π. Then

$$\begin{aligned}\Phi(x;t) &= \frac{1}{4\pi t^2}\int_{\partial B_t(x)} \varphi(y)\,dS_y,\\ \Psi(x;t) &= \frac{1}{4\pi t^2}\int_{\partial B_t(x)} \psi(y)\,dS_y.\end{aligned}$$

Therefore, we obtain formally the following expression of a solution u of (6.2.1):

$$u(x,t) = \partial_t\left(\frac{1}{4\pi t}\int_{\partial B_t(x)} \varphi(y)\,dS_y\right) + \frac{1}{4\pi t}\int_{\partial B_t(x)} \psi(y)\,dS_y, \tag{6.2.5}$$

for any $(x,t) \in \mathbb{R}^3 \times (0,\infty)$. We point out that we did not justify the compatibility condition in applying (6.1.11). Next, we prove directly that (6.2.5) is indeed a solution u of (6.2.1) under appropriate assumptions on φ and ψ.

Theorem 6.2.1. *Let $k \geq 2$ be an integer, $\varphi \in C^{k+1}(\mathbb{R}^3)$ and $\psi \in C^k(\mathbb{R}^3)$. Suppose u is defined by (6.2.5) in $\mathbb{R}^3 \times (0,\infty)$. Then $u \in C^k(\mathbb{R}^3 \times (0,\infty))$ and*

$$u_{tt} - \Delta u = 0 \quad \text{in } \mathbb{R}^3 \times (0,\infty).$$

Moreover, for any $x_0 \in \mathbb{R}^3$,

$$\lim_{(x,t)\to(x_0,0)} u(x,t) = \varphi(x_0), \qquad \lim_{(x,t)\to(x_0,0)} u_t(x,t) = \psi(x_0).$$

In fact, u can be extended to a C^k-function in $\mathbb{R}^3 \times [0,\infty)$. This can be easily seen from the proof below.

Proof. We will consider $\varphi = 0$. By (6.2.5), we have

$$u(x,t) = t\Psi(x,t),$$

where

$$\Psi(x,t) = \frac{1}{4\pi t^2}\int_{\partial B_t(x)} \psi(y)\,dS_y.$$

By the change of coordinates $y = x + \omega t$, we write

$$\Psi(x,t) = \frac{1}{4\pi}\int_{|\omega|=1} \psi(x + t\omega)\,dS_\omega.$$

In this form, $u(x,t)$ is defined for any $(x,t) \in \mathbb{R}^3 \times [0,\infty)$ and $u(\cdot,0) = 0$. Since $\psi \in C^k(\mathbb{R}^3)$, we conclude easily that $\nabla_x^i u$ exists and is continuous in $\mathbb{R}^3 \times [0,\infty)$, for $i = 0, 1, \cdots, k$. In particular,

$$\Delta u(x,t) = \frac{t}{4\pi} \int_{|\omega|=1} \Delta_x \psi(x + t\omega)\, dS_\omega.$$

For t-derivatives, we take $(x,t) \in \mathbb{R}^3 \times (0,\infty)$. Then

$$u_t = \Psi + t\Psi_t, \quad u_{tt} = 2\Psi_t + t\Psi_{tt}.$$

A simple differentiation yields

$$\Psi_t(x,t) = \frac{1}{4\pi} \int_{|\omega|=1} \frac{\partial \psi}{\partial \nu}(x + t\omega)\, dS_\omega.$$

Hence, $u_t(x,t)$ is defined for any $(x,t) \in \mathbb{R}^3 \times [0,\infty)$ and $u_t(\cdot,0) = \psi$. Moreover, $\nabla_x^i u_t$ is continuous in $\mathbb{R}^3 \times (0,\infty)$, for $i = 0, 1, \cdots, k-1$. After the change of coordinates $y = x + \omega t$ and an integration by parts, we first have

$$\Psi_t = \frac{1}{4\pi t^2} \int_{\partial B_t(x)} \frac{\partial \psi}{\partial \nu}(y)\, dS_y = \frac{1}{4\pi t^2} \int_{B_t(x)} \Delta \psi(y)\, dy.$$

Then

$$\begin{aligned}\Psi_{tt} &= -\frac{1}{2\pi t^3} \int_{B_t(x)} \Delta \psi(y)\, dy + \frac{1}{4\pi t^2} \int_{\partial B_t(x)} \Delta \psi(y)\, dS_y \\ &= -\frac{2}{t} \Psi_t(t) + \frac{1}{4\pi t^2} \int_{\partial B_t(x)} \Delta \psi(y)\, dS_y.\end{aligned}$$

By setting $y = x + \omega t$ again, we have

$$u_{tt} = \frac{t}{4\pi t^2} \int_{\partial B_t(x)} \Delta_y \psi(y)\, dS_y = \frac{t}{4\pi} \int_{|\omega|=1} \Delta_x \psi(x + t\omega)\, dS_\omega = \Delta u.$$

This implies easily that $u \in C^k(\mathbb{R}^3 \times [0,\infty))$.

A similar calculation works for $\psi = 0$. □

We point out that there are other methods to derive explicit expressions for solutions of the wave equation. Refer to Exercise 6.8 for an alternative approach to solving the three-dimensional wave equation.

By the change of variables $y = x + t\omega$ in (6.2.5), we have

$$u(x,t) = \partial_t \left(\frac{t}{4\pi} \int_{|\omega|=1} \varphi(x + t\omega)\, dS_\omega \right) + \frac{t}{4\pi} \int_{|\omega|=1} \psi(x + t\omega)\, dS_y.$$

A simple differentiation under the integral sign yields

$$u(x,t) = \frac{1}{4\pi} \int_{|\omega|=1} \big(\varphi(x + t\omega) + t\nabla \varphi(x + t\omega) \cdot \omega + t\psi(x + t\omega) \big)\, dS_\omega.$$

Hence

$$u(x,t) = \frac{1}{4\pi t^2} \int_{\partial B_t(x)} \big(\varphi(y) + \nabla_y \varphi(y) \cdot (y-x) + t\psi(y)\big)\, dS_y,$$

for any $(x,t) \in \mathbb{R}^3 \times (0,\infty)$. We note that $u(x,t)$ depends only on the initial values φ and ψ on the sphere $\partial B_t(x)$.

6.2.3. Dimension Two. We now solve initial-value problems for the wave equation in $\mathbb{R}^2 \times (0,\infty)$ by the *method of descent*. Let $\varphi \in C^2(\mathbb{R}^2)$ and $\psi \in C^1(\mathbb{R}^2)$. Suppose $u \in C^2(\mathbb{R}^2 \times (0,\infty)) \cap C^1(\mathbb{R}^2 \times [0,\infty))$ satisfies (6.2.1), i.e.,

$$\begin{aligned} u_{tt} - \Delta u &= 0 \quad \text{in } \mathbb{R}^2 \times (0,\infty), \\ u(\cdot,0) &= \varphi,\ u_t(\cdot,0) = \psi \quad \text{on } \mathbb{R}^2. \end{aligned}$$

Any solutions in $\mathbb{R}^2$ can be viewed as solutions of the same problem in $\mathbb{R}^3$, which are independent of the third space variable. Namely, by setting $\bar{x} = (x, x_3)$ for $x = (x_1, x_2) \in \mathbb{R}^2$ and

$$\bar{u}(\bar{x}, t) = u(x,t),$$

we have

$$\begin{aligned} \bar{u}_{tt} - \Delta_{\bar{x}} \bar{u} &= 0 \quad \text{in } \mathbb{R}^3 \times (0,\infty), \\ \bar{u}(\cdot,0) &= \bar{\varphi},\ \bar{u}_t(\cdot,0) = \bar{\psi} \quad \text{on } \mathbb{R}^3, \end{aligned}$$

where

$$\bar{\varphi}(\bar{x}) = \varphi(x), \quad \bar{\psi}(\bar{x}) = \psi(x).$$

By (6.2.5), we have

$$\bar{u}(\bar{x},t) = \partial_t \left(\frac{1}{4\pi t} \int_{\partial B_t(\bar{x})} \bar{\varphi}(\bar{y})\, dS_{\bar{y}} \right) + \frac{1}{4\pi t} \int_{\partial B_t(\bar{x})} \bar{\psi}(\bar{y})\, dS_{\bar{y}},$$

where $\bar{y} = (y_1, y_2, y_3) = (y, y_3)$. The integrals here are over the surface $\partial B_t(\bar{x})$ in $\mathbb{R}^3$. Now we evaluate them as integrals in $\mathbb{R}^2$ by eliminating y_3. For $x_3 = 0$, the sphere $|\bar{y} - \bar{x}| = t$ in $\mathbb{R}^3$ has two pieces given by

$$y_3 = \pm\sqrt{t^2 - |y-x|^2},$$

and its surface area element is

$$dS_{\bar{y}} = \big(1 + (\partial_{y_1} y_3)^2 + (\partial_{y_2} y_3)^2\big)^{\frac{1}{2}}\, dy_1 dy_2 = \frac{t}{\sqrt{t^2 - |y-x|^2}}\, dy.$$

Therefore, we obtain

$$
\begin{aligned}
u(x,t) &= \frac{1}{2}\partial_t\left(\frac{1}{\pi}\int_{B_t(x)}\frac{\varphi(y)}{\sqrt{t^2-|y-x|^2}}\,dy\right) \\
&\quad + \frac{1}{2}\cdot\frac{1}{\pi}\int_{B_t(x)}\frac{\psi(y)}{\sqrt{t^2-|y-x|^2}}\,dy,
\end{aligned}
\tag{6.2.6}
$$

for any $(x,t) \in \mathbb{R}^2 \times (0,\infty)$. We put the factor $1/2$ separately to emphasize that π is the area of the unit disc in $\mathbb{R}^2$.

Theorem 6.2.2. *Let $k \geq 2$ be an integer, $\varphi \in C^{k+1}(\mathbb{R}^2)$ and $\psi \in C^k(\mathbb{R}^2)$. Suppose u is defined by (6.2.6) in $\mathbb{R}^2 \times (0,\infty)$. Then $u \in C^k(\mathbb{R}^2 \times (0,\infty))$ and*

$$u_{tt} - \Delta u = 0 \quad \text{in } \mathbb{R}^2 \times (0,\infty).$$

Moreover, for any $x_0 \in \mathbb{R}^2$,

$$\lim_{(x,t)\to(x_0,0)} u(x,t) = \varphi(x_0), \qquad \lim_{(x,t)\to(x_0,0)} u_t(x,t) = \psi(x_0).$$

This follows from Theorem 6.2.1. Again, u can be extended to a C^k-function in $\mathbb{R}^2 \times [0,\infty)$.

By the change of variables $y = x + tz$ in (6.2.6), we have

$$u(x,t) = \partial_t\left(\frac{t}{2\pi}\int_{B_1}\frac{\varphi(x+tz)}{\sqrt{1-|z|^2}}\,dz\right) + \frac{t}{2\pi}\int_{B_1}\frac{\psi(x+tz)}{\sqrt{1-|z|^2}}\,dz.$$

A simple differentiation under the integral sign yields

$$u(x,t) = \frac{1}{2\pi}\int_{B_1}\frac{\varphi(x+tz) + t\nabla\varphi(x+tz)\cdot z + t\psi(x+tz)}{\sqrt{1-|z|^2}}\,dz.$$

Hence

$$u(x,t) = \frac{1}{2}\cdot\frac{1}{\pi t^2}\int_{B_t(x)}\frac{t\varphi(y) + t\nabla\varphi(y)\cdot(y-x) + t^2\psi(y)}{\sqrt{t^2-|y-x|^2}}\,dy,$$

for any $(x,t) \in \mathbb{R}^2 \times (0,\infty)$. We note that $u(x,t)$ depends on the initial values φ and ψ in the solid disc $B_t(x)$.

6.2.4. Properties of Solutions. Now we compare several formulas we obtained so far. Let u be a C^2-solution of the initial-value problem (6.2.1).

We write u_n for dimension n. Then for any $(x,t)\in\mathbb{R}^n\times(0,\infty)$,

$$u_1(x,t)=\frac{1}{2}\big(\varphi(x+t)+\varphi(x-t)\big)+\frac{1}{2}\int_{x-t}^{x+t}\psi(y)\,dy,$$

$$u_2(x,t)=\frac{1}{2\pi t^2}\int_{B_t(x)}\frac{t\varphi(y)+t\nabla\varphi(y)\cdot(y-x)+t^2\psi(y)}{\sqrt{t^2-|y-x|^2}}\,dy,$$

$$u_3(x,t)=\frac{1}{4\pi t^2}\int_{\partial B_t(x)}\big(\varphi(y)+\nabla_y\varphi(y)\cdot(y-x)+t\psi(y)\big)\,dS_y.$$

These formulas display many important properties of solutions u.

According to these expressions, the value of u at (x,t) depends on the values of φ and ψ on the interval $[x-t,x+t]$ for $n=1$ (in fact, on φ only at two endpoints), on the solid disc $B_t(x)$ of center x and radius t for $n=2$, and on the sphere $\partial B_t(x)$ of center x and radius t for $n=3$. These regions are the *domains of dependence* of solutions at (x,t) on initial values. Conversely,

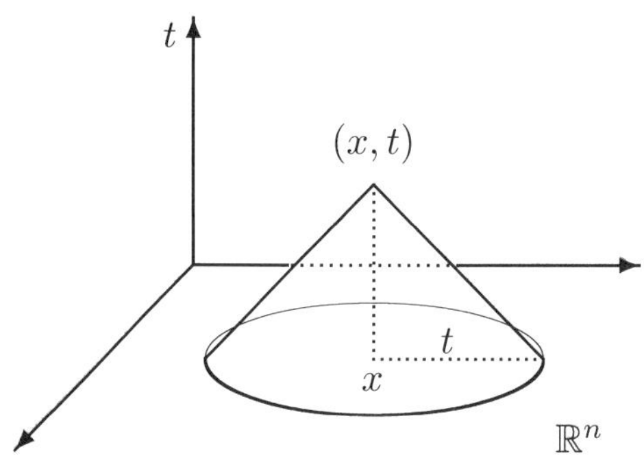

Figure 6.2.1. The domain of dependence.

the initial values φ and ψ at a point x_0 on the initial hypersurface $t=0$ *influence* u at the points (x,t) in the solid cone $|x-x_0|\le t$ for $n=2$ and only on the surface $|x-x_0|=t$ for $n=3$ at a later time t.

The central issue here is that the solution at a given point is determined by the initial values in a proper subset of the initial hypersurface. An important consequence is that the process of solving initial-value problems for the wave equation can be localized in space. Specifically, changing initial values outside the domain of dependence of a point does not change the values of solutions at this point. This is a unique property of the wave equation which distinguishes it from the heat equation.

Before exploring the difference between $n=2$ and $n=3$, we first note that it takes time (literally) for initial values to make influences. Suppose that the initial values φ,ψ have their support contained in a ball $B_r(x_0)$.

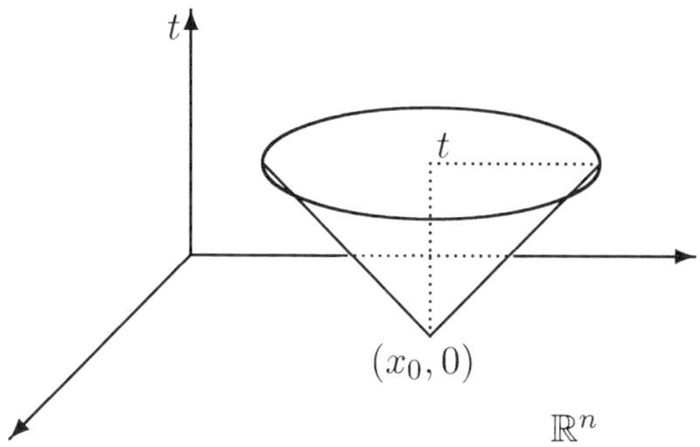

Figure 6.2.2. The range of influence.

Then at a later time t, the support of $u(\cdot, t)$ is contained in the union of all balls $B_t(\bar{x})$ for $\bar{x} \in B_r(x_0)$. It is easy to see that such a union is in fact the ball of center x_0 and radius $r+t$. The support of u spreads at a finite speed. To put it in another perspective, we fix an $x \notin B_r(x_0)$. Then $u(x,t) = 0$ for $t < |x - x_0| - r$. This is a *finite-speed propagation.*

For $n = 2$, if the supports of φ and ψ are the entire disc $B_r(x_0)$, then the support of $u(\cdot, t)$ will be the entire disc $B_{r+t}(x_0)$ in general. The influence from initial values never disappears in a finite time at any particular point, like the surface waves arising from a stone dropped into water.

For $n = 3$, the behavior of solutions is different. Again, we assume that the supports of φ and ψ are contained in a ball $B_r(x_0)$. Then at a later time t, the support of $u(\cdot, t)$ is in fact contained in the union of all spheres $\partial B_t(\bar{x})$ for $\bar{x} \in B_r(x_0)$. Such a union is the ball $B_{t+r}(x_0)$ for $t \leq r$, as in the two-dimensional case, and the annular region of center x_0 and outer and inner radii $t + r$ and $t - r$, respectively, for $t > r$. This annular region has a thickness $2r$ and spreads at a finite speed. In other words, $u(x,t)$ is not zero only if

$$t - r < |x - x_0| < t + r,$$

or

$$|x - x_0| - r < t < |x - x_0| + r.$$

Therefore, for a fixed $x \in \mathbb{R}^3$, $u(x,t) = 0$ for $t < |x - x_0| - r$ (corresponding to finite-speed propagation) and for $t > |x - x_0| + r$. So, the influence from the initial values lasts only for an interval of length $2r$ in time. This phenomenon is called *Huygens' principle* for the wave equation. (It is called the strong Huygens' principle in some literature.)

In fact, Huygens' principle holds for the wave equation in every odd space dimension n except $n = 1$ and does not hold in even space dimensions.

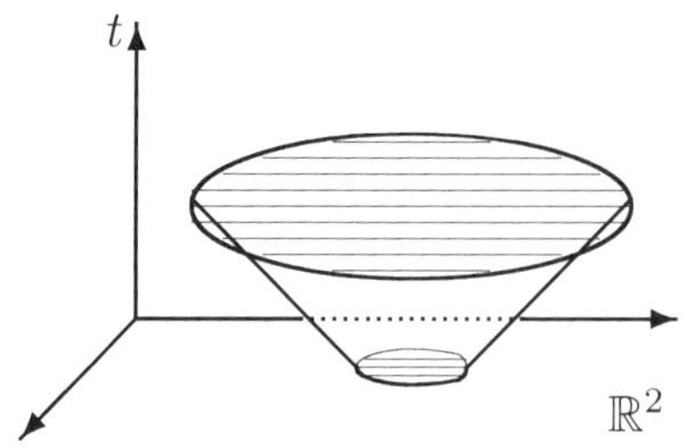

Figure 6.2.3. The range of influence for $n = 2$.

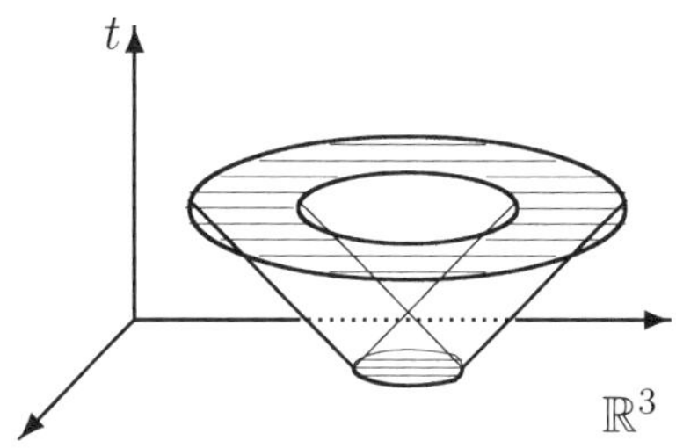

Figure 6.2.4. The range of influence for $n = 3$.

Now we compare regularity of solutions for $n = 1$ and $n = 3$. For $n = 1$, the regularity of u is clearly the same as $u(\cdot, 0)$ and one order better than $u_t(\cdot, 0)$. In other words, $u \in C^m$ and $u_t \in C^{m-1}$ initially at $t = 0$ guarantee $u \in C^m$ at a later time. However, such a result does not hold for $n = 3$. The formula for $n = 3$ indicates that u can be less regular than the initial values. There is a possible loss of one order of differentiability. Namely, $u \in C^{k+1}$ and $u_t \in C^k$ initially at $t = 0$ only guarantee $u \in C^k$ at a later time.

Example 6.2.3. We consider an initial-value problem for the wave equation in $\mathbb{R}^3$ of the form

$$\begin{aligned} u_{tt} - \Delta u = 0 \quad &\text{in } \mathbb{R}^3 \times (0, \infty), \\ u(\cdot, 0) = 0, \ u_t(\cdot, 0) = \psi \quad &\text{on } \mathbb{R}^3. \end{aligned}$$

Its solution is given by

$$u(x,t) = \frac{1}{4\pi t} \int_{\partial B_t(x)} \psi(y)\, dS_y,$$

for any $(x, t) \in \mathbb{R}^3 \times (0, \infty)$. We assume that ψ is radially symmetric, i.e., $\psi(x) = h(|x|)$ for some function h defined in $[0, \infty)$. Then

$$u(0,t) = \frac{1}{4\pi t} \int_{\partial B_t} \psi(y)\, dS_y = th(t).$$

For some integer $k \geq 3$, if $\psi(x)$ is not C^k at $|x| = 1$, then $h(t)$ is not C^k at $t = 1$. Therefore, the solution u is not C^k at $(x,t) = (0,1)$. The physical interpretation is that the singularity of initial values at $|x| = 1$ propagates along the characteristic cone and focuses at its vertex. We note that $(x,t) = (0,1)$ is the vertex of the characteristic cone $\{(x,t) : t = 1-|x|\}$ which intersects $\{t = 0\}$ at $|x| = 1$.

This example demonstrates that solutions of the higher-dimensional wave equation do not have good pointwise behavior. A loss of differentiability in the pointwise sense occurs. However, the differentiability is preserved in the L^2-sense. We will discuss the related energy estimates in the next section.

6.2.5. Arbitrary Odd Dimensions. Next, we discuss how to obtain explicit expressions for solutions of initial-value problems for the wave equation in an arbitrary dimension. For odd dimensions, we seek an appropriate combination of $U(x;r,t)$ and its derivatives to satisfy the one-dimensional wave equation and then proceed as for $n = 3$. For even dimensions, we again use the method of descent.

Let $n \geq 3$ be an odd integer. The spherical average $U(x;r,t)$ defined by (6.2.2) satisfies

$$U_{tt} = U_{rr} + \frac{n-1}{r}U_r, \tag{6.2.7}$$

for any $r > 0$ and $t > 0$. First, we write (6.2.7) as

$$U_{tt} = \frac{1}{r}\big(rU_{rr} + (n-1)U_r\big).$$

Since

$$(rU)_{rr} = rU_{rr} + 2U_r,$$

we obtain

$$U_{tt} = \frac{1}{r}\big((rU)_{rr} + (n-3)U_r\big),$$

or

$$(rU)_{tt} = (rU)_{rr} + (n-3)U_r. \tag{6.2.8}$$

If $n = 3$, then rU satisfies the one-dimensional wave equation. This is how we solved the initial-value problem for the wave equation in dimension three. By differentiating (6.2.7) with respect to r, we have

$$\begin{aligned} U_{rtt} &= U_{rrr} + \frac{n-1}{r}U_{rr} - \frac{n-1}{r^2}U_r \\ &= \frac{1}{r^2}\big(r^2U_{rrr} + (n-1)rU_{rr} - (n-1)U_r\big). \end{aligned}$$

Since

$$(r^2U_r)_{rr} = r^2U_{rrr} + 4rU_{rr} + 2U_r,$$

we obtain

$$U_{rtt} = \frac{1}{r^2}\big((r^2U)_{rr} + (n-5)rU_{rr} - (n+1)U_r\big),$$

or

$$(r^2U_r)_{tt} = (r^2U_r)_{rr} + (n-5)rU_{rr} - (n+1)U_r. \tag{6.2.9}$$

The second term in the right-hand side of (6.2.9) has a coefficient $n-5$, which is 2 less than $n-3$, the coefficient of the second term in the right-hand side of (6.2.8). Also the third term involving U_r in the right-hand side of (6.2.9) has a similar expression as the second term in the right-hand side of (6.2.8). Therefore an appropriate combination of (6.2.8) and (6.2.9) eliminates those terms involving U_r. In particular, for $n=5$, we have

$$(r^2U_r + 3rU)_{tt} = (r^2U_r + 3rU)_{rr}.$$

In other words, $r^2U_r + 3rU$ satisfies the one-dimensional wave equation. We can continue this process to obtain appropriate combinations for all odd dimensions. Next, we note that

$$r^2U_r + 3rU = \frac{1}{r}(r^3U)_r.$$

It turns out that the correct combination of U and its derivatives for arbitrary odd dimension n is given by

$$\left(\frac{1}{r}\frac{\partial}{\partial r}\right)^{\frac{n-3}{2}} (r^{n-2}U).$$

We first state a simple calculus lemma.

Lemma 6.2.4. *Let m be a positive integer and $v = v(r)$ be a C^{m+1}-function on $(0,\infty)$. Then for any $r>0$,*

(1) $$\left(\frac{d^2}{dr^2}\right)\left(\frac{1}{r}\frac{d}{dr}\right)^{m-1}\left(r^{2m-1}v(r)\right) = \left(\frac{1}{r}\frac{d}{dr}\right)^{m}\left(r^{2m}\frac{dv}{dr}(r)\right);$$

(2) $$\left(\frac{1}{r}\frac{d}{dr}\right)^{m-1}\left(r^{2m-1}v(r)\right) = \sum_{i=0}^{m-1} c_{m,i}r^{i+1}\frac{d^iv}{dr^i}(r),$$

where $c_{m,i}$ is a constant independent of v, for $i = 0, 1, \cdots, m-1$, and

$$c_{m,0} = 1\cdot 3\cdots(2m-1).$$

The proof is by induction and is omitted.

Now we let $n \geq 3$ be an odd integer and write $n = 2m+1$. Let $\varphi \in C^m(\mathbb{R}^n)$ and $\psi \in C^{m-1}(\mathbb{R}^n)$. We assume that $u \in C^{m+1}(\mathbb{R}^n \times [0,\infty))$ is a solution of the initial-value problem (6.2.1). Then U defined by (6.2.2) is

C^{m+1}, and Φ and Ψ defined by (6.2.3) are C^m and C^{m-1}, respectively. For $x \in \mathbb{R}^n$, $r > 0$ and $t \geq 0$, set

$$\tilde U(x;r,t) = \left(\frac{1}{r}\frac{\partial}{\partial r}\right)^{m-1} \big(r^{2m-1}U(x;r,t)\big), \tag{6.2.10}$$

and

$$\tilde \Phi(x;r) = \left(\frac{1}{r}\frac{\partial}{\partial r}\right)^{m-1} \big(r^{2m-1}\Phi(x;r)\big),$$
$$\tilde \Psi(x;r) = \left(\frac{1}{r}\frac{\partial}{\partial r}\right)^{m-1} \big(r^{2m-1}\Psi(x;r)\big).$$

We now claim that for each fixed $x \in \mathbb{R}^n$,

$$\begin{aligned} &\tilde U_{tt} - \tilde U_{rr} = 0 \quad \text{in } (0,\infty)\times(0,\infty),\\ &\tilde U(x;r,0) = \tilde\Phi(x;r),\ \tilde U_t(x;r,0) = \tilde\Psi(x;r) \quad \text{for } r > 0,\\ &\tilde U(x;0,t) = 0 \quad \text{for } t > 0. \end{aligned}$$

This follows by a straightforward calculation. First, in view of (6.2.4) and $n = 2m+1$, we have for any $r > 0$ and $t > 0$,

$$\begin{aligned} \frac{1}{r}\frac{\partial}{\partial r}(r^{2m}U_r) &= r^{2m-1}U_{rr} + 2mr^{2m-2}U_r \\ &= r^{2m-1}\big(U_{rr} + \frac{n-1}{r}U_r\big) = r^{2m-1}U_{tt}. \end{aligned}$$

Then by (6.2.10) and Lemma 6.2.4(1), we have

$$\begin{aligned} \tilde U_{rr} &= \left(\frac{\partial^2}{\partial r^2}\right)\left(\frac{1}{r}\frac{\partial}{\partial r}\right)^{m-1}\big(r^{2m-1}U\big) = \left(\frac{1}{r}\frac{\partial}{\partial r}\right)^{m}\big(r^{2m}U_r\big) \\ &= \left(\frac{1}{r}\frac{\partial}{\partial r}\right)^{m-1}\big(r^{2m-1}U_{tt}\big) = \tilde U_{tt}. \end{aligned}$$

The initial condition easily follows from the definition of $\tilde U$, $\tilde\Phi$ and $\tilde\Psi$. The boundary condition $\tilde U(x;0,t) = 0$ follows from Lemma 6.2.4(2).

As for $n = 3$, we have for any $t \geq r > 0$,

$$\tilde U(x;r,t) = \frac{1}{2}\big(\tilde\Phi(x;t+r) - \tilde\Phi(x;t-r)\big) + \frac{1}{2}\int_{t-r}^{t+r}\tilde\Psi(x;s)\,ds.$$

Note that by Lemma 6.2.4(2),

$$\begin{aligned} \tilde U(x;r,t) &= \left(\frac{1}{r}\frac{\partial}{\partial r}\right)^{m-1}\big(r^{2m-1}U(x;r,t)\big) \\ &= \sum_{i=0}^{m-1} c_{m,i}r^{i+1}\frac{\partial^i}{\partial r^i}U(x;r,t). \end{aligned}$$

Hence

$$\lim_{r\to 0}\frac{1}{c_{m,0}r}\tilde{U}(x;r,t)=\lim_{r\to 0}U(x;r,t)=u(x,t).$$

Therefore, we obtain

$$\begin{aligned}u(x,t)&=\frac{1}{c_{m,0}}\lim_{r\to 0}\left(\frac{1}{2r}\big(\tilde{\Phi}(x;t+r)-\tilde{\Phi}(x;t-r)\big)+\frac{1}{2r}\int_{t-r}^{r+t}\tilde{\Psi}(x;s)\,ds\right)\\&=\frac{1}{c_{m,0}}\big(\tilde{\Phi}_t(x;t)+\tilde{\Psi}(x;t)\big).\end{aligned}$$

Using $n=2m+1$, the expression for $c_{m,0}$ in Lemma 6.2.4 and the definitions of $\tilde{\Phi}$ and $\tilde{\Psi}$, we can rewrite the last formula in terms of φ and ψ. Thus, we obtain for any $x\in\mathbb{R}^n$ and $t>0$,

$$\begin{aligned}u(x,t)=\frac{1}{c_n}\Bigg[&\left(\frac{\partial}{\partial t}\right)\left(\frac{1}{t}\frac{\partial}{\partial t}\right)^{\frac{n-3}{2}}\left(\frac{1}{\omega_n t}\int_{\partial B_t(x)}\varphi\,dS\right)\\&+\left(\frac{1}{t}\frac{\partial}{\partial t}\right)^{\frac{n-3}{2}}\left(\frac{1}{\omega_n t}\int_{\partial B_t(x)}\psi\,dS\right)\Bigg],\end{aligned}\tag{6.2.11}$$

where n is an odd integer, ω_n is the surface area of the unit sphere in $\mathbb{R}^n$ and

$$c_n=1\cdot 3\cdots(n-2).\tag{6.2.12}$$

We note that $c_3=1$ and hence (6.2.11) reduces to (6.2.5) for $n=3$.

Now we check that u given by (6.2.11) indeed solves the initial-value problem (6.2.1).

Theorem 6.2.5. *Let $n\geq 3$ be an odd integer and $k\geq 2$ be an integer. Suppose $\varphi\in C^{\frac{n-1}{2}+k}(\mathbb{R}^n)$, $\psi\in C^{\frac{n-3}{2}+k}(\mathbb{R}^n)$ and u is defined in (6.2.11). Then $u\in C^k(\mathbb{R}^n\times(0,\infty))$ and*

$$u_{tt}-\Delta u=0\quad\text{in }\mathbb{R}^n\times(0,\infty).$$

Moreover, for any $x_0\in\mathbb{R}^n$,

$$\lim_{(x,t)\to(x_0,0)}u(x,t)=\varphi(x_0),\qquad\lim_{(x,t)\to(x_0,0)}u_t(x,t)=\psi(x_0).$$

In fact, u can be extended to a C^k-function in $\mathbb{R}^n\times[0,\infty)$.

Proof. The proof proceeds similarly to that of Theorem 6.2.1. We consider $\varphi=0$. Then for any $(x,t)\in\mathbb{R}^n\times(0,\infty)$,

$$u(x,t)=\frac{1}{c_n}\left(\frac{1}{t}\frac{\partial}{\partial t}\right)^{\frac{n-3}{2}}\big(t^{n-2}\Psi(x,t)\big),$$

where

$$\Psi(x,t) = \frac{1}{\omega_n t^{n-1}} \int_{\partial B_t(x)} \psi \, dS.$$

By Lemma 6.2.4(2), we have

$$u(x,t) = \frac{1}{c_n} \sum_{i=0}^{\frac{n-3}{2}} c_{\frac{n-1}{2},i} t^{i+1} \frac{\partial^i}{\partial t^i} \Psi(x,t).$$

Note that c_n in (6.2.12) is $c_{(n-1)/2,0}$ in Lemma 6.2.4. By the change of coordinates $y = x + \omega t$, we write

$$\Psi(x,t) = \frac{1}{\omega_n} \int_{|\omega|=1} \psi(x + t\omega) \, dS_\omega.$$

Therefore,

$$\frac{\partial^i}{\partial t^i} \Psi(x,t) = \frac{1}{\omega_n} \int_{|\omega|=1} \frac{\partial^i}{\partial \nu^i} \psi(x + t\omega) \, dS_\omega.$$

Hence, $u(x,t)$ is defined for any $(x,t) \in \mathbb{R}^n \times [0,\infty)$ and $u(\cdot,0) = 0$. Since $\psi \in C^{\frac{n-3}{2}+k}(\mathbb{R}^n)$, we conclude easily that $\nabla_x^i u$ exists and is continuous in $\mathbb{R}^n \times [0,\infty)$, for $i = 0, 1, \cdots, k$. For t-derivatives, we conclude similarly that $u_t(x,t)$ is defined for any $(x,t) \in \mathbb{R}^n \times [0,\infty)$ and $u_t(\cdot,0) = \psi$. Moreover, $\nabla_x^i u_t$ is continuous in $\mathbb{R}^n \times (0,\infty)$, for $i = 0, 1, \cdots, k-1$. In particular,

$$\Psi_t(x,t) = \frac{1}{\omega_n t^{n-1}} \int_{\partial B_t(x)} \frac{\partial \psi}{\partial \nu} \, dS = \frac{1}{\omega_n t^{n-1}} \int_{B_t(x)} \Delta \psi \, dy.$$

Next,

$$\begin{aligned} \Delta \Psi(x,t) &= \frac{1}{\omega_n} \int_{|\omega|=1} \Delta_x \psi(x + t\omega) \, dS_\omega \\ &= \frac{1}{\omega_n t^{n-1}} \int_{\partial B_t(x)} \Delta \psi \, dS_y. \end{aligned}$$

Hence

$$\Delta u(x,t) = \frac{1}{c_n} \left(\frac{1}{t} \frac{\partial}{\partial t} \right)^{\frac{n-3}{2}} \left(\frac{1}{\omega_n t} \int_{\partial B_t(x)} \Delta \psi \, dS_y \right).$$

On the other hand, Lemma 6.2.4(1) implies

$$u_{tt} = \frac{1}{c_n} \left(\frac{1}{t} \frac{\partial}{\partial t} \right)^{\frac{n-1}{2}} \left(t^{n-1} \Psi_t \right).$$

Hence

$$u_{tt} = \frac{1}{\omega_n c_n}\left(\frac{1}{t}\frac{\partial}{\partial t}\right)^{\frac{n-1}{2}}\left(\int_{B_t(x)} \Delta\psi\, dy\right)$$
$$= \frac{1}{\omega_n c_n}\left(\frac{1}{t}\frac{\partial}{\partial t}\right)^{\frac{n-3}{2}}\left(\frac{1}{t}\int_{\partial B_t(x)} \Delta\psi\, dS\right).$$

This implies that $u_{tt} - \Delta u = 0$ at $(x,t) \in \mathbb{R}^n \times (0,\infty)$ and then $u \in C^k(\mathbb{R}^n \times [0,\infty))$.

We can discuss the case $\psi = 0$ in a similar way. □

6.2.6. Arbitrary Even Dimensions. Let $n \geq 2$ be an even integer with $n = 2m - 2$, $\varphi \in C^m(\mathbb{R}^n)$ and $\psi \in C^{m-1}(\mathbb{R}^n)$. We assume that $u \in C^m(\mathbb{R}^n \times [0,\infty))$ is a solution of the initial-value problem (6.2.1). We will use the *method of descent* to find an explicit expression for u in terms of φ and ψ.

By setting $\bar{x} = (x, x_{n+1})$ for $x = (x_1, \cdots, x_n) \in \mathbb{R}^n$ and

$$\bar{u}(\bar{x}, t) = u(x,t),$$

we have

$$\bar{u}_{tt} - \Delta_{\bar{x}}\bar{u} = 0 \quad \text{in } \mathbb{R}^{n+1} \times (0,\infty),$$
$$\bar{u}(\cdot, 0) = \bar{\varphi},\ \bar{u}_t(\cdot,0) = \bar{\psi} \quad \text{on } \mathbb{R}^{n+1},$$

where

$$\bar{\varphi}(\bar{x}) = \varphi(x), \quad \bar{\psi}(\bar{x}) = \psi(x).$$

As $n+1$ is odd, by (6.2.11), with $n+1$ replacing n, we have

$$\bar{u}(\bar{x},t) = \frac{1}{c_{n+1}}\left[\left(\frac{\partial}{\partial t}\right)\left(\frac{1}{t}\frac{\partial}{\partial t}\right)^{\frac{n-2}{2}}\left(\frac{1}{\omega_{n+1} t}\int_{\partial B_t(\bar{x})} \bar{\varphi}(\bar{y})\, dS_{\bar{y}}\right)\right.$$
$$\left. + \left(\frac{1}{t}\frac{\partial}{\partial t}\right)^{\frac{n-2}{2}}\left(\frac{1}{\omega_{n+1} t}\int_{\partial B_t(\bar{x})} \bar{\psi}(\bar{y})\, dS_{\bar{y}}\right)\right],$$

where $\bar{y} = (y_1, \cdots, y_n, y_{n+1}) = (y, y_{n+1})$. The integrals here are over the surface $\partial B_t(\bar{x})$ in $\mathbb{R}^{n+1}$. Now we evaluate them as integrals in $\mathbb{R}^n$ by eliminating y_{n+1}. For $x_{n+1} = 0$, the sphere $|\bar{y} - \bar{x}| = t$ in $\mathbb{R}^{n+1}$ has two pieces given by

$$y_{n+1} = \pm\sqrt{t^2 - |y-x|^2},$$

and its surface area element is

$$dS_{\bar{y}} = \left(1 + |\nabla_y y_{n+1}|^2\right)^{\frac{1}{2}} dy = \frac{t}{\sqrt{t^2 - |y-x|^2}}\, dy.$$

Hence

$$\begin{aligned}\frac{1}{\omega_{n+1}t}\int_{\partial B_t(\bar{x})}\bar{\varphi}(\bar{y})\,dS_{\bar{y}} &= \frac{2}{\omega_{n+1}}\int_{B_t(x)}\frac{\varphi(y)}{\sqrt{t^2-|y-x|^2}}\,dy\\ &= \frac{2\omega_n}{n\omega_{n+1}}\cdot\frac{n}{\omega_n}\int_{B_t(x)}\frac{\varphi(y)}{\sqrt{t^2-|y-x|^2}}\,dy.\end{aligned}$$

We point out that ω_n/n is the volume of the unit ball in $\mathbb{R}^n$. A similar expression holds for ψ. By a simple substitute, we now get an expression of u in terms of φ and ψ. We need to calculate the constant in the formula. Therefore, we obtain for any $x\in\mathbb{R}^n$ and $t>0$,

$$\begin{aligned}u(x,t) = \frac{1}{c_n}\Bigg[&\left(\frac{\partial}{\partial t}\right)\left(\frac{1}{t}\frac{\partial}{\partial t}\right)^{\frac{n-2}{2}}\left(\frac{n}{\omega_n}\int_{B_t(x)}\frac{\varphi(y)}{\sqrt{t^2-|y-x|^2}}\,dy\right)\\ &+\left(\frac{1}{t}\frac{\partial}{\partial t}\right)^{\frac{n-2}{2}}\left(\frac{n}{\omega_n}\int_{B_t(x)}\frac{\psi(y)}{\sqrt{t^2-|y-x|^2}}\,dy\right)\Bigg],\end{aligned}\tag{6.2.13}$$

where n is an even integer, ω_n/n is the volume of the unit ball in $\mathbb{R}^n$ and c_n is given by

$$c_n = \frac{nc_{n+1}\omega_{n+1}}{2\omega_n}.$$

In fact, we have

$$c_n = 2\cdot 4\cdots n.$$

We note that $c_2 = 2$ and hence (6.2.13) reduces to (6.2.6) for $n=2$.

Theorem 6.2.6. *Let n be an even integer and $k\geq 2$ be an integer. Suppose $\varphi\in C^{\frac{n}{2}+k}(\mathbb{R}^n)$, $\psi\in C^{\frac{n}{2}-1+k}(\mathbb{R}^n)$ and u is defined in (6.2.13). Then $u\in C^k(\mathbb{R}^n\times(0,\infty))$ and*

$$u_{tt}-\Delta u = 0 \quad \text{in } \mathbb{R}^n\times(0,\infty).$$

Moreover, for any $x_0\in\mathbb{R}^n$,

$$\lim_{(x,t)\to(x_0,0)} u(x,t)=\varphi(x_0),\qquad \lim_{(x,t)\to(x_0,0)} u_t(x,t)=\psi(x_0).$$

This follows from Theorem 6.2.5. Again, u can be extended to a C^k-function in $\mathbb{R}^n\times[0,\infty)$.

6.2.7. Global Properties.

Next, we discuss global properties of solutions of the initial-value problem for the wave equation. First, we have the following global boundedness.

Theorem 6.2.7. *For $n\geq 2$, let ψ be a smooth function in $\mathbb{R}^n$ and u be a solution of*

$$\begin{aligned}&u_{tt}-\Delta u = 0 \quad \text{in } \mathbb{R}^n\times(0,\infty),\\ &u(\cdot,0)=0,\ u_t(\cdot,0)=\psi \quad \text{on } \mathbb{R}^n.\end{aligned}$$

Then for any $t > 0$,

$$|u(\cdot,t)|_{L^\infty(\mathbb{R}^n)} \le C \sum_{i=0}^{n-1} \|\nabla^i \psi\|_{L^1(\mathbb{R}^n)},$$

where C *is a positive constant depending only on* n.

Solutions not only are bounded globally but also decay as $t \to \infty$ for $n \geq 2$. In this aspect, there is a sharp difference between dimension 1 and higher dimensions. By d'Alembert's formula (6.1.5), it is obvious that solutions of the initial-value problem for the one-dimensional wave equation do not decay as $t \to \infty$. However, solutions in higher dimensions have a different behavior.

Theorem 6.2.8. *For* $n \geq 2$, *let* ψ *be a smooth function in* $\mathbb{R}^n$ *and* u *be a solution of*

$$\begin{aligned} u_{tt} - \Delta u &= 0 \quad \text{in } \mathbb{R}^n \times (0,\infty), \\ u(\cdot,0) = 0,\ u_t(\cdot,0) &= \psi \quad \text{on } \mathbb{R}^n. \end{aligned}$$

Then for any $t > 1$,

$$|u(\cdot,t)|_{L^\infty(\mathbb{R}^n)} \le Ct^{-\frac{n-1}{2}} \sum_{i=0}^{\left[\frac{n}{2}\right]} \|\nabla^i \psi\|_{L^1(\mathbb{R}^n)},$$

where C *is a positive constant depending only on* n.

Decay estimates in Theorem 6.2.8 are optimal for large t. They play an important role in the studies of global solutions of nonlinear wave equations. We note that decay rates vary according to dimensions.

Before presenting a proof, we demonstrate that t^{-1} is the correct decay rate for $n = 3$ by a simple geometric consideration. By (6.2.5), the solution u is given by

$$u(x,t) = \frac{1}{4\pi t} \int_{\partial B_t(x)} \psi(y)\, dS_y,$$

for any $(x,t) \in \mathbb{R}^3 \times (0,\infty)$. Suppose ψ is of compact support and $\operatorname{supp}\psi \subset B_R$ for some $R > 0$. Then

$$u(x,t) = \frac{1}{4\pi t} \int_{B_R \cap \partial B_t(x)} \psi(y)\, dS_y.$$

A simple geometric argument shows that for any $x \in \mathbb{R}^3$ and any $t > 0$,

$$\operatorname{Area}\big(B_R \cap \partial B_t(x)\big) \le CR^2,$$

where C is a constant independent of x and t. Hence,

$$|u(x,t)| \le \frac{CR^2}{t} \sup_{\mathbb{R}^3} |\psi|.$$

This clearly shows that $u(x,t)$ decays uniformly for $x \in \mathbb{R}^3$ at the rate of t^{-1} as $t \to \infty$. The drawback here is that the diameter of the support appears explicitly in the estimate. The discussion for $n = 2$ is a bit complicated and is left as an exercise. Refer to Exercise 6.7.

We now prove Theorem 6.2.7 and Theorem 6.2.8 together. The proof is based on explicit expressions for u.

Proof of Theorems 6.2.7 and 6.2.8. We first consider $n = 3$. By assuming that ψ is of compact support, we prove that for any $t > 0$,

$$|u(x,t)| \le \frac{1}{4\pi}\|\nabla^2\psi\|_{L^1(\mathbb{R}^3)},$$

and for any $t > 0$,

$$|u(x,t)| \le \frac{1}{4\pi t}\|\nabla\psi\|_{L^1(\mathbb{R}^3)}.$$

By (6.2.5), the solution u is given by

$$u(x,t) = \frac{t}{4\pi}\int_{|\omega|=1}\psi(x+t\omega)\,dS_\omega,$$

for any $(x,t) \in \mathbb{R}^3 \times (0,\infty)$. Since ψ has compact support, we have

$$\psi(x+t\omega) = -\int_t^\infty \frac{\partial}{\partial s}\psi(x+s\omega)\,ds.$$

Then

$$u(x,t) = -\frac{t}{4\pi}\int_t^\infty\int_{|\omega|=1}\frac{\partial}{\partial s}\psi(x+s\omega)\,dS_\omega ds.$$

For $s \ge t$, we have $t \le s^2/t$ and hence

$$|u(x,t)| \le \frac{1}{4\pi t}\int_t^\infty s^2\int_{|\omega|=1}|\nabla\psi(x+s\omega)|\,dS_\omega ds \le \frac{1}{4\pi t}\|\nabla\psi\|_{L^1(\mathbb{R}^3)}.$$

For the global boundedness, we first have

$$\psi(x+t\omega) = \int_t^\infty s\frac{\partial^2}{\partial s^2}\psi(x+s\omega)\,ds.$$

Then

$$u(x,t) = \frac{t}{4\pi}\int_t^\infty\int_{|\omega|=1}s\frac{\partial^2}{\partial s^2}\psi(x+s\omega)\,dS_\omega ds.$$

Hence

$$|u(x,t)| \le \frac{1}{4\pi}\int_t^\infty s^2\int_{|\omega|=1}|\nabla^2\psi(x+s\omega)|\,dS_\omega ds \le \frac{1}{4\pi}\|\nabla^2\psi\|_{L^1(\mathbb{R}^3)}.$$

We now discuss general ψ. For any $(x,t) \in \mathbb{R}^3 \times (0,\infty)$, we note that u depends on ψ only on $\partial B_t(x)$. We now take a cutoff function $\eta \in C_0^\infty(\mathbb{R}^3)$ with $\eta = 1$ in $B_{t+1}(x)$, $\eta = 0$ in $\mathbb{R}^3 \setminus B_{t+2}(x)$ and a uniform bound on $\nabla\eta$. Then in the expression for u, we may replace ψ by $\eta\psi$. We can obtain the desired estimates by repeating the argument above. We simply note that derivatives of η have uniform bounds, independent of $(x,t) \in \mathbb{R}^3 \times (0,\infty)$.

Now we consider $n=2$. By assuming that ψ is of compact support, we prove that for any $t>0$,

$$|u(x,t)| \leq \frac{1}{4}\|\nabla\psi\|_{L^1(\mathbb{R}^2)},$$

and for any $t>1$,

$$|u(x,t)| \leq \frac{1}{2\sqrt{t}}\left(\|\psi\|_{L^1(\mathbb{R}^2)} + \|\nabla\psi\|_{L^1(\mathbb{R}^2)}\right).$$

The general case follows similarly to the case of $n=3$. By (6.2.6) and a change of variables, we have

$$u(x,t) = \frac{1}{2\pi}\int_0^t \frac{r}{\sqrt{t^2-r^2}}\int_{|\omega|=1}\psi(x+r\omega)\,dS_\omega dr.$$

As in the proof for $n=3$, we have for $r>0$,

$$\int_{|\omega|=1}\psi(x+r\omega)\,dS_\omega = -\int_r^\infty\int_{|\omega|=1}\frac{\partial}{\partial s}\psi(x+s\omega)\,dS_\omega ds,$$

and hence

$$\begin{aligned}\left|r\int_{|\omega|=1}\psi(x+r\omega)\,dS_\omega\right| &\leq \int_r^\infty s\int_{|\omega|=1}|\nabla\psi(x+s\omega)|\,dS_\omega ds\\ &\leq \|\nabla\psi\|_{L^1(\mathbb{R}^2)}.\end{aligned}$$

Therefore,

$$|u(x,t)| \leq \frac{1}{2\pi}\|\nabla\psi\|_{L^1(\mathbb{R}^2)}\int_0^t\frac{1}{\sqrt{t^2-r^2}}\,dr = \frac{1}{4}\|\nabla\psi\|_{L^1(\mathbb{R}^2)}.$$

For the decay estimate, we write u as

$$u(x,t) = \frac{1}{2\pi}\left(\int_0^{t-\varepsilon} + \int_{t-\varepsilon}^t\right) = \frac{1}{2\pi}(I_1+I_2),$$

where $\varepsilon>0$ is a positive constant to be determined. We can estimate I_2 similarly to the above. In fact,

$$\begin{aligned}|I_2| &= \left|\int_{t-\varepsilon}^t\frac{1}{\sqrt{t^2-r^2}}\cdot r\int_{|\omega|=1}\psi(x+r\omega)\,dS_\omega dr\right|\\ &\leq \int_{t-\varepsilon}^t\frac{1}{\sqrt{t^2-r^2}}\,dr\cdot\|\nabla\psi\|_{L^1(\mathbb{R}^2)}.\end{aligned}$$

A simple calculation yields

$$\begin{aligned}\int_{t-\varepsilon}^{t} \frac{1}{\sqrt{t^2-r^2}}\,dr &= \int_{t-\varepsilon}^{t} \frac{1}{\sqrt{(t+r)(t-r)}}\,dr \\ &\le \frac{1}{\sqrt{t}}\int_{t-\varepsilon}^{t}\frac{1}{\sqrt{t-r}}\,dr = \frac{2\sqrt{\varepsilon}}{\sqrt{t}}.\end{aligned}$$

Hence,

$$|I_2| \le \frac{2\sqrt{\varepsilon}}{\sqrt{t}}\|\nabla\psi\|_{L^1(\mathbb{R}^2)}.$$

For I_1, we have

$$\begin{aligned}|I_1| &= \left|\int_0^{t-\varepsilon}\frac{r}{\sqrt{t^2-r^2}}\int_{|\omega|=1}\psi(x+r\omega)\,dS_\omega dr\right| \\ &\le \frac{1}{\sqrt{t^2-(t-\varepsilon)^2}}\int_0^{t-\varepsilon} r\int_{|\omega|=1}|\psi(x+r\omega)|\,dS_\omega dr \\ &\le \frac{1}{\sqrt{2\varepsilon t-\varepsilon^2}}\|\psi\|_{L^1(\mathbb{R}^2)}.\end{aligned}$$

Therefore, we obtain

$$|u(x,t)| \le \frac{1}{2\pi}\left(\frac{1}{\sqrt{2\varepsilon t-\varepsilon^2}}\|\psi\|_{L^1(\mathbb{R}^2)} + \frac{2\sqrt{\varepsilon}}{\sqrt{t}}\|\nabla\psi\|_{L^1(\mathbb{R}^2)}\right).$$

For any $t>1$, we take $\varepsilon = 1/2$ and obtain the desired result.

We leave the proof for arbitrary n as an exercise. □

6.2.8. Duhamel's Principle. We now discuss the initial-value problem for the nonhomogeneous wave equation. Let φ and ψ be C^2 and C^1 functions in $\mathbb{R}^n$, respectively, and f be a continuous function in $\mathbb{R}^n\times(0,\infty)$. Consider

$$\begin{aligned}&u_{tt}-\Delta u = f \quad \text{in } \mathbb{R}^n\times(0,\infty),\\ &u(\cdot,0)=\varphi,\ u_t(\cdot,0)=\psi \quad \text{on } \mathbb{R}^n.\end{aligned} \tag{6.2.14}$$

For $f\equiv 0$, the solution u of (6.2.14) is given by (6.2.11) for n odd and by (6.2.13) for n even. We note that there are two terms in these expressions, one being a derivative in t. This is not a coincidence.

We now decompose (6.2.14) into three problems,

$$\begin{aligned}&u_{tt}-\Delta u = 0 \quad \text{in } \mathbb{R}^n\times(0,\infty),\\ &u(\cdot,0)=\varphi,\ u_t(\cdot,0)=0 \quad \text{on } \mathbb{R}^n,\end{aligned} \tag{6.2.15}$$

$$\begin{aligned}&u_{tt}-\Delta u = 0 \quad \text{in } \mathbb{R}^n\times(0,\infty),\\ &u(\cdot,0)=0,\ u_t(\cdot,0)=\psi \quad \text{on } \mathbb{R}^n,\end{aligned} \tag{6.2.16}$$

and

$$(6.2.17)\qquad \begin{aligned} u_{tt} - \Delta u &= f \quad \text{in } \mathbb{R}^n \times (0, \infty), \\ u(\cdot, 0) = 0,\ u_t(\cdot, 0) &= 0 \quad \text{on } \mathbb{R}^n. \end{aligned}$$

Obviously, a sum of solutions of (6.2.15)–(6.2.17) yields a solution of (6.2.14).

For any $\psi \in C^{[\frac{n}{2}]+1}(\mathbb{R}^n)$, set for $(x, t) \in \mathbb{R}^n \times (0, \infty)$,

$$(6.2.18)\qquad M_\psi(x, t) = \frac{1}{c_n} \left(\frac{1}{t} \frac{\partial}{\partial t} \right)^{\frac{n-3}{2}} \left(\frac{1}{\omega_n t} \int_{\partial B_t(x)} \psi \, dS \right)$$

if $n \geq 3$ is odd, and

$$(6.2.19)\qquad M_\psi(x, t) = \frac{1}{c_n} \left(\frac{1}{t} \frac{\partial}{\partial t} \right)^{\frac{n-2}{2}} \left(\frac{n}{\omega_n} \int_{B_t(x)} \frac{\psi(y)}{\sqrt{t^2 - |y - x|^2}} \, dy \right)$$

if $n \geq 2$ is even, where ω_n is the surface area of the unit sphere in $\mathbb{R}^n$ and

$$c_n = \begin{cases} 1 \cdot 3 \cdots (n-2) & \text{for } n \geq 3 \text{ odd}, \\ 2 \cdot 4 \cdots n & \text{for } n \geq 2 \text{ even}. \end{cases}$$

We note that $[\frac{n}{2}] + 1 = \frac{n+1}{2}$ if n is odd, and $[\frac{n}{2}] + 1 = \frac{n+2}{2}$ if n is even.

Theorem 6.2.9. *Let $m \geq 2$ be an integer, $\psi \in C^{[\frac{n}{2}]+m-1}(\mathbb{R}^n)$ and set $u = M_\psi$. Then $u \in C^m(\mathbb{R}^n \times (0, \infty))$ and*

$$u_{tt} - \Delta u = 0 \quad \text{in } \mathbb{R}^n \times (0, \infty).$$

Moreover, for any $x_0 \in \mathbb{R}^n$,

$$\lim_{(x,t) \to (x_0, 0)} u(x, t) = 0, \qquad \lim_{(x,t) \to (x_0, 0)} u_t(x, t) = \psi(x_0).$$

Proof. This follows easily from Theorem 6.2.5 and Theorem 6.2.6 for $\varphi = 0$. As we have seen, u is in fact C^m in $\mathbb{R}^n \times [0, \infty)$. □

We now prove that solutions of (6.2.15) can be obtained directly from those of (6.2.16).

Theorem 6.2.10. *Let $m \geq 2$ be an integer, $\varphi \in C^{[\frac{n}{2}]+m}(\mathbb{R}^n)$ and set $u = \partial_t M_\varphi$. Then $u \in C^m(\mathbb{R}^n \times (0, \infty))$ and*

$$u_{tt} - \Delta u = 0 \quad \text{in } \mathbb{R}^n \times (0, \infty).$$

Moreover, for any $x_0 \in \mathbb{R}^n$,

$$\lim_{(x,t) \to (x_0, 0)} u(x, t) = \varphi(x_0), \qquad \lim_{(x,t) \to (x_0, 0)} u_t(x, t) = 0.$$

Proof. The proof is based on straightforward calculations. We point out that u is C^m in $\mathbb{R}^n \times [0, \infty)$. By the definition of $M_\varphi(x,t)$, we have

$$\begin{aligned}&\partial_{tt}M_\varphi - \Delta M_\varphi = 0 \quad \text{in } \mathbb{R}^n \times (0,\infty),\\&M_\varphi(\cdot,0) = 0, \ \partial_t M_\varphi(\cdot,0) = \varphi \quad \text{on } \mathbb{R}^n.\end{aligned}$$

Then

$$\partial_{tt}u - \Delta u = (\partial_{tt} - \Delta)\partial_t M_\varphi = \partial_t(\partial_{tt}M_\varphi - \Delta M_\varphi) = 0 \quad \text{in } \mathbb{R}^n \times (0,\infty),$$

and

$$\begin{aligned}&u(\cdot,0) = \partial_t M_\varphi(x,t)(\cdot,0) = \varphi \quad \text{on } \mathbb{R}^n,\\&\partial_t u(\cdot,0) = \partial_{tt}M_\varphi(\cdot,0) = \Delta M_\varphi(\cdot,0) = 0 \quad \text{on } \mathbb{R}^n.\end{aligned}$$

We have the desired result. □

The next result is referred to as *Duhamel's principle.*

Theorem 6.2.11. *Let $m \geq 2$ be an integer, $f \in C^{[\frac{n}{2}]+m-1}(\mathbb{R}^n \times [0,\infty))$ and u be defined by*

$$u(x,t) = \int_0^t M_{f_\tau}(x, t-\tau)\,d\tau,$$

where $f_\tau = f(\cdot,\tau)$. Then $u \in C^m(\mathbb{R}^n \times (0,\infty))$ and

$$u_{tt} - \Delta u = f \quad \text{in } \mathbb{R}^n \times (0,\infty).$$

Moreover, for any $x_0 \in \mathbb{R}^n$,

$$\lim_{(x,t)\to(x_0,0)} u(x,t) = 0, \qquad \lim_{(x,t)\to(x_0,0)} u_t(x,t) = 0.$$

Proof. The regularity of u easily follows from Theorem 6.2.9. We will verify that u satisfies $u_{tt} - \Delta u = f$ and the initial conditions. For each fixed $\tau > 0$, $w(x,t) = M_{f_\tau}(x,t-\tau)$ satisfies

$$\begin{aligned}&w_{tt} - \Delta w = 0 \quad \text{in } \mathbb{R}^n \times (\tau,\infty),\\&w(\cdot,\tau) = 0, \ \partial_t w(\cdot,\tau) = f(\cdot,\tau) \quad \text{on } \mathbb{R}^n.\end{aligned}$$

We note that the initial conditions here are prescribed on $\{t = \tau\}$. Then

$$\begin{aligned}u_t &= M_{f_\tau}(x,t-\tau)|_{\tau=t} + \int_0^t \partial_t M_{f_\tau}(x,t-\tau)\,d\tau\\&= \int_0^t \partial_t M_{f_\tau}(x,t-\tau)\,d\tau,\end{aligned}$$

and

$$\begin{aligned} u_{tt} &= \partial_t M_{f_\tau}(x, t-\tau)|_{\tau=t} + \int_0^t \partial_{tt} M_{f_\tau}(x, t-\tau)\, d\tau \\ &= f(x,t) + \int_0^t \Delta M_{f_\tau}(x, t-\tau)\, d\tau \\ &= f(x,t) + \Delta \int_0^t M_{f_\tau}(x, t-\tau)\, d\tau \\ &= f(x,t) + \Delta u. \end{aligned}$$

Hence $u_{tt} - \Delta u = f$ in $\mathbb{R}^n \times (0, \infty)$ and $u(\cdot, 0) = 0$, $u_t(\cdot, 0) = 0$ in $\mathbb{R}^n$. $\square$

As an application of Theorem 6.2.11, we consider the initial-value problem (6.2.17) for $n = 3$. Let u be a C^2-solution of

$$\begin{aligned} u_{tt} - \Delta u = f \quad &\text{in } \mathbb{R}^3 \times (0, \infty), \\ u(\cdot, 0) = 0, \ u_t(\cdot, 0) = 0 \quad &\text{on } \mathbb{R}^3. \end{aligned}$$

By (6.2.18) for $n = 3$, we have for any $\psi \in C^2(\mathbb{R}^3)$,

$$M_\psi(x, t) = \frac{1}{4\pi t} \int_{\partial B_t(x)} \psi(y)\, dS_y.$$

Then, by Theorem 6.2.11,

$$u(x,t) = \int_0^t M_{f_\tau}(x, t-\tau)\, d\tau = \frac{1}{4\pi} \int_0^t \frac{1}{t-\tau} \int_{\partial B_{t-\tau}(x)} f(y, \tau)\, dS_y d\tau.$$

By the change of variables $\tau = t - s$, we have

$$u(x,t) = \frac{1}{4\pi} \int_0^t \int_{\partial B_s(x)} \frac{f(y, t-s)}{s}\, dS_y ds.$$

Therefore,

$$u(x,t) = \frac{1}{4\pi} \int_{B_t(x)} \frac{f(y, t - |y-x|)}{|y-x|}\, dy, \tag{6.2.20}$$

for any $(x,t) \in \mathbb{R}^3 \times (0, \infty)$. We note that the value of the solution u at (x,t) depends on the values of f only at the points (y, s) with

$$|y - x| = t - s, \quad 0 < s < t.$$

This is exactly the backward characteristic cone with vertex (x,t).

Theorem 6.2.12. *Let $m \geq 2$ be an integer, $f \in C^m(\mathbb{R}^3 \times [0, \infty))$ and u be defined by (6.2.20). Then $u \in C^m(\mathbb{R}^3 \times (0, \infty))$ and*

$$u_{tt} - \Delta u = f \quad \text{in } \mathbb{R}^3 \times (0, \infty).$$

Moreover, for any $x_0 \in \mathbb{R}^3$,

$$\lim_{(x,t)\to(x_0,0)} u(x,t) = 0, \qquad \lim_{(x,t)\to(x_0,0)} u_t(x,t) = 0.$$

6.3. Energy Estimates

In this section, we derive energy estimates of solutions of initial-value problems for a class of hyperbolic equations slightly more general than the wave equation.

Before we start, we demonstrate by a simple case what is involved. Suppose u is a C^2-solution of

$$u_{tt} - \Delta u = 0 \quad \text{in } \mathbb{R}^n \times (0, \infty).$$

We assume that $u(\cdot, 0)$ and $u_t(\cdot, 0)$ have compact support. By finite-speed propagation, $u(\cdot, t)$ also has compact support for any $t > 0$. We multiply the wave equation by u_t and integrate in $B_R \times (0, t)$. Here we choose R sufficiently large such that B_R contains the support of $u(\cdot, s)$, for any $s \in (0, t)$. Note that

$$u_t u_{tt} - u_t \Delta u = \frac{1}{2}(u_t^2 + |\nabla_x u|^2)_t - \sum_{i=1}^n (u_t u_{x_i})_{x_i}.$$

Then a simple integration in $B_R \times (0, t)$ yields

$$\frac{1}{2}\int_{\mathbb{R}^n \times \{t\}} (u_t^2 + |\nabla_x u|^2)\, dx = \frac{1}{2}\int_{\mathbb{R}^n \times \{0\}} (u_t^2 + |\nabla_x u|^2)\, dx.$$

This is *conservation of energy*: the L^2-norm of derivatives at each time slice is a constant independent of time. For general hyperbolic equations, conservation of energy is not expected. However, we have the *energy estimates*: the energy at later time is controlled by the initial energy.

Let a, c and f be continuous functions in $\mathbb{R}^n \times [0, \infty)$ and φ and ψ be continuous functions in $\mathbb{R}^n$. We consider the initial-value problem

$$\begin{aligned} &u_{tt} - a\Delta u + cu = f \quad \text{in } \mathbb{R}^n \times (0, \infty), \\ &u(\cdot, 0) = \varphi,\ u_t(\cdot, 0) = \psi \quad \text{in } \mathbb{R}^n. \end{aligned} \tag{6.3.1}$$

We assume that a is a positive function satisfying

$$\lambda \le a(x, t) \le \Lambda \quad \text{for any } (x, t) \in \mathbb{R}^n \times [0, \infty), \tag{6.3.2}$$

for some positive constants λ and Λ. For the wave equation, we have $a = 1$ and $c = 0$ and hence we can choose $\lambda = \Lambda = 1$ in (6.3.2).

In the following, we set

$$\kappa = \frac{1}{\sqrt{\Lambda}}.$$

For any point $P = (X, T) \in \mathbb{R}^n \times (0, \infty)$, consider the cone $C_\kappa(P)$ (opening downward) with vertex at P defined by

$$C_\kappa(P) = \{(x, t) :\ 0 < t < T,\ \kappa|x - X| < T - t\}.$$

As in Section 2.3, we denote by $\partial_s C_\kappa(P)$ and $\partial_- C_\kappa(P)$ the side and bottom of the boundary, respectively, i.e.,

$$\partial_s C_\kappa(P) = \{(x,t) : \ 0 < t \le T, \ \kappa|x - X| = T - t\},$$
$$\partial_- C_\kappa(P) = \{(x,0) : \ \kappa|x - X| \le T\}.$$

We note that $\partial_- C_\kappa(P)$ is simply the closed ball in $\mathbb{R}^n \times \{0\}$ centered at $(X, 0)$ with radius T/κ.

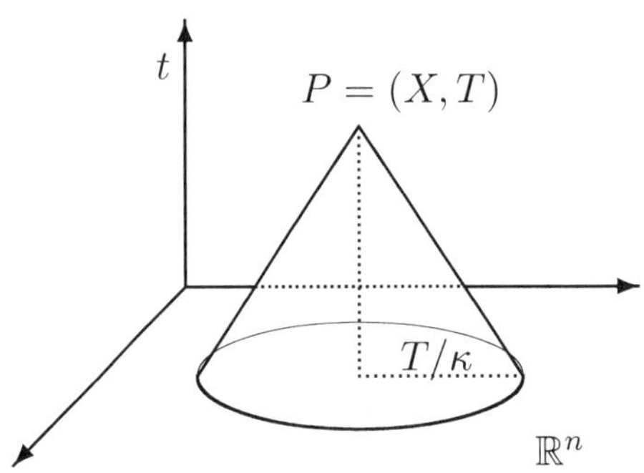

Figure 6.3.1. The cone $C_\kappa(P)$.

Theorem 6.3.1. *Let a be C^1, c and f be continuous in $\mathbb{R}^n \times [0, \infty)$, and let φ be C^1 and ψ be continuous in $\mathbb{R}^n$. Suppose (6.3.2) holds and $u \in C^2(\mathbb{R}^n \times (0,\infty)) \cap C^1(\mathbb{R}^n \times [0,\infty))$ is a solution of (6.3.1). Then for any point $P = (X, T) \in \mathbb{R}^n \times (0, \infty)$ and any $\eta > \eta_0$,*

$$\begin{aligned}(\eta - \eta_0) &\int_{C_\kappa(P)} e^{-\eta t}(u^2 + u_t^2 + a|\nabla u|^2)\,dxdt \\ &\le \int_{\partial_- C_\kappa(P)} (\varphi^2 + \psi^2 + a|\nabla\varphi|^2)\,dx + \int_{C_\kappa(P)} e^{-\eta t} f^2\,dxdt,\end{aligned}$$

where η_0 is a positive constant depending only on n, λ, the C^1-norm of a and the L^∞-norm of c in $C_\kappa(P)$.

Proof. We multiply the equation in (6.3.1) by $2e^{-\eta t}u_t$ and integrate in $C_\kappa(P)$, for a nonnegative constant η to be determined. First, we note that

$$2e^{-\eta t}u_t u_{tt} = (e^{-\eta t}u_t^2)_t + \eta e^{-\eta t}u_t^2,$$

and

$$-2e^{-\eta t}au_t\Delta u = -2e^{-\eta t}au_t\sum_{i=1}^n u_{x_ix_i}$$
$$= \sum_{i=1}^n\big(-2(e^{-\eta t}au_tu_{x_i})_{x_i} + 2e^{-\eta t}au_{x_i}u_{tx_i} + 2e^{-\eta t}a_{x_i}u_tu_{x_i}\big)$$
$$= \sum_{i=1}^n\big(-2(e^{-\eta t}au_tu_{x_i})_{x_i} + (e^{-\eta t}au_{x_i}^2)_t + 2e^{-\eta t}a_{x_i}u_tu_{x_i}$$
$$+ \eta e^{-\eta t}au_{x_i}^2 - e^{-\eta t}a_tu_{x_i}^2\big),$$

where we used $2u_{x_i}u_{tx_i} = (u_{x_i}^2)_t$. Therefore, we obtain

$$(e^{-\eta t}u_t^2 + e^{-\eta t}a|\nabla u|^2)_t - 2\sum_{i=1}^n(e^{-\eta t}au_tu_{x_i})_{x_i} + \eta e^{-\eta t}(u_t^2 + a|\nabla u|^2)$$
$$+ \sum_{i=1}^n 2e^{-\eta t}a_{x_i}u_tu_{x_i} - e^{-\eta t}a_t|\nabla u|^2 + 2e^{-\eta t}cuu_t$$
$$= 2e^{-\eta t}u_tf.$$

We note that the first two terms in the left-hand side are derivatives of quadratic expressions in $\nabla_x u$ and u_t and that the next three terms are quadratic in $\nabla_x u$ and u_t. In particular, the third term is a positive quadratic form. The final term in the left-hand side involves u itself. To control this term, we note that

$$(e^{-\eta t}u^2)_t + \eta e^{-\eta t}u^2 - 2e^{-\eta t}uu_t = 0.$$

Then a simple addition yields

$$\big(e^{-\eta t}(u^2+u_t^2 + a|\nabla u|^2)\big)_t - \sum_{i=1}^n 2(e^{-\eta t}au_tu_{x_i})_{x_i}$$
$$+ \eta e^{-\eta t}(u^2 + u_t^2 + a|\nabla u|^2) = RHS,$$

where

$$RHS = -2e^{-\eta t}\sum_{i=1}^n a_{x_i}u_tu_{x_i} + e^{-\eta t}a_t|\nabla u|^2 - 2e^{-\eta t}(c-1)uu_t + 2e^{-\eta t}u_tf.$$

The first three terms in RHS are quadratic in u_t, u_{x_i} and u. Now by (6.3.2) and the Cauchy inequality, we have

$$2|a_{x_i}u_tu_{x_i}| \le |a_{x_i}|(u_t^2 + u_{x_i}^2) \le |a_{x_i}|\left(u_t^2 + \frac{1}{\lambda}au_{x_i}^2\right),$$

and similar estimates for other three terms in RHS. Hence

$$RHS \le \eta_0e^{-\eta t}(u^2 + u_t^2 + a|\nabla u|^2) + e^{-\eta t}f^2,$$

where η_0 is a positive constant which can be taken as

$$\eta_0 = \frac{1}{\lambda}\sup_{C_\kappa(P)}|a_t| + \left(n+\frac{1}{\lambda}\right)\sup_{C_\kappa(P)}|\nabla_x a| + \sup_{C_\kappa(P)}|c| + 2.$$

Then a simple substitution yields

$$\begin{aligned}\left(e^{-\eta t}(u^2+u_t^2+a|\nabla u|^2)\right)_t &- 2\sum_{i=1}^n (e^{-\eta t}au_tu_{x_i})_{x_i}\\ &+(\eta-\eta_0)e^{-\eta t}(u^2+u_t^2+a|\nabla u|^2) \le e^{-\eta t}f^2.\end{aligned}$$

Upon integrating over $C_\kappa(P)$, we obtain

$$\begin{aligned}(\eta-\eta_0)&\int_{C_\kappa(P)} e^{-\eta t}(u^2+u_t^2+a|\nabla u|^2)\,dxdt\\ &+\int_{\partial_s C_\kappa(P)} e^{-\eta t}\left((u^2+u_t^2+a|\nabla u|^2)\nu_t - 2\sum_{i=1}^n au_tu_{x_i}\nu_i\right)dS\\ &\le \int_{\partial_- C_\kappa(P)}(u^2+u_t^2+a|\nabla u|^2)\,dx + \int_{C_\kappa(P)} e^{-\eta t}f^2\,dxdt,\end{aligned}$$

where the unit exterior normal vector on $\partial_s C_\kappa(P)$ is given by

$$(\nu_1,\cdots,\nu_n,\nu_t) = \frac{1}{\sqrt{1+\kappa^2}}\left(\kappa\frac{x-X}{|x-X|},1\right).$$

We need only prove that the integrand for $\partial_s C_\kappa(P)$ is nonnegative. We claim that

$$BI \equiv (u_t^2+a|\nabla u|^2)\nu_t - 2\sum_{i=1}^n au_tu_{x_i}\nu_i \ge 0 \quad \text{on } \partial_s C_\kappa(P).$$

To prove this, we first note that, by the Cauchy inequality,

$$\left|\sum_{i=1}^n u_{x_i}\nu_i\right| \le \left(\sum_{i=1}^n u_{x_i}^2\right)^{\frac{1}{2}}\cdot\left(\sum_{i=1}^n \nu_i^2\right)^{\frac{1}{2}} = |\nabla u|\sqrt{1-\nu_t^2}.$$

With $\nu_t = 1/\sqrt{1+\kappa^2}$, we have

$$BI \ge \frac{1}{\sqrt{1+\kappa^2}}\big(u_t^2+a|\nabla u|^2 - 2\kappa a|u_t|\cdot|\nabla u|\big).$$

By (6.3.2) and $\kappa = 1/\sqrt{\Lambda}$, we have $\kappa\sqrt{a}\le 1$. Hence

$$BI \ge \frac{1}{\sqrt{1+\kappa^2}}\big(u_t^2+a|\nabla u|^2 - 2\sqrt{a}|u_t|\cdot|\nabla u|\big) \ge 0.$$

Therefore, the boundary integral on $\partial_s C_\kappa(P)$ is nonnegative and can be discarded. □

A consequence of Theorem 6.3.1 is the uniqueness of solutions of (6.3.1). We can also discuss the domain of dependence and the range of influence as in the previous section.

We note that the cone $C_\kappa(P)$ in Theorem 6.3.1 plays the same role as the cone in Theorem 2.3.4. The constant κ is chosen so that the boundary integral over $\partial_s C_\kappa(P)$ is nonnegative and hence can be dropped from the estimate.

Similar to Theorem 2.3.5, we have the following result.

Theorem 6.3.2. *Let a be C^1, c and f be continuous in $\mathbb{R}^n \times [0, \infty)$, and let φ be C^1 and ψ be continuous in $\mathbb{R}^n$. Suppose (6.3.2) holds and $u \in C^2(\mathbb{R}^n \times (0, \infty)) \cap C^1(\mathbb{R}^n \times [0, \infty))$ is a solution of (6.3.1). For a fixed $T > 0$, if $f \in L^2(\mathbb{R}^n \times (0, T))$ and $\varphi, \nabla_x\varphi, \psi \in L^2(\mathbb{R}^n)$, then for any $\eta > \eta_0$,*

$$\begin{aligned}\int_{\mathbb{R}^n\times\{T\}} & e^{-\eta t}(u^2 + u_t^2 + a|\nabla u|^2)\,dx \\ &+ (\eta - \eta_0)\int_{\mathbb{R}^n\times(0,T)} e^{-\eta t}(u^2 + u_t^2 + a|\nabla u|^2)\,dxdt \\ &\le \int_{\mathbb{R}^n}(\varphi^2 + \psi^2 + a|\nabla\varphi|^2)\,dx + \int_{\mathbb{R}^n\times(0,T)} e^{-\eta t} f^2\,dxdt,\end{aligned}$$

where η_0 is a positive constant depending only on n, λ, the C^1-norm of a and the L^∞-norm of c in $\mathbb{R}^n \times [0, T]$.

Usually, we call $u_t^2 + a|\nabla u|^2$ the *energy density* and its integral over $\mathbb{R}^n \times \{t\}$ the *energy* at time t. Then Theorem 6.3.2 asserts, in the case of $c = 0$ and $f = 0$, that the initial energy (the energy at $t = 0$) controls the energy at later time.

Next, we consider the initial-value problem in general domains. Let Ω be a bounded domain in $\mathbb{R}^n$ and h_- and h_+ be two piecewise C^1-functions in Ω with $h_- < h_+$ in Ω and $h_- = h_+$ on $\partial\Omega$. Set

$$D = \{(x, t) :\ h_-(x) < t < h_+(x),\ x \in \Omega\}.$$

Let a be C^1 and c be continuous in $\bar{D}$. We assume that

$$\lambda \le a \le \Lambda \quad \text{in } D.$$

We now consider

$$u_{tt} - a\Delta u + cu = f \quad \text{in } D. \tag{6.3.3}$$

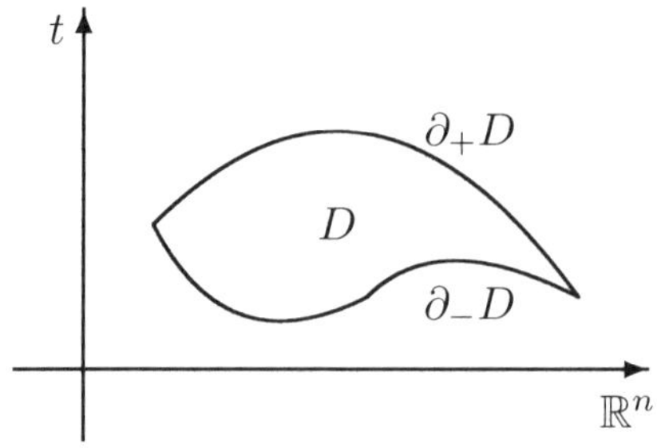

Figure 6.3.2. A general domain.

We can perform a similar integration in D as in the proof of Theorem 6.3.1 and obtain

$$\begin{aligned}
&\int_{\partial_+ D} e^{-\eta t}\left((u^2+u_t^2+a|\nabla u|^2)\nu_{+t} - 2\sum_{i=1}^n a u_t u_{x_i}\nu_{+i}\right) dS \\
&\qquad + (\eta-\eta_0)\int_D e^{-\eta t}(u^2+u_t^2+a|\nabla u|^2)\,dxdt \\
&\le \int_{\partial_- D} e^{-\eta t}\left((u^2+u_t^2+a|\nabla u|^2)\nu_{-t} - 2\sum_{i=1}^n a u_t u_{x_i}\nu_{-i}\right) dS \\
&\qquad + \int_D e^{-\eta t} f^2\,dxdt,
\end{aligned}$$

where $\nu_\pm = (\nu_{\pm 1}, \cdots, \nu_{\pm n}, \nu_{\pm t})$ are unit normal vectors pointing in the positive t-direction along $\partial_\pm D$. We are interested in whether the integrand for $\partial_+ D$ is nonnegative. As in the proof of Theorem 6.3.1, we have, by the Cauchy inequality,

$$\left|\sum_{i=1}^n u_{x_i}\nu_{+i}\right| \le \left(\sum_{i=1}^n u_{x_i}^2\right)^{\frac12}\cdot\left(\sum_{i=1}^n \nu_{+i}^2\right)^{\frac12} = |\nabla u|\sqrt{1-\nu_{+t}^2}.$$

Then it is easy to see that

$$\begin{aligned}
&(u_t^2+a|\nabla u|^2)\nu_{+t} - 2\sum_{i=1}^n a u_t u_{x_i}\nu_{+i} \\
&\qquad \ge (u_t^2+a|\nabla u|^2)\nu_{+t} - 2\sqrt{a(1-\nu_{+t}^2)}\cdot\sqrt{a}|u_t|\cdot|\nabla u| \ge 0 \quad \text{on } \partial_+ D
\end{aligned}$$

if

$$\nu_{+t} \ge \sqrt{a(1-\nu_{+t}^2)}.$$

This condition can be written as

$$\nu_{+t} \ge \sqrt{\frac{a}{1+a}} \quad \text{on } \partial_+ D. \tag{6.3.4}$$

In conclusion, under the condition (6.3.4), we obtain

$$\int_{\partial_+ D} e^{-\eta t} u^2 \nu_{+t}\, dS + (\eta - \eta_0) \int_D e^{-\eta t}(u^2 + u_t^2 + a|\nabla u|^2)\, dxdt$$
$$\leq \int_{\partial_- D} e^{-\eta t} \left((u^2 + u_t^2 + a|\nabla u|^2)\nu_{-t} - 2\sum_{i=1}^n a u_t u_{x_i} \nu_{-i} \right) dS$$
$$+ \int_D e^{-\eta t} f^2\, dxdt.$$

If we prescribe u and u_t on $\partial_- D$, then $\nabla_x u$ can be calculated on $\partial_- D$ in terms of u and u_t. Hence, the expressions in the right-hand side are known. In particular, if $u = u_t = 0$ on $\partial_- D$ and $f = 0$ in D, then $u = 0$ in D.

Now we introduce the notion of space-like and time-like surfaces.

Definition 6.3.3. Let Σ be a C^1-hypersurface in $\mathbb{R}^n \times \mathbb{R}_+$ and $\nu = (\nu_x, \nu_t)$ be a unit normal vector field on Σ with $\nu_t \geq 0$. Then Σ is *space-like* at (x, t) for (6.3.3) if

$$\nu_t(x,t) > \sqrt{\frac{a(x,t)}{1 + a(x,t)}};$$

Σ is *time-like* at (x, t) if

$$\nu_t(x,t) < \sqrt{\frac{a(x,t)}{1 + a(x,t)}}.$$

If the hypersurface Σ is given by $t = t(x)$, it is easy to check that Σ is space-like at $(x, t(x))$ if

$$|\nabla t(x)| < \frac{1}{\sqrt{a(x, t(x))}}.$$

Now we consider the wave equation

$$u_{tt} - \Delta u = f. \tag{6.3.5}$$

With $a = 1$, the hypersurface Σ is space-like at (x, t) if $\nu_t(x, t) > 1/\sqrt{2}$. If Σ is given by $t = t(x)$, then Σ is space-like at $(x, t(x))$ if

$$|\nabla t(x)| < 1.$$

In the following, we demonstrate the importance of space-like hypersurfaces by the wave equation.

Let Σ be a space-like hypersurface for the wave equation. Then for any $(x_0, t_0) \in \Sigma$, the range of influence of (x_0, t_0) is given by the cone $\{(x, t) :\ t - t_0 > |x - x_0|\}$ and hence is always above Σ. This suggests that prescribing initial values on space-like surfaces yields a well-posed problem.

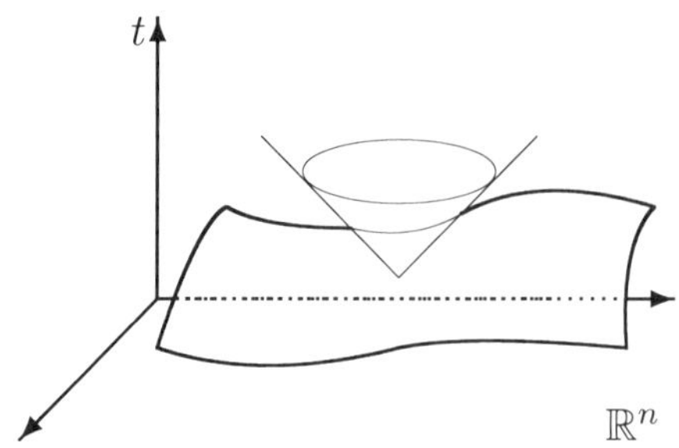

Figure 6.3.3. A space-like hypersurface.

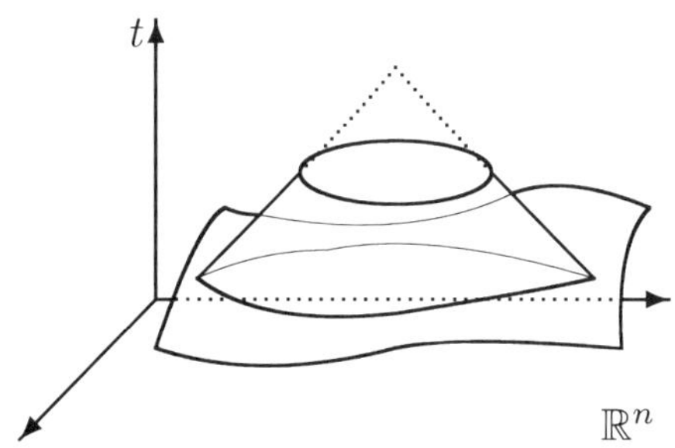

Figure 6.3.4. An integral domain for space-like initial hypersurfaces.

In fact, domains of integration for energy estimates can be constructed accordingly.

Next, we discuss briefly initial-value problems with initial values prescribed on a time-like hypersurface. Consider

$$u_{tt} = u_{xx} + u_{yy} \quad \text{for } x > 0 \text{ and } y, t \in \mathbb{R},$$

$$u = \frac{1}{m^2} \sin my, \ \frac{\partial u}{\partial x} = \frac{1}{m} \sin my \quad \text{on } \{x = 0\}.$$

Here we treat $\{x = 0\}$ as the initial hypersurface, which is time-like for the wave equation. A solution is given by

$$u_m(x, y) = \frac{1}{m^2} e^{mx} \sin my.$$

Note that

$$u_m \to 0, \ \frac{\partial u_m}{\partial x} \to 0 \quad \text{on } \{x = 0\} \text{ as } m \to \infty.$$

Meanwhile, for any $x > 0$,

$$\sup_{\mathbb{R}^2} |u_m(x, \cdot)| = \frac{1}{m^2} e^{mx} \to \infty \quad \text{as } m \to \infty.$$

Therefore, there is no continuous dependence on the initial values.

To conclude this section, we discuss a consequence of Theorem 6.3.2. In Subsection 2.3.3, we proved in Theorem 2.3.7 the existence of weak solutions of the initial-value problem for the first-order linear PDEs with the help of estimates in Theorem 2.3.5. By a similar process, we can prove the existence of weak solutions of (6.3.1) using Theorem 6.3.2. However, there is a significant difference. The weak solutions in Definition 2.3.6 are in L^2 because an estimate of the L^2-norms of solutions is established in Theorem 2.3.5. In the present situation, Theorem 6.3.2 establishes an estimate of the L^2-norms of solutions and *their derivatives*. This naturally leads to a new norm defined by

$$\|u\|_{H^1(\mathbb{R}^n\times(0,T))} = \left(\int_{\mathbb{R}^n\times(0,T)} (u^2 + u_t^2 + |\nabla_x u|^2)\, dxdt\right)^{\frac{1}{2}}.$$

The superscript 1 in H^1 indicates the order of derivatives. With such a norm, we can define the Sobolev space $H^1(\mathbb{R}^n \times (0,T))$ as the completion of smooth functions of finite H^1-norms with respect to the H^1-norm. Obviously, $H^1(\mathbb{R}^n\times(0,T))$ defined in this way is complete. In fact, it is a Hilbert space, since the H^1-norm is naturally induced by an H^1-inner product given by

$$(u,v)_{H^1(\mathbb{R}^n\times(0,T))} = \int_{\mathbb{R}^n\times(0,T)} (uv + u_t v_t + \nabla_x u \cdot \nabla_x v)\, dxdt.$$

Then we can prove that (6.3.1) admits a weak H^1-solution in $\mathbb{R}^n \times (0,T)$ if $\varphi = \psi = 0$. We will not provide the details here. The purpose of this short discussion is to demonstrate the importance of Sobolev spaces in PDEs. We refer to Subsection 4.4.2 for a discussion of weak solutions of the Poisson equation.

6.4. Exercises

Exercise 6.1. Let l be a positive constant, $\varphi \in C^2([0,l])$ and $\psi \in C^1([0,l])$. Consider

$$\begin{aligned} &u_{tt} - u_{xx} = 0 \quad \text{in } (0,l)\times(0,\infty),\\ &u(\cdot,0) = \varphi,\ u_t(\cdot,0) = \psi \quad \text{in } [0,l],\\ &u(0,t) = 0,\ u_x(l,t) = 0 \quad \text{for } t > 0. \end{aligned}$$

Find a compatibility condition and prove the existence of a C^2-solution under such a condition.

Exercise 6.2. Let φ_1 and φ_2 be C^2-functions in $\{x < 0\}$ and $\{x > 0\}$, respectively. Consider the characteristic initial-value problem

$$\begin{aligned} u_{tt} - u_{xx} &= 0 \quad \text{for } t > |x|, \\ u(x, -x) &= \varphi_1(x) \quad \text{for } x < 0, \\ u(x, x) &= \varphi_2(x) \quad \text{for } x > 0. \end{aligned}$$

Solve this problem and find the domain of dependence for any point (x, t) with $t > |x|$.

Exercise 6.3. Let φ_1 and φ_2 be C^2-functions in $\{x > 0\}$. Consider the *Goursat problem*

$$\begin{aligned} &u_{tt} - u_{xx} = 0 \quad \text{for } 0 < t < x, \\ &u(x, 0) = \varphi_1(x), \ \ u(x, x) = \varphi_2(x) \quad \text{for } x > 0. \end{aligned}$$

Solve this problem and find the domain of dependence for any point (x, t) with $0 < t < x$.

Exercise 6.4. Let α be a constant and φ and ψ be C^2-functions on $(0, \infty)$ which vanish near $x = 0$. Consider

$$\begin{aligned} &u_{tt} - u_{xx} = 0 \quad \text{for } x > 0, \ t > 0, \\ &u(x, 0) = \varphi(x), \ u_t(x, 0) = \psi(x) \quad \text{for } x > 0, \\ &u_t(0, t) = \alpha u_x(0, t) \quad \text{for } t > 0. \end{aligned}$$

Find a solution for $\alpha \neq -1$ and prove that in general there exist no solutions for $\alpha = -1$.

Exercise 6.5. Let a be a constant with $|a| < 1$. Prove that the wave equation

$$u_{tt} - \Delta_x u = 0 \quad \text{in } \mathbb{R}^3 \times \mathbb{R}$$

is preserved by a *Lorentz transformation*, i.e., a change of variables given by

$$\begin{aligned} s &= \frac{t - ax_1}{\sqrt{1 - a^2}}, \\ y_1 &= \frac{x_1 - at}{\sqrt{1 - a^2}}, \\ y_i &= x_i \quad \text{for } i = 2, 3. \end{aligned}$$

Exercise 6.6. Let λ be a positive constant and $\psi \in C^2(\mathbb{R}^2)$. Solve the following initial-value problems by the method of descent:

$$\begin{aligned} &u_{tt} = \Delta u + \lambda^2 u \quad \text{in } \mathbb{R}^2 \times (0, \infty), \\ &u(\cdot, 0) = 0, \ u_t(\cdot, 0) = \psi \quad \text{on } \mathbb{R}^2, \end{aligned}$$

and

$$\begin{aligned} u_{tt} &= \Delta u - \lambda^2 u \quad \text{in } \mathbb{R}^2 \times (0,\infty), \\ u(\cdot,0) &= 0, \ u_t(\cdot,0) = \psi \quad \text{on } \mathbb{R}^2. \end{aligned}$$

Hint: Use complex functions temporarily to solve the second problem.

Exercise 6.7. Let ψ be a bounded function defined in $\mathbb{R}^2$ with $\psi = 0$ in $\mathbb{R}^2 \setminus B_1$. For any $(x,t) \in \mathbb{R}^2 \times (0,\infty)$, define

$$u(x,t) = \frac{1}{2\pi} \int_{B_t(x)} \frac{\psi(y)}{\sqrt{t^2 - |y-x|^2}}\, dy.$$

(1) For any $\alpha \in (0,1)$, prove

$$\sup_{B_{\alpha t}} |u(\cdot,t)| \le \frac{C}{t} \sup_{\mathbb{R}^2} |\psi| \quad \text{for any } t > 1,$$

where C is a positive constant depending only on α.

(2) Assume, in addition, that $\psi = 1$ in B_1. For any unit vector $e \in \mathbb{R}^2$, find the decay rate of $u(te,t)$ as $t \to \infty$.

Exercise 6.8. Let $\varphi \in C^2(\mathbb{R}^3)$ and $\psi \in C^1(\mathbb{R}^3)$. Suppose that $u \in C^2(\mathbb{R}^3 \times [0,\infty))$ is a solution of the initial-value problem

$$\begin{aligned} u_{tt} - \Delta u &= 0 \quad \text{in } \mathbb{R}^3 \times (0,\infty), \\ u(\cdot,0) &= \varphi, \ u_t(\cdot,0) = \psi \quad \text{on } \mathbb{R}^3. \end{aligned}$$

(1) For any fixed $(x_0,t_0) \in \mathbb{R}^3 \times (0,\infty)$, set for any $x \in B_{t_0}(x_0) \setminus \{x_0\}$,

$$\mathbf{v}(x) = \left(\frac{\nabla_x u(x,t)}{|x-x_0|} + \frac{x-x_0}{|x-x_0|^3} u(x,t) + \frac{x-x_0}{|x-x_0|^2} u_t(x,t) \right)\Bigg|_{t=t_0-|x-x_0|}.$$

Prove that $\operatorname{div} \mathbf{v} = 0$.

(2) Derive an expression of $u(x_0,t_0)$ in terms of φ and ψ by integrating $\operatorname{div} \mathbf{v}$ in $B_{t_0}(x_0) \setminus B_\varepsilon(x_0)$ and then letting $\varepsilon \to 0$.

Remark: This exercise gives an alternative approach to solving the initial-value problem for the three-dimensional wave equation.

Exercise 6.9. Let a be a positive constant and u be a C^2-solution of the characteristic initial-value problem

$$\begin{aligned} &u_{tt} - \Delta u = 0 \quad \text{in } \{(x,t) \in \mathbb{R}^3 \times (0,\infty) : \ t > |x| > a\}, \\ &u(x,|x|) = 0 \quad \text{for } |x| > a. \end{aligned}$$

(1) For any fixed $(x_0,t_0) \in \mathbb{R}^3 \times \mathbb{R}_+$ with $t_0 > |x_0| > a$, integrate $\operatorname{div} \mathbf{v}$ (introduced in Exercise 6.8) in the region bounded by $|x-x_0|+|x| = t_0$, $|x| = a$ and $|x-x_0| = \varepsilon$. By letting $\varepsilon \to 0$, express $u(x_0,t_0)$ in terms of an integral over ∂B_a.

(2) For any $\omega \in \mathbb{S}^2$ and $\tau > 0$, prove that the limit

$$\lim_{r\to\infty} \big(ru(r\omega, r+\tau)\big)$$

exists and the convergence is uniform for $\omega \in \mathbb{S}^2$ and $\tau \in (0, \tau_0]$, for any fixed $\tau_0 > 0$.

Remark: The limit in (2) is called the *radiation field.* [1]

Exercise 6.10. Prove Theorem 6.2.7 and Theorem 6.2.8 for $n \geq 2$.

Exercise 6.11. Set $Q_T = \{(x,t) : \ 0 < x < l, \ 0 < t < T\}$. Consider the equation

$$Lu \equiv 2u_{tt} + 3u_{tx} + u_{xx} = 0.$$

(1) Give a correct presentation of the boundary-value problem in Q_T.
(2) Find an explicit expression of a solution with prescribed boundary values.
(3) Derive an estimate of the integral of $u_x^2 + u_t^2$ in Q_T.

Hint: For (2), divide Q_T into three regions separated by characteristic curves from $(0,0)$. For (3), integrate an appropriate linear combination of $u_t Lu$ and $u_x Lu$ to make integrands on $[0,l] \times \{t\}$ and $\{l\} \times [0,t]$ positive definite.

Exercise 6.12. For some constant $a > 0$, let f be a C^1-function in $a < |x| < t + a$, φ a C^1-function on $r_0 < |x| = t - a$ and ψ a C^1-function on $|x| = a$ and $t > 0$. Consider the *characteristic initial-value problem* for the wave equation

$$\begin{aligned} u_{tt} - \Delta u &= f(x,t) \quad \text{in } a < |x| < t + a, \\ u &= \varphi(x,t) \quad \text{on } |x| > a, \ t = |x| - a, \\ u &= \psi(x,t) \quad \text{on } |x| = a, \ t > 0. \end{aligned}$$

Derive an energy estimate in an appropriate domain in $a < |x| < t + a$.

[1] F. G. Friedlander, *On the radiation field of pulse solutions of the wave equation*, Proc. Roy. Soc. A, **269** (1962), 53–65.

Chapter 7

First-Order Differential Systems

In this chapter, we discuss partial differential systems of the first order and focus on local existence of solutions.

In Section 7.1, we introduce the notion of noncharacteristic hypersurfaces for initial-value problems. We proceed here for linear partial differential equations and partial differential systems of arbitrary order similarly to how we did for first-order linear PDEs in Section 2.1 and second-order linear PDEs in Section 3.1. We show that we can compute all derivatives of solutions on initial hypersurfaces if initial values are prescribed on noncharacteristic initial hypersurfaces. We also demonstrate that partial differential systems of arbitrary order can always be transformed to those of the first order.

In Section 7.2, we discuss analytic solutions of the initial-value problem for first-order linear differential systems. The main result is the Cauchy-Kovalevskaya theorem, which asserts the local existence of analytic solutions if the coefficient matrices and the nonhomogeneous terms are analytic and the initial values are analytic on analytic noncharacteristic hypersurfaces. The proof is based on the convergence of the formal power series of solutions. In this section, we also prove a uniqueness result due to Holmgren, which asserts that the solutions in the Cauchy-Kovalevskaya theorem are the only solutions in the C^∞-category.

In Section 7.3, we construct a first-order linear differential system in $\mathbb{R}^3$ that does not admit smooth solutions in any subsets of $\mathbb{R}^3$. In this system, the coefficient matrices are analytic and the nonhomogeneous term is a suitably chosen smooth function. (For analytic nonhomogeneous terms there would always be solutions by the Cauchy-Kovalevskaya theorem). We need to point out that such a nonhomogeneous term is proved to exist by a contradiction argument. An important role is played by the Baire category theorem.

7.1. Noncharacteristic Hypersurfaces

The main focus in this section is on linear partial differential systems of arbitrary order.

7.1.1. Linear Partial Differential Equations. We start with linear partial differential equations of arbitrary order and proceed here as in Sections 2.1 and 3.1.

Let Ω be a domain in $\mathbb{R}^n$ containing the origin, m be a positive integer and a_α be a continuous function in Ω, for $\alpha \in \mathbb{Z}_+^n$ with $|\alpha| \leq m$. Consider an mth-order linear differential operator L defined by

$$Lu = \sum_{|\alpha| \leq m} a_\alpha(x)\partial^\alpha u \quad \text{in } \Omega. \tag{7.1.1}$$

Here, a_α is called the *coefficient* of $\partial^\alpha u$.

Definition 7.1.1. Let L be a linear differential operator of order m as in (7.1.1) defined in $\Omega \subset \mathbb{R}^n$. The *principal part* L_0 and the *principal symbol* p of L are defined by

$$L_0 u = \sum_{|\alpha| = m} a_\alpha(x)\partial^\alpha u \quad \text{in } \Omega,$$

and

$$p(x;\xi) = \sum_{|\alpha| = m} a_\alpha(x)\xi^\alpha,$$

for any $x \in \Omega$ and $\xi \in \mathbb{R}^n$.

The principal part L_0 is a differential operator consisting of terms involving derivatives of order m in L, and the principal symbol is a homogeneous polynomial of degree m with coefficients given by the coefficients of L_0. Principal symbols play an important role in discussions of differential operators.

We discussed first-order and second-order linear differential operators in Chapter 2 and Chapter 3, respectively. Usually, they are written in the forms

$$Lu = \sum_{i=1}^n a_i(x)u_{x_i} + b(x)u \quad \text{in } \Omega,$$

and

$$Lu = \sum_{i,j=1}^{n} a_{ij}(x)u_{x_i x_j} + \sum_{i=1}^{n} b_i(x)u_{x_i} + c(x)u \quad \text{in } \Omega.$$

Their principal symbols are given by

$$p(x;\xi) = \sum_{i=1}^{n} a_i(x)\xi_i,$$

and

$$p(x;\xi) = \sum_{i,j=1}^{n} a_{ij}(x)\xi_i\xi_j,$$

for any $x \in \Omega$ and $\xi \in \mathbb{R}^n$. For second-order differential operators, we usually assume that (a_{ij}) is a symmetric matrix in Ω.

Let f be a continuous function in Ω. We consider the equation

$$Lu = f(x) \quad \text{in } \Omega. \tag{7.1.2}$$

The function f is called the *nonhomogeneous term* of the equation.

Let Σ be a smooth hypersurface in Ω with a unit normal vector field $\nu = (\nu_1, \cdots, \nu_n)$. For any integer $j \geq 1$, any point $x_0 \in \Sigma$ and any C^j-function u defined in a neighborhood of x_0, the *jth normal derivative* of u at x_0 is defined by

$$\frac{\partial^j u}{\partial \nu^j} = \sum_{|\alpha|=j} \nu^\alpha \partial^\alpha u = \sum_{\alpha_1+\cdots+\alpha_n=j} \nu_1^{\alpha_1}\cdots\nu_n^{\alpha_n}\partial_{x_1}^{\alpha_1}\cdots\partial_{x_n}^{\alpha_n} u.$$

We now prescribe the values of u and its normal derivatives on Σ so that we can find a solution u of (7.1.2) in Ω. Let $u_0, u_1, \cdots, u_{m-1}$ be continuous functions defined on Σ. We set

$$u = u_0, \; \frac{\partial u}{\partial \nu} = u_1, \; \cdots, \; \frac{\partial^{m-1} u}{\partial \nu^{m-1}} = u_{m-1} \quad \text{on } \Gamma. \tag{7.1.3}$$

We call Σ the *initial hypersurface* and $u_0, \cdots, u_{m-1}$ the *initial values* or *Cauchy values*. The problem of solving (7.1.2) together with (7.1.3) is called the *initial-value problem* or *Cauchy problem*.

We note that there are m functions $u_0, u_1, \cdots, u_{m-1}$ in (7.1.3). This reflects the general fact that m conditions are needed for initial-value problems for PDEs of order m.

As the first step in discussing the solvability of the initial-value problem (7.1.2)–(7.1.3), we intend to find all derivatives of u on Σ. We consider a special case where Σ is the hyperplane $\{x_n = 0\}$. In this case, we can take $\nu = e_n$ and the initial condition (7.1.3) has the form

$$u(\cdot,0) = u_0, \; \partial_{x_n} u(\cdot,0) = u_1, \; \cdots, \; \partial_{x_n}^{m-1} u(\cdot,0) = u_{m-1} \quad \text{on } \mathbb{R}^{n-1}. \tag{7.1.4}$$

Let $u_0, u_1, \cdots, u_{m-1}$ be smooth functions on $\mathbb{R}^{n-1}$ and u be a smooth solution of (7.1.2) and (7.1.4) in a neighborhood of the origin. In the following, we investigate whether we can compute all derivatives of u at the origin in terms of the equation and initial values. We write $x = (x', x_n)$ for $x' \in \mathbb{R}^{n-1}$. First, we can find all x'-derivatives of u at the origin in terms of those of u_0. Next, we can find all x'-derivatives of u_{x_n} at the origin in terms of those of u_1. By continuing this process, we can find all x'-derivatives of u, u_{x_n}, $\cdots$, $\partial_{x_n}^{m-1}u$ at the origin in terms of those of u_0, u_1, $\cdots$, u_{m-1}. In particular, for derivatives up to order m, we find all except $\partial_{x_n}^m u$. To find $\partial_{x_n}^m u(0)$, we need to use the equation. We note that $a_{(0,\cdots,0,m)}$ is the coefficient of $\partial_{x_n}^m u$ in (7.1.2). If we assume

$$a_{(0,\cdots,0,m)}(0) \neq 0, \tag{7.1.5}$$

then by (7.1.2),

$$\partial_{x_n}^m u(0) = -\frac{1}{a_{(0,\cdots,0,m)}(0)} \left(\sum_{\alpha \neq (0,\cdots,0,m)} a_\alpha(0)\partial^\alpha u(0) - f(0) \right).$$

Hence, we can compute all derivatives up to order m at 0 in terms of the coefficients and nonhomogeneous term in (7.1.2) and the initial values $u_0, u_1, \cdots, u_{m-1}$ in (7.1.4). In fact, we can compute the derivatives of u of arbitrary order at the origin. For an illustration, we find the derivatives of u of order $m+1$. By (7.1.5), $a_{(0,\cdots,0,m)}$ is not zero in a neighborhood of the origin. Hence, by (7.1.2),

$$\partial_{x_n}^m u = -\frac{1}{a_{(0,\cdots,0,m)}} \left(\sum_{\alpha \neq (0,\cdots,0,m)} a_\alpha \partial^\alpha u - f \right).$$

By evaluating at $x \in \mathbb{R}^{n-1} \times \{0\}$ close to the origin, we find $\partial_{x_n}^m u(x)$ for $x \in \mathbb{R}^{n-1} \times \{0\}$ sufficiently small. As before, we can find all x'-derivatives of $\partial_{x_n}^m u$ at the origin. Hence for derivatives up to order $m+1$, we find all except $\partial_{x_n}^{m+1}u$. To find $\partial_{x_n}^{m+1}u(0)$, we again need to use the equation. By differentiating (7.1.2) with respect to x_n, we obtain

$$a_{(0,\cdots,0,m)}\partial_{x_n}^{m+1}u + \cdots = f_{x_n},$$

where the dots denote a linear combination of derivatives of u whose values on $\mathbb{R}^{n-1} \times \{0\}$ are already calculated in terms of the derivatives of u_0, u_1, $\cdots$, u_{m-1}, f and the coefficients in the equation. By (7.1.5) and the above equation, we can find $\partial_{x_n}^{m+1}u(0)$. We can continue this process for derivatives of arbitrary order. In summary, we can find all derivatives of u of any order at the origin under the condition (7.1.5), which will be defined as the noncharacteristic condition later on.

In general, consider the hypersurface Σ given by $\{\varphi = 0\}$ for a smooth function φ in a neighborhood of the origin with $\nabla\varphi \neq 0$. We note that the vector field $\nabla\varphi$ is normal to the hypersurface Σ at each point of Σ. We take a point on Σ, say the origin. Then $\varphi(0) = 0$. Without loss of generality, we assume that $\varphi_{x_n}(0) \neq 0$. Then by the implicit function theorem, we can solve $\varphi = 0$ for $x_n = \psi(x_1, \cdots, x_{n-1})$ in a neighborhood of the origin. Consider the change of variables

$$x \mapsto y = (x_1, \cdots, x_{n-1}, \varphi(x)).$$

This is a well-defined transformation with a nonsingular Jacobian in a neighborhood of the origin. With

$$u_{x_i} = \sum_{k=1}^{n} y_{k,x_i} u_{y_k} = \varphi_{x_i} u_{y_n} + \text{ terms not involving } u_{y_n},$$

and in general, for any $\alpha \in \mathbb{Z}_+^n$ with $|\alpha| = m$,

$$\partial_x^\alpha u = \partial_{x_1}^{\alpha_1} \cdots \partial_{x_n}^{\alpha_n} u = \varphi_{x_1}^{\alpha_1} \cdots \varphi_{x_n}^{\alpha_n} \partial_{y_n}^m u + \text{terms not involving } \partial_{y_n}^m u,$$

we can write the operator L in the y-coordinates as

$$Lu = \sum_{|\alpha|=m} a_\alpha\big(x(y)\big) \varphi_{x_1}^{\alpha_1} \cdots \varphi_{x_n}^{\alpha_n} \partial_{y_n}^m u + \text{ terms not involving } \partial_{y_n}^m u.$$

The initial hypersurface Σ is given by $\{y_n = 0\}$ in the y-coordinates. With $y_n = \varphi$, the coefficient of $\partial_{y_n}^m u$ is given by

$$\sum_{|\alpha|=m} a_\alpha(x) \varphi_{x_1}^{\alpha_1} \cdots \varphi_{x_n}^{\alpha_n}.$$

This is the principal symbol $p(x;\xi)$ evaluated at $\xi = \nabla\varphi(x)$.

Definition 7.1.2. Let L be a linear differential operator of order m defined as in (7.1.1) in a neighborhood of $x_0 \in \mathbb{R}^n$ and Σ be a smooth hypersurface containing x_0. Then Σ is *noncharacteristic* at x_0 if

$$p(x_0; \nu) = \sum_{|\alpha|=m} a_\alpha(x_0)\nu^\alpha \neq 0, \tag{7.1.6}$$

where $\nu = (\nu_1, \cdots, \nu_n)$ is normal to Σ at x_0. Otherwise, Σ is *characteristic* at x_0.

A hypersurface is noncharacteristic if it is noncharacteristic at every point. Strictly speaking, a hypersurface is characteristic if it is not noncharacteristic, i.e., if it is characteristic at some point. In this book, we will abuse this terminology. When we say a hypersurface is characteristic, we mean it is characteristic *everywhere*. This should cause no confusion. In $\mathbb{R}^2$, hypersurfaces are curves, so we shall speak of characteristic curves and noncharacteristic curves.

When the hypersurface Σ is given by $\{\varphi = 0\}$ with $\nabla\varphi \neq 0$, its normal vector field is given by $\nabla\varphi = (\varphi_{x_1}, \cdots, \varphi_{x_n})$. Hence we may take $\nu = \nabla\varphi(x_0)$ in (7.1.6). We note that the condition (7.1.6) is preserved under C^m-changes of coordinates. By this condition, we can find successively the values of all derivatives of u at x_0, as far as they exist. Then, we could write *formal* power series at x_0 for solutions of initial-value problems. If the initial hypersurface is analytic and the coefficients, nonhomogeneous terms and initial values are analytic, then this formal power series converges to an analytic solution. This is the content of the Cauchy-Kovalevskaya theorem, which we will discuss in Section 7.2.

Now we introduce a special class of linear differential operators.

Definition 7.1.3. Let L be a linear differential operator of order m defined as in (7.1.1) in a neighborhood of $x_0 \in \mathbb{R}^n$. Then L is *elliptic* at x_0 if

$$p(x_0; \xi) = \sum_{|\alpha|=m} a_\alpha(x_0)\xi^\alpha \neq 0,$$

for any $\xi \in \mathbb{R}^n \setminus \{0\}$.

A linear differential operator defined in Ω is called elliptic in Ω if it is elliptic at every point in Ω.

According to Definition 7.1.3, linear differential operators are elliptic if every hypersurface is noncharacteristic.

Consider a first-order linear differential operator of the form

$$Lu = \sum_{i=1}^n a_i(x)u_{x_i} + b(x)u \quad \text{in } \Omega \subset \mathbb{R}^n.$$

Its principal symbol is given by

$$p(x; \xi) = \sum_{i=1}^n a_i(x)\xi_i,$$

for any $x \in \Omega$ and any $\xi \in \mathbb{R}^n$. Hence first-order linear differential equations with real coefficients are never elliptic. Complex coefficients may yield elliptic equations. For example, take $a_1 = 1/2$ and $a_2 = i/2$ in $\mathbb{R}^2$. Then $\partial_z = (\partial_{x_1} + i\partial_{x_2})/2$ is elliptic.

The notion of ellipticity was introduced in Definition 3.1.2 for second-order linear differential operators of the form

$$Lu = \sum_{i,j=1}^n a_{ij}(x)u_{x_i x_j} + \sum_{i=1}^n b_i(x)u_{x_i} + c(x)u \quad \text{in } \Omega \subset \mathbb{R}^n.$$

The principal symbol of L is given by

$$p(x;\xi) = \sum_{i,j=1}^{n} a_{ij}(x)\xi_i\xi_j,$$

for any $x \in \Omega$ and any $\xi \in \mathbb{R}^n$. Then L is elliptic at $x \in \Omega$ if

$$\sum_{i,j=1}^{n} a_{ij}(x)\xi_i\xi_j \neq 0 \quad \text{for any } \xi \in \mathbb{R}^n \setminus \{0\}.$$

If $(a_{ij}(x))$ is a real-valued $n \times n$ symmetric matrix, L is elliptic at x if $(a_{ij}(x))$ is a definite matrix at x, positive definite or negative definite.

7.1.2. Linear Partial Differential Systems. The concept of noncharacteristics can be generalized to linear partial differential equations for vector-valued functions. Let $m, N \geq 1$ be integers and $\Omega \subset \mathbb{R}^n$ be a domain. A smooth $N \times N$ matrix A in Ω is an $N \times N$ matrix whose components are smooth functions in Ω. Similarly, a smooth N-vector u is a vector of N components which are smooth functions in Ω. Alternatively, we may call them a smooth $N \times N$ matrix-valued function and a smooth N-vector-valued function, or a smooth $\mathbb{R}^N$-valued function, respectively. In the following, a function may mean a scalar-valued function, a vector-valued function, or a matrix-valued function. This should cause no confusion. Throughout this chapter, all vectors are in the form of column vectors.

Let A_α be a smooth $N \times N$ matrix in Ω, for each $\alpha \in \mathbb{Z}_+^n$ with $|\alpha| \leq m$. Consider a linear partial differential operator of the form

$$Lu = \sum_{|\alpha| \leq m} A_\alpha(x)\partial^\alpha u \quad \text{in } \Omega, \tag{7.1.7}$$

where u is a smooth N-vector in Ω. Here, A_α is called the *coefficient matrix* of $\partial^\alpha u$.

We define principal parts, principal symbols and noncharacteristic hypersurfaces similarly to those for single differential equations.

Definition 7.1.4. Let L be a linear differential operator defined in $\Omega \subset \mathbb{R}^n$ as in (7.1.7). The *principal part* L_0 and the *principal symbol* p of L are defined by

$$L_0 u = \sum_{|\alpha| = m} A_\alpha(x)\partial^\alpha u \quad \text{in } \Omega,$$

and

$$p(x;\xi) = \det\left(\sum_{|\alpha|=m} A_\alpha(x)\xi^\alpha\right),$$

for any $x \in \Omega$ and $\xi \in \mathbb{R}^n$.

Definition 7.1.5. Let L be a linear differential operator defined in a neighborhood of $x_0 \in \mathbb{R}^n$ as in (7.1.7) and Σ be a smooth hypersurface containing x_0. Then Σ is *noncharacteristic* at x_0 if

$$p(x_0;\nu) = \det\left(\sum_{|\alpha|=m} A_\alpha(x_0)\nu^\alpha\right) \neq 0,$$

where $\nu = (\nu_1, \cdots, \nu_n)$ is normal to Σ at x_0. Otherwise, Σ is *characteristic* at x_0.

Let f be a smooth N-vector in Ω. We consider the linear differential equation

$$Lu = f(x) \quad \text{in } \Omega. \tag{7.1.8}$$

The function f is called the *nonhomogeneous term* of the equation. We often call (7.1.8) a *partial differential system*, treating (7.1.8) as a collection of partial differential equations for the components of u.

Let Σ be a smooth hypersurface in Ω with a normal vector field ν and let $u_0, u_1, \cdots, u_{m-1}$ be smooth N-vectors on Σ. We prescribe

$$u = u_0, \ \frac{\partial u}{\partial \nu} = u_1, \ \cdots, \ \frac{\partial^{m-1} u}{\partial \nu^{m-1}} = u_{m-1} \quad \text{on } \Sigma. \tag{7.1.9}$$

We call Σ the *initial hypersurface* and $u_0, \cdots, u_{m-1}$ the *initial values* or *Cauchy values*. The problem of solving (7.1.8) together with (7.1.9) is called the *initial-value problem* or *Cauchy problem*.

We now examine first-order linear partial differential systems. Let A_1, $\cdots$, A_n and B be smooth $N \times N$ matrices in a neighborhood of $x_0 \in \mathbb{R}^n$. Consider a first-order linear differential operator

$$Lu = \sum_{i=1}^n A_i u_{x_i} + Bu.$$

A hypersurface Σ containing x_0 is *noncharacteristic* at x_0 if

$$\det\left(\sum_{i=1}^n \nu_i A_i(x_0)\right) \neq 0,$$

where $\nu = (\nu_1, \cdots, \nu_n)$ is normal to Σ at the x_0.

We now demonstrate that we can always reduce the order of differential systems to 1 by increasing the number of equations and the number of components of solution vectors.

Proposition 7.1.6. *Let L be a linear differential operator defined in a neighborhood of $x_0 \in \mathbb{R}^n$ as in (7.1.7), Σ be a smooth hypersurface containing x_0 which is noncharacteristic at x_0 for the operator L, and $u_0, u_1, \cdots, u_{m-1}$*

be smooth on Σ. Then the initial-value problem (7.1.8)–(7.1.9) in a neighborhood of x_0 is equivalent to an initial-value problem for a first-order differential system with appropriate initial values prescribed on Σ, and Σ is a noncharacteristic hypersurface at x_0 for the new first-order differential system.

Proof. We assume that x_0 is the origin. In the following, we write $x = (x', x_n) \in \mathbb{R}^n$ and $\alpha = (\alpha', \alpha_n) \in \mathbb{Z}_+^n$.

Step 1. Straightening initial hypersurfaces. We assume that Σ is given by $\{\varphi = 0\}$ for a smooth function φ in a neighborhood of the origin with $\varphi_{x_n} \neq 0$. Then we introduce a change of coordinates $x = (x', x_n) \mapsto (x', \varphi(x))$. In the new coordinates, still denoted by x, the hypersurface Σ is given by $\{x_n = 0\}$ and the initial condition (7.1.9) is given by

$$\partial_{x_n}^j u(x', 0) = u_j(x') \quad \text{for } j = 0, 1, \cdots, m-1.$$

Step 2. Reductions to canonical forms and zero initial values. In the new coordinates, $\{x_n = 0\}$ is noncharacteristic at 0. Then, the coefficient matrix $A_{(0,\cdots,0,m)}$ is nonsingular at the origin and hence also in a neighborhood of the origin. Multiplying the partial differential system (7.1.8) by the inverse of this matrix, we may assume that $A_{(0,\cdots,0,m)}$ is the identity matrix in a neighborhood of the origin. Next, we may assume

$$u_j(x') = 0 \quad \text{for } j = 0, 1, \cdots, m-1.$$

To see this, we introduce a function v such that

$$u(x) = v(x) + \sum_{j=0}^{m-1} \frac{1}{j!} u_j(x') x_n^j.$$

Then the differential system for v is the same as that for u with f replaced by

$$f(x) - \sum_{j=0}^{m-1} \sum_{|\alpha| \le m} A_\alpha(x) \partial^\alpha \left(\frac{1}{j!} u_j(x') x_n^j \right).$$

Moreover,

$$\partial_{x_n}^j v(x', 0) = 0 \quad \text{for } j = 0, 1, \cdots, m-1.$$

With Step 1 and Step 2 done, we assume that (7.1.8) and (7.1.9) have the form

$$\partial_{x_n}^m u + \sum_{\alpha_n=0}^{m-1} \sum_{|\alpha'| \le m - \alpha_n} A_\alpha \partial^\alpha u = f,$$

with

$$\partial_{x_n}^j u(x', 0) = 0 \quad \text{for } j = 0, 1, \cdots, m-1.$$

Step 3. Lowering the order. We now change this differential system to an equivalent system of order $m-1$. Introduce new functions

$$U_0 = u, \quad U_i = u_{x_i} \text{ for } i = 1, \cdots, n,$$

and

$$U = (U_0^T, U_1^T, \cdots, U_n^T)^T, \tag{7.1.10}$$

where T indicates the transpose. We note that U is a column vector of $(n+1)N$ components. Then

$$U_{0,x_n} = U_n, \quad U_{i,x_n} = U_{n,x_i} \text{ for } i = 1, \cdots, n-1.$$

Hence

$$\partial_{x_n}^{m-1} U_0 - \partial_{x_n}^{m-2} U_n = 0, \tag{7.1.11}$$

$$\partial_{x_n}^{m-1} U_i - \partial_{x_i} \partial_{x_n}^{m-2} U_n = 0 \quad \text{for } i = 1, \cdots, n-1. \tag{7.1.12}$$

To get an $(m-1)$th-order differential equation for U_n, we write the equation for u as

$$\partial_{x_n}^m u + \sum_{\alpha_n=1}^{m-1} \sum_{|\alpha'| \le m-\alpha_n} A_\alpha \partial^\alpha u + \sum_{|\alpha'| \le m} A_{(\alpha',0)} \partial^{(\alpha',0)} u = f.$$

We substitute $U_n = u_{x_n}$ in the first two terms in the left-hand side to get

$$\partial_{x_n}^{m-1} U_n + \sum_{\alpha_n=0}^{m-2} \sum_{|\alpha'| \le m-\alpha_n-1} A_\alpha \partial^\alpha U_n + \sum_{|\alpha'| \le m} A_{(\alpha',0)} \partial^{(\alpha',0)} u = f. \tag{7.1.13}$$

In the last summation in the left-hand side, any mth-order derivative of u can be changed to an $(m-1)$th-order derivative of U_i for some $i = 1, \cdots, n-1$, since no derivatives with respect to x_n are involved. Now we can write a differential system for U in the form

$$\partial_{x_n}^{m-1} U + \sum_{\alpha_n=0}^{m-2} \sum_{|\alpha'| \le m-\alpha_n-1} A_\alpha^{(1)} \partial^\alpha U = F^{(1)}. \tag{7.1.14}$$

The initial value for U is given by

$$\partial_{x_n}^j U(x', 0) = 0 \quad \text{for } j = 0, 1, \cdots, m-2.$$

Hence, we reduce the original initial-value problem for a differential system of order m to an initial-value problem for the differential system of the form (7.1.14) of order $m-1$.

Now let U be a solution of (7.1.14) with zero initial values. By writing U as in (7.1.10), we prove that U_0 is a solution of the initial-value problem for the original differential system of order m. To see this, we first prove that $U_i = U_{0,x_i}$, for $i = 1, \cdots, n$. By (7.1.11) and the initial conditions for U, we have

$$\partial_{x_n}^{m-2}(U_n - U_{0,x_n}) = 0,$$

and on $\{x_n = 0\}$,

$$\partial_{x_n}^{j}(U_n - U_{0,x_n}) = 0 \quad \text{for } j = 0, \cdots, m-3.$$

This easily implies $U_n = U_{0,x_n}$. Next, for $i = 1, \cdots, n-1$,

$$\partial_{x_n}^{m-1}U_i - \partial_{x_i}\partial_{x_n}^{m-2}U_n = \partial_{x_n}^{m-1}U_i - \partial_{x_i}\partial_{x_n}^{m-1}U_0 = \partial_{x_n}^{m-1}(U_i - \partial_{x_i}U_0).$$

By (7.1.12) and the initial conditions, we have

$$\partial_{x_n}^{m-1}(U_i - \partial_{x_i}U_0) = 0,$$

and on $\{x_n = 0\}$

$$\partial_{x_n}^{j}(U_i - \partial_{x_i}U_0) = 0 \quad \text{for } j = 0, \cdots, m-2.$$

Hence, $U_i = U_{0,x_i}$, for $i = 1, \cdots, n-1$. Substituting $U_i = U_{0,x_i}$, for $i = 1, \cdots, n$, in (7.1.13), we conclude that U_0 is a solution for the original mth-order differential system.

Now, we can repeat the procedure to reduce m to 1. □

We point out that straightening initial hypersurfaces and reducing initial values to zero are frequently used techniques in discussions of initial-value problems.

7.2. Analytic Solutions

For a given first-order linear partial differential system in a neighborhood of $x_0 \in \mathbb{R}^n$ and an initial value u_0 prescribed on a hypersurface Σ containing x_0, we first intend to find a solution u formally. To this end, we need to determine all derivatives of u at x_0, in terms of the derivatives of the initial value u_0 and of the coefficients and the nonhomogeneous term in the equation. Obviously, all tangential derivatives (with respect to Σ) of u are given by derivatives of u_0. In order to find the derivatives of u involving the normal direction, we need help from the equation. It has been established that, if Σ is noncharacteristic at x_0, the initial-value problem leads to evaluations of all derivatives of u at x_0. This is clearly a necessary first step to the determination of a solution of the initial-value problem. If the coefficient matrices and initial values are analytic, a Taylor series solution could be developed for u. The Cauchy-Kovalevskaya theorem asserts the convergence of this Taylor series in a neighborhood of x_0.

To motivate our discussion, we study an example of first-order partial differential systems which may admit no solutions in any neighborhood of the origin, unless the initial values prescribed on analytic noncharacteristic hypersurfaces are analytic.

Example 7.2.1. Let $g = g(x)$ be a real-valued function in $\mathbb{R}$. Consider the partial differential system in $\mathbb{R}^2_+ = \{(x,y) : \ y > 0\}$,

$$\begin{aligned} u_y + v_x &= 0, \\ u_x - v_y &= 0, \end{aligned} \tag{7.2.1}$$

with initial values given by

$$u = g(x), \ v = 0 \quad \text{on } \{y = 0\}.$$

We point out that (7.2.1) is simply the Cauchy-Riemann equation in $\mathbb{C} = \mathbb{R}^2$. It can be written in the matrix form

$$\begin{pmatrix} 1 & 0 \\ 0 & 1 \end{pmatrix} \begin{pmatrix} u \\ v \end{pmatrix}_y + \begin{pmatrix} 0 & 1 \\ -1 & 0 \end{pmatrix} \begin{pmatrix} u \\ v \end{pmatrix}_x = \begin{pmatrix} 0 \\ 0 \end{pmatrix}.$$

Note that $\{y = 0\}$ is noncharacteristic. In fact, there are no characteristic curves. To see this, we need to calculate the principal symbol. By taking $\xi = (\xi_1, \xi_2) \in \mathbb{R}^2$, we have

$$\xi_2 \begin{pmatrix} 1 & 0 \\ 0 & 1 \end{pmatrix} + \xi_1 \begin{pmatrix} 0 & 1 \\ -1 & 0 \end{pmatrix} = \begin{pmatrix} \xi_2 & \xi_1 \\ -\xi_1 & \xi_2 \end{pmatrix}.$$

The determinant of this matrix is $\xi_1^2 + \xi_2^2$, which is not zero for any $\xi \neq 0$. Therefore, there are no characteristic curves. We now write (7.2.1) in a complex form. Suppose we have a solution (u, v) for (7.2.1) with the given initial values and let $w = u + iv$. Then

$$\begin{aligned} w_x + i w_y &= 0 \quad \text{in } \mathbb{R}^2_+, \\ w(\cdot, 0) &= g \quad \text{on } \mathbb{R}. \end{aligned}$$

Therefore, w is (complex) analytic in the upper half-plane and its imaginary part is zero on the x-axis. By the Schwartz reflection principle, w can be extended across $y = 0$ to an analytic function in $\mathbb{C} = \mathbb{R}^2$. This implies in particular that g is (real) analytic since $w(\cdot, 0) = g$. We conclude that (7.2.1) admits no solutions with the given initial value g on $\{y = 0\}$ unless g is real analytic.

Example 7.2.1 naturally leads to discussions of analytic solutions.

7.2.1. Real Analytic Functions. We introduced real analytic functions in Section 4.2. We now discuss this subject in detail.

For (real) analytic functions, we need to study convergence of infinite series of the form

$$\sum_\alpha c_\alpha,$$

where the c_α are real numbers defined for all multi-indices $\alpha \in \mathbb{Z}_+^n$. Throughout this section, the term *convergence* always refers to *absolute convergence*. Hence, a series $\sum_\alpha c_\alpha$ is convergent if and only if $\sum_\alpha |c_\alpha| < \infty$. Here, the summation is over all multi-indices $\alpha \in \mathbb{Z}_+^n$.

Definition 7.2.2. A function $u : \mathbb{R}^n \to \mathbb{R}$ is called *analytic* near $x_0 \in \mathbb{R}^n$ if there exist an $r > 0$ and constants $\{u_\alpha\}$ such that

$$u(x) = \sum_\alpha u_\alpha (x - x_0)^\alpha \quad \text{for } x \in B_r(x_0).$$

If u is analytic near x_0, then u is smooth near x_0. Moreover, the constants u_α are given by

$$u_\alpha = \frac{1}{\alpha!} \partial^\alpha u(x_0) \quad \text{for } \alpha \in \mathbb{Z}_+^n.$$

Thus u is equal to its Taylor series about x_0, i.e.,

$$u(x) = \sum_\alpha \frac{1}{\alpha!} \partial^\alpha u(x_0)(x - x_0)^\alpha \quad \text{for } x \in B_r(x_0).$$

For brevity, we will take $x_0 = 0$.

Now we discuss an important analytic function.

Example 7.2.3. For $r > 0$, set

$$u(x) = \frac{r}{r - (x_1 + \cdots + x_n)} \quad \text{for } x \in B_{\frac{r}{\sqrt{n}}}.$$

Then

$$\begin{aligned} u(x) &= \left(1 - \frac{x_1 + \cdots + x_n}{r}\right)^{-1} = \sum_{k=0}^\infty \left(\frac{x_1 + \cdots + x_n}{r}\right)^k \\ &= \sum_{k=0}^\infty \frac{1}{r^k} \sum_{|\alpha|=k} \binom{k}{\alpha} x^\alpha = \sum_\alpha \frac{|\alpha|!}{r^{|\alpha|}\alpha!} x^\alpha. \end{aligned}$$

This power series is absolutely convergent for $|x| < r/\sqrt{n}$ since

$$\sum_\alpha \frac{|\alpha|!}{r^{|\alpha|}\alpha!} |x^\alpha| = \sum_{k=0}^\infty \left(\frac{|x_1| + \cdots + |x_n|}{r}\right)^k < \infty,$$

for $|x_1| + \cdots + |x_n| \le |x|\sqrt{n} < r$. We also note that

$$\partial^\alpha u(0) = \frac{|\alpha|!}{r^{|\alpha|}} \quad \text{for } \alpha \in \mathbb{Z}_+^n.$$

We point out that all derivatives of u at 0 are positive.

An effective method to prove analyticity of functions is to control their derivatives by the derivatives of functions known to be analytic. For this, we introduce the following terminology.

Definition 7.2.4. Let u and v be smooth functions defined in $B_r \subset \mathbb{R}^n$, for some $r > 0$. Then v *majorizes* u in B_r, denoted by $v \gg u$ or $u \ll v$, if

$$\partial^\alpha v(0) \geq |\partial^\alpha u(0)| \quad \text{for any } \alpha \in \mathbb{Z}_+^n.$$

We also call v a *majorant* of u in B_r.

The following simple result concerns the convergence of Taylor series.

Lemma 7.2.5. *Let u and v be smooth functions in B_r. If $v \gg u$ and the Taylor series of v about the origin converges in B_r, then the Taylor series of u about the origin converges in B_r.*

Proof. We simply note that

$$\sum_\alpha \frac{1}{\alpha!}|\partial^\alpha u(0)x^\alpha| \leq \sum_\alpha \frac{1}{\alpha!}\partial^\alpha v(0)|x^\alpha| < \infty \quad \text{for } x \in B_r.$$

Hence we have the desired convergence for u. □

Next, we prove that every analytic function has a majorant.

Lemma 7.2.6. *If the Taylor series of u about the origin is convergent to u in B_r and $0 < s\sqrt{n} < r$, then u has an analytic majorant in $B_{s/\sqrt{n}}$.*

Proof. Set $y = s(1, \cdots, 1)$. Then, $|y| = s\sqrt{n} < r$ and

$$\sum_\alpha \frac{1}{\alpha!}\partial^\alpha u(0)y^\alpha$$

is a convergent series. There exists a constant C such that for any $\alpha \in \mathbb{Z}_+^n$,

$$\left|\frac{1}{\alpha!}\partial^\alpha u(0)y^\alpha\right| \leq C,$$

and in particular,

$$\left|\frac{1}{\alpha!}\partial^\alpha u(0)\right| \leq \frac{C}{y_1^{\alpha_1}\cdots y_n^{\alpha_n}} \leq C\frac{|\alpha|!}{s^{|\alpha|}\alpha!}.$$

Now set

$$v(x) \equiv \frac{Cs}{s - (x_1 + \cdots + x_n)} = C\sum_\alpha \frac{|\alpha|!}{s^{|\alpha|}\alpha!}x^\alpha.$$

Then v is analytic in $B_{s/\sqrt{n}}$ and majorizes u. □

So far, our discussions are limited to scalar-valued functions. All definitions and results can be generalized to vector-valued functions easily. For example, a vector-valued function $u = (u_1, \cdots, u_N)$ is analytic if each of its components is analytic. For vector-valued functions $u = (u_1, \cdots, u_N)$ and $v = (v_1, \cdots, v_N)$, we say $u \ll v$ if $u_i \ll v_i$ for $i = 1, \cdots, N$.

We have the following results for compositions of functions.

Lemma 7.2.7. *Let u, v be smooth functions in a neighborhood of $0 \in \mathbb{R}^n$ with range in $\mathbb{R}^m$ and f, g be smooth functions in a neighborhood of $0 \in \mathbb{R}^m$ with range in $\mathbb{R}^N$, with $u(0) = 0$, $f(0) = 0$, $u \ll v$ and $f \ll g$. Then $f \circ u \ll g \circ v$.*

Lemma 7.2.8. *Let u be an analytic function near $0 \in \mathbb{R}^n$ with range in $\mathbb{R}^m$ and f be an analytic function near $u(0) \in \mathbb{R}^m$ with range in $\mathbb{R}^N$. Then $f \circ u$ is analytic near $0 \in \mathbb{R}^n$.*

We leave the proofs as exercises.

7.2.2. Cauchy-Kovalevskaya Theorem. Now we are ready to discuss real analytic solutions of initial-value problems. We study first-order quasilinear partial differential systems of N equations for N unknowns in $\mathbb{R}^{n+1} = \{(x, t)\}$ with initial values prescribed on the noncharacteristic hyperplane $\{t = 0\}$.

Let $A_1, \cdots, A_n$ be smooth $N \times N$ matrices in $\mathbb{R}^{n+1+N}$, F be a smooth N-vector in $\mathbb{R}^{n+1+N}$ and u_0 be a smooth N-vector in $\mathbb{R}^n$. Consider

$$u_t = \sum_{j=1}^{n} A_j(x, t, u) u_{x_j} + F(x, t, u), \tag{7.2.2}$$

with

$$u(\cdot, 0) = u_0. \tag{7.2.3}$$

We assume that $A_1, \cdots, A_n$, F and u_0 are analytic in their arguments and seek an analytic solution u. We point out that $\{t = 0\}$ is noncharacteristic for (7.2.2). Noncharacteristics was defined for linear differential systems in Section 7.1 and can be generalized easily to quasilinear differential systems. We refer to Section 2.1 for such a generalization for single quasilinear differential equations.

The next result is referred to as the *Cauchy-Kovalevskaya theorem.*

Theorem 7.2.9. *Let u_0 be an analytic N-vector near $0 \in \mathbb{R}^n$, and let A_1, $\cdots$, A_n be analytic $N \times N$ matrices and F be an analytic N-vector near $(0, 0, u_0(0)) \in \mathbb{R}^{n+1+N}$. Then the problem* (7.2.2)–(7.2.3) *admits an analytic solution u near $0 \in \mathbb{R}^{n+1}$.*

Proof. Without loss of generality, we assume $u_0 = 0$. To this end, we introduce v by $v(x, t) = u(x, t) - u_0(x)$. Then the differential system for v is similar to that for u. Next, we add t as an additional component of u by introducing u_{N+1} such that $u_{N+1,t} = 1$ and $u_{N+1}(\cdot, 0) = 0$. This increases the number of equations and the number of components of the solution vector in (7.2.2) by 1 and at the same time deletes t from $A_1, \cdots, A_n$ and

F. For brevity, we still denote by N the number of equations and the number of components of solution vectors.

In the following, we study

$$u_t = \sum_{j=1}^{n} A_j(x, u)u_{x_j} + F(x, u), \tag{7.2.4}$$

with

$$u(\cdot, 0) = 0, \tag{7.2.5}$$

where $A_1, \cdots, A_n$ are analytic $N \times N$ matrices and F is an analytic N-vector in a neighborhood of the origin in $\mathbb{R}^{n+N}$. We seek an analytic solution u in a neighborhood of the origin in $\mathbb{R}^{n+1}$. To this end, we will compute derivatives of u at $0 \in \mathbb{R}^{n+1}$ in terms of derivatives of $A_1, \cdots, A_n$ and F at $(0, 0) \in \mathbb{R}^{n+N}$ and then prove that the Taylor series of u at 0 converges in a neighborhood of $0 \in \mathbb{R}^{n+1}$. We note that t does not appear explicitly in the right hand side of (7.2.4).

Since $u = 0$ on $\{t = 0\}$, we have

$$\partial_x^\alpha u(0) = 0 \quad \text{for any } \alpha \in \mathbb{Z}_+^n.$$

For any $i = 1, \cdots, n$, by differentiating (7.2.4) with respect to x_i, we get

$$u_{x_i t} = \sum_{j=1}^{n} (A_j u_{x_i x_j} + A_{j,x_i} u_{x_j} + A_{j,u} u_{x_i} u_{x_j}) + F_u u_{x_i} + F_{x_i}.$$

In view of (7.2.5), we have

$$u_{x_i t}(0) = F_{x_i}(0, 0).$$

More generally, we obtain by induction

$$\partial_x^\alpha \partial_t u(0) = \partial_x^\alpha F(0, 0) \quad \text{for any } \alpha \in \mathbb{Z}_+^n.$$

Next, for any $\alpha \in \mathbb{Z}_+^n$, we have

$$\begin{aligned}
\partial_x^\alpha \partial_t^2 u = \partial_x^\alpha \partial_t u_t &= \partial_x^\alpha \partial_t \left(\sum_{j=1}^{n} A_j u_{x_j} + F \right) \\
&= \partial_x^\alpha \left(\sum_{j=1}^{n} (A_j u_{x_j t} + A_{j,u} u_t u_{x_j}) + F_u u_t \right).
\end{aligned}$$

Here we used the fact that A_j and F are independent of t. Thus,

$$\partial_x^\alpha \partial_t^2 u(0) = \partial_x^\alpha \left(\sum_{j=1}^{n} (A_j u_{x_j t} + A_{j,u} u_t u_{x_j}) + F_u u_t \right) \Bigg|_{(x,t,u)=0}.$$

The expression in the right-hand side can be worked out to be a polynomial with nonnegative coefficients in various derivatives of $A_1, \cdots, A_n$ and F and the derivatives $\partial_x^\beta \partial_t^l u$ with $|\beta| + l \leq |\alpha| + 2$ and $l \leq 1$.

More generally, for any $\alpha \in \mathbb{Z}_+^n$ and $k \geq 0$, we have

$$\partial_x^\alpha \partial_t^k u(0) = p_{\alpha,k}(\partial_x^\eta \partial_u^\gamma A_1, \cdots, \partial_x^\eta \partial_u^\gamma A_n, \partial_x^\eta \partial_u^\gamma F, \partial_x^\beta \partial_t^l u)\big|_{(x,t,u)=0}, \tag{7.2.6}$$

where $p_{\alpha,k}$ is a polynomial with nonnegative coefficients and the indices η, γ, β, l range over $\eta, \beta \in \mathbb{Z}_+^n$, $\gamma \in \mathbb{Z}_+^N$ and $l \in \mathbb{Z}_+$ with $|\eta| + |\gamma| \leq |\alpha| + k - 1$, $|\beta| + l \leq |\alpha| + k$ and $l \leq k - 1$.

We point out that $p_{\alpha,k}(\partial_x^\eta \partial_u^\gamma A_1, \cdots)$ is considered as a polynomial in the components of $\partial_x^\eta \partial_u^\gamma A_1, \cdots$. We denote by $p_{\alpha,k}(|\partial_x^\eta \partial_u^\gamma A_1|, \cdots)$ the value of $p_{\alpha,k}$ when all components $\partial_x^\eta \partial_u^\gamma A_1, \cdots$ are replaced by their absolute values. Since $p_{\alpha,k}$ has nonnegative coefficients, we conclude that

$$\begin{aligned} &\big|p_{\alpha,k}(\partial_x^\eta \partial_u^\gamma A_1, \cdots, \partial_x^\eta \partial_u^\gamma A_n, \partial_x^\eta \partial_u^\gamma F, \partial_x^\beta \partial_t^l u)\big|_{(x,t,u)=0}\big| \\ &\quad \leq p_{\alpha,k}(|\partial_x^\eta \partial_u^\gamma A_1|, \cdots, |\partial_x^\eta \partial_u^\gamma A_n|, |\partial_x^\eta \partial_u^\gamma F|, |\partial_x^\beta \partial_t^l u|)\big|_{(x,t,u)=0}. \end{aligned} \tag{7.2.7}$$

We now consider a new differential system

$$\begin{aligned} v_t &= \sum_{j=1}^n B_j(x,v) v_{x_i} + G(x,v), \\ v(\cdot, 0) &= 0, \end{aligned} \tag{7.2.8}$$

where $B_1, \cdots, B_n$ are analytic $N \times N$ matrices and G is an analytic N-vector in a neighborhood of the origin in $\mathbb{R}^{n+N}$. We will choose $B_1, \cdots, B_n$ and G such that

$$B_j \gg A_j \quad \text{for } j = 1, \cdots, n \quad \text{and} \quad G \gg F. \tag{7.2.9}$$

Hence, for any $(\eta, \gamma) \in \mathbb{Z}_+^{n+N}$,

$$\partial_x^\eta \partial_u^\gamma B_j(0) \geq |\partial_x^\eta \partial_u^\gamma A_j(0)| \quad \text{for } j = 1, \cdots, n,$$

and

$$\partial_x^\eta \partial_u^\gamma G(0) \geq |\partial_x^\eta \partial_u^\gamma F(0)|.$$

The above inequalities should be understood as holding componentwise.

Let v be a solution of (7.2.8). We now claim that

$$|\partial_x^\alpha \partial_t^k u(0)| \leq \partial_x^\alpha \partial_t^k v(0) \quad \text{for any } (\alpha, k) \in \mathbb{Z}_+^{n+1}.$$

The proof is by induction on the order of t-derivatives. The general step follows since

$$\begin{aligned}|\partial_x^\alpha\partial_t^k u(0)| &= \left|p_{\alpha,k}(\partial_x^\eta\partial_u^\gamma A_1,\cdots,\partial_x^\eta\partial_u^\gamma A_n,\partial_x^\eta\partial_u^\gamma F,\partial_x^\beta\partial_t^l u)\big|_{(x,t,u)=0}\right|\\ &\le p_{\alpha,k}(|\partial_x^\eta\partial_u^\gamma A_1|,\cdots,|\partial_x^\eta\partial_u^\gamma A_n|,|\partial_x^\eta\partial_u^\gamma F|,|\partial_x^\beta\partial_t^l u|)\big|_{(x,t,u)=0}\\ &\le p_{\alpha,k}(\partial_x^\eta\partial_u^\gamma B_1,\cdots,\partial_x^\eta\partial_u^\gamma B_n,\partial_x^\eta\partial_u^\gamma G,\partial_x^\beta\partial_t^l v)\big|_{(x,t,u)=0}\\ &= \partial_x^\alpha\partial_t^k v(0),\end{aligned}$$

where we used (7.2.6), (7.2.7) and the fact that $p_{\alpha,k}$ has nonnegative coefficients. Thus

$$v \gg u. \tag{7.2.10}$$

It remains to prove that the Taylor series of v at 0 converges in a neighborhood of $0 \in \mathbb{R}^{n+1}$.

To this end, we consider

$$B_1 = \cdots = B_n = \frac{Cr}{r-(x_1+\cdots+x_n+v_1+\cdots+v_N)}\begin{pmatrix}1 & \cdots & 1\\ \vdots & \ddots & \vdots\\ 1 & \cdots & 1\end{pmatrix}$$

and

$$G = \frac{Cr}{r-(x_1+\cdots+x_n+v_1+\cdots+v_N)}\begin{pmatrix}1\\ \vdots\\ 1\end{pmatrix},$$

for positive constants C and r, with $|x|+|v| < r/\sqrt{n+N}$. As demonstrated in the proof of Lemma 7.2.6, we may choose C sufficiently large and r sufficiently small such that (7.2.9) holds.

Set

$$v = w\begin{pmatrix}1\\ \vdots\\ 1\end{pmatrix},$$

for some scalar-valued function w in a neighborhood of $0 \in \mathbb{R}^{n+1}$. Then (7.2.8) is reduced to

$$\begin{aligned}w_t &= \frac{Cr}{r-(x_1+\cdots+x_n+Nw)}\left(N\sum_{i=1}^n w_{x_i}+1\right),\\ w(\cdot,0) &= 0.\end{aligned}$$

This is a (single) first-order quasilinear partial differential equation. We now seek a solution w of the form

$$w(x_1,\cdots,x_n,t) = \tilde{w}(x_1+\cdots+x_n,t).$$

Then $\tilde{w} = \tilde{w}(z,t)$ satisfies

$$\tilde{w}_t = \frac{Cr}{r - z - N\tilde{w}}(nN\tilde{w}_z + 1),$$
$$\tilde{w}(\cdot, 0) = 0.$$

By using the method of characteristics as in Section 2.2, we have an explicit solution

$$\tilde{w}(z,t) = \frac{1}{(n+1)N}\{r - z - [(r-z)^2 - 2Cr(n+1)Nt]^{\frac{1}{2}}\},$$

and hence

$$w(x,t) = \frac{1}{(n+1)N}\left\{r - \sum_{i=1}^n x_i - \left[\left(r - \sum_{i=1}^n x_i\right)^2 - 2Cr(n+1)Nt\right]^{\frac{1}{2}}\right\}.$$

This function is analytic near the origin and its Taylor series about the origin is convergent for $|(x,t)| < s$, for sufficiently small $s > 0$. Hence, the corresponding solution v of (7.2.8) is analytic and its Taylor series about the origin is convergent for $|(x,t)| < s$. By Lemma 7.2.5 and (7.2.10), the Taylor series of u about the origin is convergent and hence defines an analytic function for $|(x,t)| < s$, which we denote by u. Since the Taylor series of the analytic functions u_t and $\sum_{j=1}^n A_j(x,u)u_{x_j} + F(x,u)$ have the same coefficients at the origin, they agree throughout the region $|(x,t)| < s$. $\square$

At the beginning of the proof, we introduced an extra component for the solution vector to get rid of t in the coefficient matrices of the differential system. Had we chosen to preserve t, we would have to solve the initial-value problem

$$\tilde{w}_t = \frac{Cr}{r - z - t - Nw}(nN\tilde{w}_z + 1),$$
$$\tilde{w}(\cdot, 0) = 0.$$

It is difficult, if not impossible, to find an explicit expression of the solution $\tilde{w}$.

7.2.3. The Uniqueness Theorem of Holmgren. The solution given in Theorem 7.2.9 is the only analytic solution since all derivatives of the solution are computed at the origin and they uniquely determine the analytic solution. A natural question is whether there are other solutions, which are not analytic.

Let $A_0, A_1, \cdots, A_n$ and B be analytic $N \times N$ matrices, and let F be an analytic N-vector in a neighborhood of the origin in $\mathbb{R}^{n+1}$ and u_0 be an analytic N-vector in a neighborhood of the origin in $\mathbb{R}^n$. We consider the initial-value problem for linear differential systems of the form

$$\begin{aligned} A_0(x,t)u_t + \sum_{j=1}^{n} A_j(x,t)u_{x_j} + B(x,t)u = F(x,t), \\ u(x,0) = u_0(x). \end{aligned} \tag{7.2.11}$$

The next result is referred to as the *local Holmgren uniqueness theorem*. It asserts that there do not exist nonanalytic solutions.

Theorem 7.2.10. *Let $A_0, A_1, \cdots, A_n$ and B be analytic $N \times N$ matrices and F be an analytic N-vector near the origin in $\mathbb{R}^{n+1}$ and u_0 be an analytic N-vector near the origin in $\mathbb{R}^n$. If $\{t = 0\}$ is noncharacteristic at the origin, then any C^1-solution of (7.2.11) is analytic in a sufficiently small neighborhood of the origin in $\mathbb{R}^{n+1}$.*

For the proof, we need to introduce adjoint operators. Let L be a differential operator defined by

$$Lu = A_0(x,t)u_t + \sum_{i=1}^{n} A_i(x,t)u_{x_i} + B(x,t)u.$$

For any N-vectors u and v, we write

$$v^T Lu = (v^T A_0 u)_t + \sum_{i=1}^{n}(v^T A_i u)_{x_i} - \left((A_0^T v)_t + \sum_{i=1}^{n}(A_i^T v)_{x_i} - B^T v \right)^T u.$$

We define the *adjoint operator* L^* of L by

$$\begin{aligned} L^* v &= -\,(A_0^T v)_t - \sum_{i=1}^{n}(A_i^T v)_{x_i} + B^T v \\ &= -\,A_0^T v_t - \sum_{i=1}^{n} A_i^T v_{x_i} + \left(B^T - A_{0,t}^T - \sum_{i=1}^{n} A_{i,x_i}^T \right) v. \end{aligned}$$

Then

$$v^T Lu = (v^T A_0 u)_t + \sum_{i=1}^{n}(v^T A_i u)_{x_i} + (L^* v)^T u.$$

Proof of Theorem 7.2.10. We will prove that any C^1-solution u of $Lu = 0$ with a zero initial value on $\{t = 0\}$ is in fact zero. We introduce an analytic change of coordinates so that the initial hypersurface $\{t = 0\}$ becomes a paraboloid

$$t = |x|^2.$$

For any $\varepsilon > 0$, we set

$$\Omega_\varepsilon = \{(x,t) : \ |x|^2 < t < \varepsilon\}.$$

We will prove that $u = 0$ in Ω_ε for a sufficiently small ε. In the following, we denote by $\partial_+\Omega_\varepsilon$ and $\partial_-\Omega_\varepsilon$ the upper and lower boundary of Ω_ε, respectively, i.e.,

$$\partial_+\Omega_\varepsilon = \{(x,t) : \ |x|^2 < t = \varepsilon\},$$
$$\partial_-\Omega_\varepsilon = \{(x,t) : \ |x|^2 = t < \varepsilon\}.$$

We note that $\det(A_0(0)) \neq 0$ since Σ is noncharacteristic at the origin. Hence A_0 is nonsingular in a neighborhood of the origin. By multiplying the equation in (7.2.11) by A_0^{-1}, we assume $A_0 = I$.

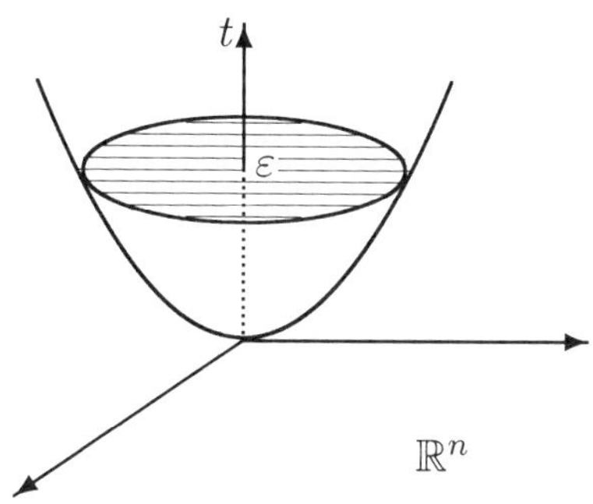

Figure 7.2.1. A parabola.

For any N-vector v defined in a neighborhood of the origin containing Ω_ε, we have

$$0 = \int_{\Omega_\varepsilon} v^T Lu \, dxdt = \int_{\Omega_\varepsilon} u^T L^* v \, dxdt + \int_{\partial_+\Omega_\varepsilon} uv^T \, dx.$$

There is no boundary integral over $\partial_-\Omega_\varepsilon$ since $u = 0$ there. Let $P_k = P_k(x)$ be an arbitrary polynomial in $\mathbb{R}^n$, $k = 1, \cdots, N$, and form $P = (P_1, \cdots, P_N)$. We consider the initial-value problem

$$L^*v = 0 \quad \text{in } B_r,$$
$$v = P \quad \text{on } B_r \cap \{t = \varepsilon\},$$

where B_r is the ball in $\mathbb{R}^{n+1}$ with center at the origin and radius r. The principal part of L^* is the same as that of L, except a different sign and a transpose. We fix r so that $\{t = \varepsilon\} \cap B_r$ is noncharacteristic for L^*, for each small ε. By Theorem 7.2.9, an analytic solution v exists in B_r for ε small. We need to point out that the domain of convergence of v is independent of P, whose components are polynomials. We choose ε small such that $\Omega_\varepsilon \subset B_r$. Then we have

$$\int_{\partial_+\Omega_\varepsilon} u \cdot P \, dx = 0.$$

By the Weierstrass approximation theorem, any continuous function in a compact domain can be approximated in the L^∞-norm by a sequence of polynomials. Hence,

$$\int_{\partial_+\Omega_\varepsilon} u \cdot w \, dx = 0,$$

for any continuous function w on $\partial_+\Omega_\varepsilon \cap \bar{B}_r$. Therefore, $u = 0$ on $\partial_+\Omega_\varepsilon$ for any small ε and hence in Ω_ε. □

Theorem 7.2.9 guarantees the existence of solutions of initial-value problems in the analytic setting. As the next example shows, we do not expect any estimates of solutions in terms of initial values.

Example 7.2.11. In $\mathbb{R}^2$, consider the first-order homogeneous linear differential system (7.2.1),

$$\begin{aligned} u_y + v_x &= 0, \\ u_x - v_y &= 0. \end{aligned}$$

Note that all coefficients are constant. As shown in Example 7.2.1, $\{y = 0\}$ is noncharacteristic. For any integer $k \geq 1$, consider

$$u_k(x, y) = \sin(kx)e^{ky}, \ v_k(x, y) = \cos(kx)e^{ky} \quad \text{for any } (x, y) \in \mathbb{R}^2.$$

Then (u_k, v_k) satisfies (7.2.1) and on $\{y = 0\}$,

$$u_k(x, 0) = \sin(kx), \ v_k(x, 0) = \cos(kx) \quad \text{for any } x \in \mathbb{R}.$$

Obviously,

$$u_k^2(x, 0) + v_k^2(x, 0) = 1 \quad \text{for any } x \in \mathbb{R},$$

and for any $y > 0$,

$$\sup_{x \in \mathbb{R}} \left(u_k^2(x, y) + v_k^2(x, y)\right) = e^{2ky} \to \infty \ \text{ as } k \to \infty.$$

Therefore, there is no continuous dependence on initial values.

7.3. Nonexistence of Smooth Solutions

In this section, we construct a linear differential equation which does not admit smooth solutions *anywhere*, due to Lewy. In this equation, the coefficients are complex-valued analytic functions and the nonhomogeneous term is a suitably chosen complex-valued smooth function. We need to point out that such a nonhomogeneous term is proved to exist by a contradiction argument. This single equation with complex coefficients for a complex-valued solution is equivalent to a system of two differential equations with real coefficients for two real-valued functions.

Define a linear differential operator L in $\mathbb{R}^3 = \{(x, y, z)\}$ by

$$Lu = u_x + iu_y - 2i(x + iy)u_z. \tag{7.3.1}$$

We point out that L acts on complex-valued functions.

The main result in this section is the following theorem.

Theorem 7.3.1. *Let L be the linear differential operator in $\mathbb{R}^3$ defined in (7.3.1). Then there exists an $f \in C^\infty(\mathbb{R}^3)$ such that $Lu = f$ has no C^2-solutions in any open subset of $\mathbb{R}^3$.*

Before we prove Theorem 7.3.1, we rewrite L as a differential system of two equations with real coefficients for two real-valued functions. By writing $u = v + iw$ for real-valued functions v and w, we can write L as a differential operator acting on vectors $(v, w)^T$. Hence

$$L\begin{pmatrix} v \\ w \end{pmatrix} = \begin{pmatrix} v_x - w_y + 2(yv_z + xw_z) \\ w_x + v_y + 2(yw_z - xv_z) \end{pmatrix}.$$

In the matrix form, we have

$$L\begin{pmatrix} v \\ w \end{pmatrix} = \begin{pmatrix} v \\ w \end{pmatrix}_x + \begin{pmatrix} 0 & -1 \\ 1 & 0 \end{pmatrix}\begin{pmatrix} v \\ w \end{pmatrix}_y + \begin{pmatrix} 2y & 2x \\ -2x & 2y \end{pmatrix}\begin{pmatrix} v \\ w \end{pmatrix}_z.$$

By a straightforward calculation, the principal symbol is given by

$$p(P; \xi) = (\xi_1 + 2y\xi_3)^2 + (\xi_2 - 2x\xi_3)^2,$$

for any $P = (x, y, z) \in \mathbb{R}^3$ and $(\xi_1, \xi_2, \xi_3) \in \mathbb{R}^3$. For any fixed $P \in \mathbb{R}^3$, $p(P; \cdot)$ is a nontrivial quadratic polynomial in $\mathbb{R}^3$. Therefore, if f is an analytic function near P, we can always find an analytic solution of $Lu = f$ near P. In fact, we can always find an analytic hypersurface containing P which is noncharacteristic at P. Then by prescribing analytic initial values on this hypersurface, we can solve $Lu = f$ by the Cauchy-Kovalevskaya theorem. Theorem 7.3.1 illustrates that the analyticity of the nonhomogeneous term f is necessary in solving $Lu = f$ even for local solutions.

We first construct a differential equation which does not admit solutions near a given point.

Lemma 7.3.2. *Let (x_0, y_0, z_0) be a point in $\mathbb{R}^3$ and L be the differential operator defined in (7.3.1). Suppose $h = h(z)$ is a real-valued smooth function in $z \in \mathbb{R}$ that is not analytic at z_0. Then there exist no C^1-solutions of the equation*

$$Lu = h'(z - 2y_0x + 2x_0y)$$

in any neighborhood of (x_0, y_0, z_0).

Proof. We first consider the special case $x_0 = y_0 = 0$ and prove it by contradiction. Suppose there exists a C^1-solution u of

$$Lu = h'(z)$$

in a neighborhood of $(0, 0, z_0)$, say

$$\Omega = B_{\sqrt{R}}(0) \times (z_0 - R, z_0 + R) \subset \mathbb{R}^2 \times \mathbb{R},$$

for some $R > 0$. Set

$$v(r, \theta, z) = e^{i\theta}\sqrt{r}u(\sqrt{r}\cos\theta, \sqrt{r}\sin\theta, z).$$

As a function of (r, θ, z), v is C^1 in $(0, R) \times \mathbb{R} \times (z_0 - R, z_0 + R)$ and is continuous at $r = 0$ with $v(0, \theta, z) = 0$. Moreover, v is 2π-periodic in θ. A straightforward calculation yields

$$Lu = 2v_r + \frac{i}{r}v_\theta - 2iv_z = h'(z).$$

Consider the function

$$V(r, z) = \int_0^{2\pi} v(r, \theta, z)\, d\theta.$$

Then V is C^1 in $(r, z) \in (0, R) \times (z_0 - R, z_0 + R)$, is continuous up to $r = 0$ with $V(0, z) = 0$, and satisfies

$$V_z + iV_r = \frac{i}{2}\int_0^{2\pi}\left(2v_r + \frac{i}{r}v_\theta - 2iv_z\right) d\theta = i\pi h'(z).$$

Define

$$W = V(r, z) - i\pi h(z).$$

Then W is C^1 in $(0, R) \times (z_0 - R, z_0 + R)$, is continuous up to $r = 0$, and satisfies $W_z + iW_r = 0$. Thus W is an analytic function of $z + ir$ for $(r, z) \in (0, R) \times (z_0 - R, z_0 + R)$, continuous at $r = 0$, and has a vanishing real part there. Hence we can extend W as an analytic function of $z + ir$ to $(r, z) \in (-R, R) \times (z_0 - R, z_0 + R)$. Hence $-\pi h(z)$, the imaginary part of $W(0, z)$, is real analytic for $z \in (z_0 - R, z_0 + R)$.

Now we consider the general case. Set

$$\tilde{x} = x - x_0, \quad \tilde{y} = y - y_0, \quad \tilde{z} = z - 2y_0\tilde{x} + 2x_0\tilde{y},$$

and

$$\tilde{u}(\tilde{x}, \tilde{y}, \tilde{z}) = u(x, y, z).$$

Then $\tilde{u}(\tilde{x}, \tilde{y}, \tilde{z})$ is C^1 in a neighborhood of $(0, 0, z_0)$. A straightforward calculation yields

$$\tilde{u}_{\tilde{x}} + i\tilde{u}_{\tilde{y}} - 2i(\tilde{x} + i\tilde{y})\tilde{u}_{\tilde{z}} = h'(\tilde{z}).$$

We now apply the special case we have just proved to $\tilde{u}$. □

In the following, we let $h = h(z)$ be a real-valued periodic smooth function in $\mathbb{R}$ which is not real analytic at any $z \in \mathbb{R}$. We take a sequence of points $P_k = (x_k, y_k, z_k) \in \mathbb{R}^3$ which is dense in $\mathbb{R}^3$ and set

$$\rho_k = 2(|x_k| + |y_k|),$$

and

$$c_k = 2^{-k} e^{-\rho_k}.$$

We also denote by ℓ^∞ the collection of bounded infinite sequences $\tau = (a_1, a_2, \cdots)$ of real numbers a_i. This is a Banach space with respect to the norm

$$\|\tau\|_{\ell^\infty} = \sup_k |a_k|.$$

For any $\tau = (a_1, a_2, \cdots) \in \ell^\infty$, we set

$$f_\tau(x, y, z) = \sum_{k=1}^{\infty} a_k c_k h'(z - 2y_k x + 2x_k y) \quad \text{in } \mathbb{R}^3. \tag{7.3.2}$$

We note that f_τ depends on τ linearly. This fact will be needed later on.

Lemma 7.3.3. *Let f_τ be defined as in (7.3.2) for some $\tau \in \ell^\infty$. Then $f_\tau \in C^\infty(\mathbb{R}^3)$. Moreover, for any $\alpha \in \mathbb{Z}_+^3$,*

$$\sup_{\mathbb{R}^3} |\partial^\alpha f_\tau| \leq \left(\frac{|\alpha|}{e}\right)^{|\alpha|} \|\tau\|_{\ell^\infty} \sup_{\mathbb{R}} |h^{(|\alpha|+1)}|.$$

Proof. We need to prove that all formal derivatives of f_τ converge uniformly in $\mathbb{R}^3$. Set

$$M_k = \sup_{z \in \mathbb{R}} |h^{(k)}(z)|.$$

Then $M_k < \infty$ since h is periodic. Hence for any $\alpha \in \mathbb{Z}_+^3$ with $|\alpha| = m$,

$$\begin{aligned}
|a_k c_k \partial^\alpha h'(z - 2y_k x + 2x_k y)| &\leq \|\tau\|_{\ell^\infty} c_k M_{m+1} \rho_k^m \\
&\leq 2^{-k} \|\tau\|_{\ell^\infty} M_{m+1} \rho_k^{|\alpha|} e^{-\rho_k} \leq 2^{-k} \|\tau\|_{\ell^\infty} M_{m+1} \left(\frac{m}{e}\right)^m.
\end{aligned}$$

In the last inequality, we used the fact that the function $f(r) = r^m e^{-r}$ in $[0, \infty)$ has a maximum $m^m e^{-m}$ at $r = m$. This implies the uniform convergence of the series for $\partial^\alpha f_\tau$. $\Box$

We introduce a Hölder space which will be needed in the next result. Let $\mu \in (0, 1)$ be a constant and $\Omega \subset \mathbb{R}^n$ be a domain. We define $C^{1,\mu}(\Omega)$ as the collection of functions $u \in C^1(\Omega)$ with

$$|\nabla u(x) - \nabla u(y)| \leq C|x - y|^\mu \quad \text{for any } x, y \in \Omega,$$

where C is a positive constant. We define the $C^{1,\mu}$-norm in Ω by

$$|u|_{C^{1,\mu}(\Omega)} = \sup_\Omega |u| + \sup_\Omega |\nabla u| + \sup_{x,y \in \Omega, x \neq y} \frac{|\nabla u(x) - \nabla u(y)|}{|x - y|^\mu}.$$

We will need the following important compactness property.

Lemma 7.3.4. *Let Ω be a domain in $\mathbb{R}^n$, and $\mu \in (0,1)$ and $M > 0$ be constants. Suppose $\{u_k\}$ is a sequence of functions in $C^{1,\mu}(\Omega)$ with $\|u_k\|_{C^{1,\mu}(\Omega)} \leq M$ for any k. Then there exist a function $u \in C^{1,\mu}(\Omega)$ and a subsequence $\{u_{k'}\}$ such that $u_{k'} \to u$ in $C^1(\Omega')$ for any bounded subset Ω' with $\bar{\Omega}' \subset \Omega$ and $\|u\|_{C^{1,\mu}(\Omega)} \leq M$.*

Proof. We note that a uniform bound of $C^{1,\mu}$-norms of u_k implies that u_k and their first derivatives are equibounded and equicontinuous in Ω. Hence, the desired result follows easily from Arzelà's theorem. □

We point out that the limit is a $C^{1,\mu}$-function, although the convergence is only in C^1.

Next, we set

$$B_{k,m} = B_{\frac{1}{\sqrt{m}}}(P_k).$$

We fix a constant $\mu \in (0,1)$.

Definition 7.3.5. For positive integers m and k, we denote by $\mathcal{E}_{k,m}$ the collection of $\tau \in \ell^\infty$ such that there exists a solution $u \in C^{1,\mu}(B_{k,m})$ of

$$Lu = f_\tau \quad \text{in } B_{k,m},$$

with

$$u(P_k) = 0, \quad |u|_{C^{1,\mu}(B_{k,m})} \leq m,$$

where f_τ is the function defined in (7.3.2).

We have the following result concerning $\mathcal{E}_{k,m}$.

Lemma 7.3.6. *For any positive integers k and m, $\mathcal{E}_{k,m}$ is closed and nowhere dense in ℓ^∞.*

We recall that a subset is nowhere dense if it has no interior points.

Proof. We first prove that $\mathcal{E}_{k,m}$ is closed. Take any $\tau_1, \tau_2, \cdots \in \mathcal{E}_{k,m}$ and $\tau \in \ell^\infty$ such that

$$\lim_{j\to\infty} \|\tau_j - \tau\|_{\ell^\infty} = 0.$$

By Lemma 7.3.3, we have

$$\sup_{\mathbb{R}^3} |f_{\tau_j} - f_\tau| \leq \|\tau_j - \tau\|_{\ell^\infty} \sup_{\mathbb{R}} |h'|.$$

For each j, let $u_j \in C^{1,\mu}(B_{k,m})$ be as in Definition 7.3.5 for f_{τ_j}, i.e., $Lu_j = f_{\tau_j}$ in $B_{k,m}$, $u_j(P_k) = 0$ and

$$|u_j|_{C^{1,\mu}(B_{k,m})} \leq m.$$

By Lemma 7.3.4, there exist a $u \in C^{1,\mu}(B_{k,m})$ and a subsequence $\{u_{j'}\}$ such that $u_{j'}$ converges uniformly to u together with its first derivatives in any compact subset of $B_{k,m}$. Then, $Lu = f_\tau$ in $B_{k,m}$, $u(P_k) = 0$ and

$$|u|_{C^{1,\mu}(B_{k,m})} \leq m.$$

Hence $\tau \in \mathcal{E}_{k,m}$. This shows that $\mathcal{E}_{k,m}$ is closed.

Next, we prove that $\mathcal{E}_{k,m}$ has no interior points. To do this, we first denote by $\eta \in \ell^\infty$ the bounded sequence all of whose elements are zero, except the kth element, which is given by $1/c_k$. By (7.3.2), we have $f_\eta = h'(z - 2y_k x + 2x_k y)$. By Lemma 7.3.2, there exist no C^1-solutions of $Lu = f_\eta$ in any neighborhood of P_k.

For any $\tau \in \mathcal{E}_{k,m}$, we claim that

$$\tau + \varepsilon\eta \notin \mathcal{E}_{k,m},$$

for any ε. We will prove this by contradiction. Suppose $\tau + \varepsilon\eta \in \mathcal{E}_{k,m}$ for some ε. Set $\tilde{\tau} = \tau + \varepsilon\eta$ and let u and $\tilde{u}$ be solutions of $Lu = f_\tau$ and $L\tilde{u} = f_{\tilde{\tau}}$, respectively, as in Definition 7.3.5. Set $v = (\tilde{u} - u)/\varepsilon$. Then v is a $C^{1,\mu}$-solution of $Lv = f_\eta$ in $B_{k,m}$. This leads to a contradiction, for $|\varepsilon|$ can be arbitrarily small. □

Now we are ready to prove Theorem 7.3.1.

Proof of Theorem 7.3.1. Let $\mu \in (0,1)$ be the constant as in the definition of $\mathcal{E}_{k,m}$. We will prove that for some $\tau \in \ell^\infty$, the equation $Lu = f_\tau$ admits no $C^{1,\mu}$-solutions in any domain $\Omega \subset \mathbb{R}^3$. If not, then for every $\tau \in \ell^\infty$ there exist an open set $\Omega_\tau \subset \mathbb{R}^3$ and a $u \in C^{1,\mu}(\Omega_\tau)$ such that

$$Lu = f_\tau \quad \text{in } \Omega_\tau.$$

By the density of $\{P_k\}$ in $\mathbb{R}^3$, there exists a $P_k \in \Omega_\tau$ for some $k \geq 1$. Then $B_{k,m} \subset \Omega_\tau$ for all sufficiently large m. Next, we may assume $u(P_k) = 0$. Otherwise, we replace u by $u - u(P_k)$. Then, for m sufficiently large, we have

$$|u|_{C^{1,\mu}(B_{k,m})} \leq m.$$

This implies $\tau \in \mathcal{E}_{k,m}$. Hence

$$\ell^\infty = \bigcup_{k,m=1}^{\infty} \mathcal{E}_{k,m}.$$

Therefore, the Banach space ℓ^∞ is a union of a countable set of closed nowhere dense subsets. This contradicts the Baire category theorem. □

7.4. Exercises

Exercise 7.1. Classify the following 4th-order equation in $\mathbb{R}^3$:

$$2\partial_x^4 u + 2\partial_x^2\partial_y^2 u + \partial_y^4 u - 2\partial_x^2\partial_z^2 u + \partial_z^4 u = f.$$

Exercise 7.2. Prove Lemma 7.2.7 and Lemma 7.2.8.

Exercise 7.3. Consider the initial-value problem

$$\begin{aligned} &u_{tt} - u_{xx} - u = 0 \quad \text{in } \mathbb{R}\times(0,\infty),\\ &u(x,0) = x,\ u_t(x,0) = -x. \end{aligned}$$

Find a solution as a power series expansion about the origin and identify this solution.

Exercise 7.4. Let A be an $N\times N$ *diagonal* C^1-matrix on $\mathbb{R}\times(0,T)$ and $f:\mathbb{R}\times(0,T)\times\mathbb{R}^N\to\mathbb{R}^N$ be a C^2-function. Consider the initial-value problem for $u:\mathbb{R}\times(0,T)\to\mathbb{R}^N$ of the form

$$u_t + A(x,t)u_x = f(x,t,u) \quad \text{in } \mathbb{R}\times(0,T),$$

with

$$u(\cdot,0) = 0 \quad \text{on } \mathbb{R}.$$

Under appropriate conditions on f, prove that the above initial-value problem admits a C^1-solution by using the contraction mapping principle.

Hint: It may be helpful to write it as a system of equations instead of using a matrix form.

Exercise 7.5. Set $D = \{(x,t):\ x>0, t>0\}\subset\mathbb{R}^2$ and let a be C^1, b_{ij} be continuous in D, and φ,ψ be continuous in $[0,\infty)$ with $\varphi(0)=\psi(0)$. Suppose $(u,v)\in C^1(D)\cap C(\bar{D})$ is a solution of the problem

$$\begin{aligned} u_t + au_x + b_{11}u + b_{12}v &= f,\\ v_x + b_{12}u + b_{22}v &= g, \end{aligned}$$

with

$$u(x,0) = \varphi(x) \text{ for } x>0 \text{ and } v(0,t) = \psi(t) \text{ for } t>0.$$

(1) Assume $a(0,t)\le 0$ for any $t>0$. Derive an energy estimate for (u,v) in an appropriate domain in D.

(2) Assume $a(0,t)\le 0$ for any $t>0$. For any $T>0$, derive an estimate for $\sup_{[0,T]}|u(0,\cdot)|$ in terms of sup-norms of f,g,φ and ψ.

(3) Discuss whether similar estimates can be derived if $a(0,t)$ is positive for some $t>0$.

Exercise 7.6. Let a, b_{ij} be analytic in a neighborhood of $0 \in \mathbb{R}^2$ and φ, ψ be analytic in a neighborhood of $0 \in \mathbb{R}$. In a neighborhood of the origin in $\mathbb{R}^2 = \{(x,t)\}$, consider

$$\begin{aligned} u_t + au_x + b_{11}u + b_{12}v &= f, \\ v_x + b_{12}u + b_{22}v &= g, \end{aligned}$$

with the condition

$$u(x,0) = \varphi(x) \text{ and } v(0,t) = \psi(t).$$

(1) Let (u,v) be a smooth solution in a neighborhood of the origin. Prove that all derivatives of u and v at 0 can be expressed in terms of those of a, b_{ij}, f, g, φ and ψ at 0.

(2) Prove that there exists an analytic solution (u,v) in a neighborhood of $0 \in \mathbb{R}^2$.

Chapter 8

Epilogue

In the final chapter of this book, we present a list of differential equations we expect to study in more advanced PDE courses. Discussions in this chapter will be brief. We mention several function spaces, including Sobolev spaces and Hölder spaces, without rigorously defining them.

In Section 8.1, we talk about several basic linear differential equations of the second order, including elliptic, parabolic and hyperbolic equations, and linear symmetric hyperbolic differential systems of the first order. These equations appear frequently in many applications. We introduce the appropriate boundary-value problems and initial-value problems and discuss the correct function spaces to study these problems.

In Section 8.2, we discuss more specialized differential equations. We introduce several important nonlinear equations and focus on the background of these equations. Discussions in this section are extremely brief.

8.1. Basic Linear Differential Equations

In this section, we discuss several important linear differential equations. We will focus on elliptic, parabolic and hyperbolic differential equations of the second order and symmetric hyperbolic differential systems of the first order.

8.1.1. Linear Elliptic Differential Equations. Let Ω be a domain in $\mathbb{R}^n$ and a_{ij}, b_i and c be continuous functions in Ω. Linear elliptic differential equations of the second order are given in the form

$$\sum_{i,j=1}^{n} a_{ij}u_{x_ix_j} + \sum_{i=1}^{n} b_iu_{x_i} + cu = f \quad \text{in } \Omega, \tag{8.1.1}$$

where the a_{ij} satisfy

$$\sum_{i,j=1}^{n} a_{ij}(x)\xi_i\xi_j \geq \lambda|\xi|^2 \quad \text{for any } x \in \Omega \text{ and } \xi \in \mathbb{R}^n,$$

for some positive constant λ. The equation (8.1.1) reduces to the Poisson equation if $a_{ij} = \delta_{ij}$ and $b_i = c = 0$. In many cases, it is advantageous to write (8.1.1) in the form

$$\sum_{i,j=1}^{n} (a_{ij}u_{x_i})_{x_j} + \sum_{i=1}^{n} b_i u_{x_i} + cu = f \quad \text{in } \Omega, \tag{8.1.2}$$

by renaming the coefficients b_i. The equation (8.1.2) is said to be in the *divergence form*. For comparison, the equation (8.1.1) is said to be in the *nondivergence form*.

Naturally associated with the elliptic differential equations are boundary-value problems. There are several important classes of boundary-value problems. In the Dirichlet problem, the values of solutions are prescribed on the boundary, while in the Neumann problem, the normal derivatives of solutions are prescribed.

In solving boundary-value problems for elliptic differential equations, we work in Hölder spaces $C^{k,\alpha}$ and Sobolev spaces $W^{k,p}$. Here, k is a nonnegative integer, $p > 1$ and $\alpha \in (0,1)$ are constants. For elliptic equations in the divergence form, it is advantageous to work in Sobolev spaces $H^k = W^{k,2}$ due to their Hilbert space structure.

8.1.2. Linear Parabolic Differential Equations. We denote by (x,t) points in $\mathbb{R}^n \times \mathbb{R}$. Let D be a domain in $\mathbb{R}^n \times \mathbb{R}$ and a_{ij}, b_i and c be continuous functions in D. Linear parabolic differential equations of the second order are given in the form

$$u_t - \sum_{i,j=1}^{n} a_{ij}u_{x_ix_j} + \sum_{i=1}^{n} b_i u_{x_i} + cu = f \quad \text{in } D, \tag{8.1.3}$$

where the a_{ij} satisfy

$$\sum_{i,j=1}^{n} a_{ij}(x,t)\xi_i\xi_j \geq \lambda|\xi|^2 \quad \text{for any } (x,t) \in D \text{ and } \xi \in \mathbb{R}^n,$$

for some positive constant λ. The equation (8.1.3) reduces to the heat equation if $a_{ij} = \delta_{ij}$ and $b_i = c = 0$.

Naturally associated with the parabolic differential equations are initial-value problems and initial/boundary-value problems. In initial-value problems, $D = \mathbb{R}^n \times (0,\infty)$ and the values of solutions are prescribed on $\mathbb{R}^n \times \{0\}$. In initial/boundary-value problems, D has the form $\Omega \times (0,\infty)$, where Ω is a bounded domain in $\mathbb{R}^n$, appropriate boundary values are prescribed on $\partial\Omega \times (0, \infty)$ and the values of solutions are prescribed on $\Omega \times \{0\}$. Many results for elliptic equations have their counterparts for parabolic equations.

8.1.3. Linear Hyperbolic Differential Equations. We denote by (x, t) points in $\mathbb{R}^n \times \mathbb{R}$. Let D be a domain in $\mathbb{R}^n \times \mathbb{R}$ and a_{ij}, b_i and c be continuous functions in D. Linear hyperbolic differential equations of the second order are given in the form

$$u_{tt} - \sum_{i,j=1}^{n} a_{ij} u_{x_i x_j} + \sum_{i=1}^{n} b_i u_{x_i} + cu = f \quad \text{in } D, \tag{8.1.4}$$

where the a_{ij} satisfy

$$\sum_{i,j=1}^{n} a_{ij}(x,t)\xi_i\xi_j \geq \lambda|\xi|^2 \quad \text{for any } (x,t) \in D \text{ and } \xi \in \mathbb{R}^n,$$

for some positive constant λ. The equation (8.1.4) reduces to the wave equation if $a_{ij} = \delta_{ij}$ and $b_i = c = 0$.

Naturally associated with the hyperbolic differential equations are initial-value problems. We note that $\{t = 0\}$ is a noncharacteristic hypersurface for (8.1.4). In initial-value problems, $D = \mathbb{R}^n \times (0, \infty)$ and the values of solutions together with their first t-derivatives are prescribed on $\mathbb{R}^n \times \{0\}$. Solutions can be proved to exist in Sobolev spaces under appropriate assumptions. Energy estimates play fundamental roles in hyperbolic differential equations.

8.1.4. Linear Symmetric Hyperbolic Differential Systems. We denote by (x, t) points in $\mathbb{R}^n \times \mathbb{R}$. Let N be a positive integer, $A_0, A_1, \cdots, A_n$ and B be continuous $N \times N$ matrices and f be continuous N-vector in $\mathbb{R}^n \times \mathbb{R}$. We consider a first-order linear differential system in $\mathbb{R}^n \times \mathbb{R}$ of the form

$$A_0 u_t + \sum_{k=1}^{n} A_k u_{x_k} + Bu = f. \tag{8.1.5}$$

We always assume that $A_0(x, t)$ is nonsigular for any (x, t), i.e.,

$$\det(A_0(x,t)) \neq 0.$$

Hence, the hypersurface $\{t = 0\}$ is noncharacteristic. Naturally associated with (8.1.5) are initial-value problems.

If $N = 1$, the system (8.1.5) is reduced to a differential equation for a scalar-valued function u, and the initial-value problem for (8.1.5) can be solved by the method of characteristics. For $N > 1$, extra conditions are needed.

The differential system (8.1.5) is *symmetric hyperbolic* at (x,t) if $A_0(x,t)$, $A_1(x,t)$, $\cdots$, $A_n(x,t)$ are symmetric and $A_0(x,t)$ is positive definite. It is symmetric hyperbolic in $\mathbb{R}^n \times \mathbb{R}$ if it is symmetric hyperbolic at every point in $\mathbb{R}^n \times \mathbb{R}$.

For $N > 1$, the symmetry plays an essential role in solving initial-value problems for (8.1.5). Symmetric hyperbolic differential systems in general dimensions behave like single differential equations of a similar form. We can derive energy estimates and then prove the existence of solutions of the initial-value problems for (8.1.5) in appropriate Sobolev spaces.

We need to point out that hyperbolic differential equations of the second order can be transformed to symmetric hyperbolic differential systems of the first order.

8.2. Examples of Nonlinear Differential Equations

In this section, we introduce some nonlinear differential equations and systems and discuss briefly their background. The aim of this section is to illustrate the diversity of nonlinear partial differential equations. We have no intention of including here *all* important nonlinear PDEs of mathematics and physics.

8.2.1. Nonlinear Differential Equations. We first introduce some important nonlinear differential equations.

The *Hamilton-Jacobi equation* is a first-order nonlinear PDE for a function $u = u(x,t)$,

$$u_t + H(Du, x) = 0.$$

This equation is derived from Hamiltonian mechanics by treating u as the generating function for a canonical transformation of the classical Hamiltonian $H = H(p,x)$. The Hamilton-Jacobi equation is important in identifying conserved quantities for mechanical systems. A part of its characteristic ODE is given by

$$\begin{aligned}\dot{x}_i &= H_{p_i}(p,x),\\ \dot{p}_i &= -H_{x_i}(p,x).\end{aligned}$$

This is referred to as *Hamilton's ODE*, which arises in the classical calculus of variations and in mechanics.

In continuum physics, a *conservation law* states that a particular measurable property of an isolated physical system does not change as the system evolves. In mathematics, a *scalar conservation law* is a first-order nonlinear PDE

$$u_t + \big(F(u)\big)_x = 0.$$

Here, F is a given function in $\mathbb{R}$ and $u = u(x, t)$ is the unknown function in $\mathbb{R} \times \mathbb{R}$. It reduces to the inviscid Burgers' equation if $F(u) = u^2/2$. In general, global smooth solutions do not exist for initial-value problems. Even for smooth initial values, solutions may develop discontinuities, which are referred to as *shocks*.

Minimal surfaces are defined as surfaces with zero mean curvature. The *minimal surface equation* is a second-order PDE for $u = u(x)$ of the form

$$\operatorname{div}\left(\frac{\nabla u}{\sqrt{1+|\nabla u|^2}}\right) = 0.$$

This is a quasilinear elliptic differential equation. Let Ω be a domain in $\mathbb{R}^n$. For any function u defined in Ω, the area of the graph of u is given by

$$A(u) = \int_\Omega \sqrt{1+|\nabla u|^2}\, dx.$$

The minimal surface equation is the Euler-Lagrange equation of the area functional A.

A *Monge-Ampère equation* is a nonlinear second-order PDE for a function $u = u(x)$ of the form

$$\det(\nabla^2 u) = f(x),$$

where f is a given function defined in $\mathbb{R}^n$. This is an elliptic equation if u is strictly convex. Monge-Ampère equations arise naturally from many problems in Riemannian geometry and conformal geometry. One of the simplest of these problems is the problem of prescribed Gauss curvature. Suppose that Ω is a bounded domain in $\mathbb{R}^n$ and that K is a function defined in Ω. In the problem of prescribed Gauss curvature, we seek a hypersurface of $\mathbb{R}^{n+1}$ as a graph $y = u(x)$ over $x \in \Omega$ so that at each point $(x, u(x))$ of the surface, the Gauss curvature is given by $K(x)$. The resulting partial differential equation is

$$\det(\nabla^2 u) = K(x)(1 + |Du|^2)^{\frac{n+2}{2}}.$$

Scalar reaction-diffusion equations are second-order semilinear parabolic differential equations of the form

$$u_t - a\Delta u = f(u),$$

where $u = u(x, t)$ represents the concentration of a substance, a is the diffusion coefficient and f accounts for all local reactions. They model changes of the concentration of substances under the influence of two processes: local chemical reactions, in which the substances are transformed into each other, and diffusion, which causes the substances to spread out in space. They have a wide range of applications in chemistry as well as biology, ecology and physics.

In quantum mechanics, the *Schrödinger equation* describes how the quantum state of a physical system changes in time. It is as central to quantum mechanics as Newton's laws are to classical mechanics. The Schrödinger equation takes several different forms, depending on physical situations. For a single particle, the Schrödinger equation takes the form

$$iu_t = -\Delta u + Vu,$$

where $u = u(x, t)$ is the probability amplitude for the particle to be found at position x at time t, and V is the potential energy. We allow u to be complex-valued. In forming this equation, we rescale position and time so that the Planck constant and the mass of the particle are absent. The nonlinear Schrödinger equation has the form

$$iu_t = -\Delta u + \kappa |u|^2 u,$$

where κ is a constant.

The *Korteweg-de Vries equation* (KdV equation for short) is a mathematical model of waves on shallow water surfaces. The KdV equation is a nonlinear, dispersive PDE for a function $u = u(x, t)$ of two real variables, space x and time t, in the form

$$u_t + uu_x + u_{xxx} = 0.$$

It admits solutions of the form $v(x - ct)$, which represent waves traveling to the right at speed c. These are called *soliton solutions.*

8.2.2. Nonlinear Differential Systems. Next, we introduce some nonlinear differential systems.

In fluid dynamics, the *Euler equations* govern inviscid flow. They are usually written in the conservation form to emphasize the conservation of mass, momentum and energy. The Euler equations are a system of first-order PDEs given by

$$\begin{aligned}
&\rho_t + \nabla \cdot (\rho \mathbf{u}) = 0, \\
&(\rho \mathbf{u})_t + \nabla \cdot (\mathbf{u} \otimes (\rho \mathbf{u})) + \nabla p = 0, \\
&(\rho E)_t + \nabla \cdot (\mathbf{u}(\rho E + p)) = 0,
\end{aligned}$$

where ρ is the fluid *mass density,* $\mathbf{u}$ is the fluid *velocity vector*, p is the *pressure* and E is the *energy* per unit volume. We assume

$$E = e + \frac{1}{2}|\mathbf{u}|^2,$$

where e is the internal energy per unit mass and the second term corresponds to the *kinetic energy* per unit mass. When the flow is incompressible,

$$\nabla \cdot \mathbf{u} = 0.$$

If the flow is further assumed to be homogeneous, the density ρ is constant and does not change with respect to space. The Euler equations for incompressible flow have the form

$$\begin{aligned} \mathbf{u}_t + \mathbf{u} \cdot \nabla \mathbf{u} &= -\nabla p, \\ \nabla \cdot \mathbf{u} &= 0. \end{aligned}$$

In forming these equations, we take the density ρ to be 1 and neglect the equation for E.

The *Navier-Stokes equations* describe the motion of incompressible and homogeneous fluid substances when viscosity is present. These equations arise from applying Newton's second law to fluid motion under appropriate assumptions on the fluid stress. With the same notation for the Euler equations, the Navier-Stokes equations have the form

$$\begin{aligned} \mathbf{u}_t + \mathbf{u} \cdot \nabla \mathbf{u} &= \nu \Delta \mathbf{u} - \nabla p, \\ \nabla \cdot \mathbf{u} &= 0, \end{aligned}$$

where ν is the viscosity constant. We note that (incompressible) Euler equations correspond to the (incompressible) Navier-Stokes equations with zero viscosity. It is a Millennium Prize Problem to prove the existence and smoothness of solutions of the initial-value problem for Navier-Stokes equations.

In differential geometry, a *geometric flow* is the gradient flow associated with a functional on a manifold which has a geometric interpretation, usually associated with some extrinsic or intrinsic curvature. A geometric flow is also called a *geometric evolution equation*.

The *mean curvature flow* is a geometric flow of hypersurfaces in Euclidean space or, more generally, in a Riemannian manifold. In mean curvature flows, a family of surfaces evolves with the velocity at each point on the surface given by the mean curvature of the surface. For closed hypersurfaces in Euclidean space $\mathbb{R}^{n+1}$, the mean curvature flow is the geometric evolution equation of the form

$$\mathbf{F}_t = H\nu,$$

where $\mathbf{F}(t) : M \to \mathbb{R}^{n+1}$ is an embedding with an inner normal vector field ν and the mean curvature H. We can rewrite this equation as

$$\mathbf{F}_t = \Delta_{g(t)} \mathbf{F},$$

where $g(t)$ is the induced metric of the evolving hypersurface $\mathbf{F}(t)$. When expressed in an appropriate coordinate system, the mean curvature flow forms a second-order nonlinear parabolic system of PDEs for the components of $\mathbf{F}$.

The *Ricci flow* is an intrinsic geometric flow in differential geometry which deforms the metric of a Riemannian manifold. For any metric g on a Riemannian manifold M, we denote by Ric its Ricci curvature tensor. The Ricci flow is the geometric evolution equation of the form

$$\partial_t g = -2Ric.$$

Here we view the metric tensor and its associated Ricci tensor as functions of a variable $x \in M$ and an extra variable t, which is interpreted as *time*. In local coordinate systems, the components R_{ij} of the Ricci curvature tensor can be expressed in terms of the components g_{ij} of the metric tensor g and their derivatives up to order 2. When expressed in an appropriate coordinate system, the Ricci flow forms a second-order quasilinear parabolic system of PDEs for g_{ij}. The Ricci flow plays an essential role in the solution of the Poincaré conjecture, a Millennium Prize Problem.

In general relativity, the *Einstein field equations* describe how the curvature of spacetime is related to the matter/energy content of the universe. They are given by

$$G = T,$$

where G is the *Einstein tensor* of a Lorentzian manifold (M, g), or *spacetime*, and T is the *stress-energy tensor*. The Einstein tensor is defined by

$$G = Ric - \frac{1}{2}Sg,$$

where Ric is the Ricci curvature tensor and S is the scalar curvature of (M, g). While the Einstein tensor is a type of curvature, and as such relates to gravity, the stress-energy tensor contains all the information concerning the matter fields. Thus, the Einstein field equations exhibit how matter acts as a source for gravity. When expressed in an appropriate gauge (coordinate system), the Einstein field equations form a second-order quasilinear hyperbolic system of PDEs for components g_{ij} of the metric tensor g. In general, the stress-energy tensor T depends on the metric g and its first derivatives. If T is zero, then the Einstein field equations are referred to as the *Einstein vacuum field equations*, and are equivalent to the vanishing of the Ricci curvature.

Yang-Mills theory, also known as non-Abelian gauge theory, was formulated by Yang and Mills in 1954 in an effort to extend the original concept of gauge theory for an Abelian group to the case of a non-Abelian group and has great impact on physics. It explains the electromagnetic and the strong and weak nuclear interactions. It also succeeds in studying the topology of smooth 4-manifolds in mathematics. Let M be a Riemannian manifold and P a principal G-bundle over M, where G is a compact Lie group, referred to as the gauge group. Let A be a connection on P and F be its curvature. Then the *Yang-Mills functional* is defined by

$$\int_M |F|^2 dV_g.$$

The *Yang-Mills equations* are the Euler-Lagrange equations for this functional and can be written as

$$d_A^* F = 0,$$

where d_A^* is the adjoint of d_A, the gauge-covariant extension of the exterior derivative. We point out that F also satisfies

$$d_A F = 0.$$

This is the Bianchi identity, which follows from the exterior differentiation of F. In general, Yang-Mills equations are nonlinear. It is a Millennium Prize Problem to prove that a nontrivial Yang-Mills theory exists on $\mathbb{R}^4$ and has a positive mass gap for any compact simple gauge group G.

8.2.3. Variational Problems. Last, we introduce some variational problems with elliptic characters. As we know, harmonic functions in an arbitrary domain $\Omega \subset \mathbb{R}^n$ can be regarded as minimizers or critical points of the *Dirichlet energy*

$$\int_\Omega |\nabla u|^2 \, dx.$$

This is probably the simplest variational problem.

There are several ways to generalize such a problem. We may take a function $F : \mathbb{R}^n \to \mathbb{R}$ and consider

$$\int_\Omega F(\nabla u) \, dx.$$

It is the Dirichlet energy if $F(p) = |p|^2$ for any $p \in \mathbb{R}^n$. When $F(p) = \sqrt{1 + |p|^2}$, the integral above is the area of the hypersurface of the graph $y = u(x)$ in $\mathbb{R}^n \times \mathbb{R}$. This corresponds to the *minimal surface equation* we have introduced earlier.

Another generalization is to consider the Dirichlet energy,

$$\int_\Omega |\nabla u|^2 \, dx,$$

for vector-valued functions $u : \Omega \subset \mathbb{R}^n \to \mathbb{R}^m$ with an extra requirement that the image $u(\Omega)$ lies in a given submanifold of $\mathbb{R}^m$. For example, we may take this submanifold to be the unit sphere in $\mathbb{R}^m$. Minimizers of such a variational problem are called *minimizing harmonic maps*. In general, minimizing harmonic maps are not smooth. They are smooth away from a subset Σ, referred to as a *singular set.* The study of singular sets and behavior of minimizing harmonic maps near singular sets constitutes an important subject.

One more way to generalize is to consider the Dirichlet energy,

$$\int_\Omega |\nabla u|^2\, dx,$$

for scalar-valued functions $u : \Omega \subset \mathbb{R}^n \to \mathbb{R}$ with an extra requirement that $u \geq \psi$ in Ω for a given function ψ. This is the simplest *obstacle problem* or *free boundary problem*, where ψ is an *obstacle.* Let u be a minimizer and set $\Lambda = \{x \in \Omega; u(x) > \psi(x)\}$. It can be proved that u is harmonic in Λ. The set $\partial\Lambda$ in Ω is called the *free boundary.* It is important to study the regularity of free boundaries.

Bibliography

[1] Alinhac, S., *Hyperbolic Partial Differential Equations*, Springer, 2009.

[2] Carlson, J., Jaffe, A., Wiles, A. (Editors), *The Millennium Prize Problems*, Clay Math. Institute, 2006.

[3] Chen, Y.-Z., Wu, L.-C., *Second Order Elliptic Equations and Elliptic Systems*, Amer. Math. Soc., 1998.

[4] Courant, R., Hilbert, D., *Methods of Mathematical Physics*, Vol. II, Interscience Publishers, 1962.

[5] DiBenedetto, E., *Partial Differential Equations*, Birkhäuser, 1995.

[6] Evans, L., *Partial Differential Equations*, Amer. Math. Soc., 1998.

[7] Folland, G., *Introduction to Partial Differential Equations*, Princeton University Press, 1976.

[8] Friedman, A., *Partial Differential Equations,* Holt, Rinehart, Winston, 1969.

[9] Friedman, A., *Partial Differential Equations of Parabolic Type,* Prentice-Hall, 1964.

[10] Garabedian, P., *Partial Differential Equations*, Wiley, 1964.

[11] Gilbarg, D., Trudinger, N., *Elliptic Partial Differential Equations of Second Order* (2nd ed.), Springer, 1983.

[12] Han, Q., Lin, F.-H., *Elliptic Partial Differential Equations*, Amer. Math. Soc., 2000.

[13] Hörmander, L., *Lectures on Nonlinear Hyperbolic Differential Equation,* Springer, 1996.

[14] Hörmander, L., *The Analysis of Linear Partial Differential Operators,* Vols. 1–4, Springer, 1983–85.

[15] John, F., *Partial Differential Equations* (4th ed.), Springer, 1991.

[16] Lax, P., *Hyperbolic Partial Differential Equations*, Amer. Math. Soc., 2006.

[17] Lieberman, G. M., *Second Order Parabolic Partial Differential Equations*, World Scientific, 1996.

[18] MacRobert, T. M., *Spherical Harmonics, An Elementary Treatise on Harmonic Functions with Applications*, Pergamon Press, 1967.

[19] Protter, M., Weinberger, H., *Maximum Principles in Differential Equations*, Prentice-Hall, 1967.

[20] Rauch, J., *Partial Differential Equations*, Springer, 1992.

[21] Schoen, R., Yau, S.-T., *Lectures on Differential Geometry*, International Press, 1994.

[22] Shatah, J., Struwe M., *Geometric Wave Equations*, Amer. Math. Soc., 1998.

[23] Smoller, J., *Shock Waves and Reaction-Diffusion Equations*, Springer, 1983.

[24] Strauss, W., *Partial Differential Equations: An Introduction*, Wiley, 1992.

[25] Taylor, M., *Partial Differential Equations*, Vols. I–III, Springer, 1996.

Index